Concurrent Engineering Design

Integrating the Best Practices for Process Improvement

By
Landon C. G. Miller

Published by the
Society of Manufacturing Engineers
Publications Development Department
Reference Publications Division
One SME Drive
P.O. Box 930
Dearborn, Michigan 48121

Concurrent Engineering Design

Integrating the Best Practices for Process Improvement

Society of Manufacturing Engineers
Dearborn, Michigan 48121

First Edition
Second Printing

Library of Congress Catalog Card Number: 92-085526
International Standard Book Number: 0-87263-433-7
Manufactured in the United States of America

Foreword

This book introduces Concurrent Engineering Design to all parts of the organization. Concurrent Engineering Design crosses many traditional functional elements of a manufacturing organization. Because of its broad scope and impact, this book is intended for a wide group of audiences. The book, as a whole, is directed at management.

This first section introduces Concurrent Engineering Design (CE Design). CE Design is described from an Executive Summary perspective in Chapter 1. In Chapter 2, the dimensions, or new and advantageous characteristics of this "paradigm shift" creating concept are discussed.

Interest in CE Design is accelerating in impact for a variety of reasons; these reasons are discussed in Chapter 3. World class manufacturing, and CE Design's relationship to it, as well as CE Design's enabling characteristics for world class manufacturing are discussed in Chapter 4.

Section II describes the business, technical and managerial processes within which CE Design's activities occur. This section is important to engineering and production manufacturing management. Section III focuses on the Computing architecture necessary for CE Design and an overall implementation plan for CE Design. This section should be of interest to all management. The Appendices contain self-assessment guides, suggestions for other self assessment materials, and a glossary.

Acknowledgments

I wish to thank and acknowledge the assistance of Mr. Thomas R. Currie, P.E., M.B.A., Senior Engineer and Manager, Engineering Computing Architecture, Boeing Commercial Airplane Group, The Boeing Company, who over a five-year association provided many valuable insights as well as assistance in the preparation of this book. I would also like to thank Mr. James M. Maloney, M.S., M.B.A., Manager, Integrated Systems, Boeing Computer Services, The Boeing Company; Mr. Lew Espinosa, Director, Engineering Operations, deHavilland Aircraft Company; and Mr. David W. Hunter, Executive Vice President, Vertical Systems, Inc., who all reviewed this book for me. I must also thank the many clients with whom I have worked over the past 15 years whose trust gave me the opportunity to assist in improving their organization while gaining the insights which lead to the formation of the concepts and experiences described in this book.

Table of Contents

SECTION I

The Business Environment Surrounding Concurrent Engineering Design

This initial section highlights the advantages gained by a manufacturing facility after incorporating Concurrent Engineering Design throughout its organization. Both the basic concept of CE Design as well as its benefits are detailed. Giving graphic evidence of CE Design's advantages, *Figure 1-4* details advantages gained through CE Design, giving percentages of improvement over traditional methods.

An essential element of CE Design is that it must be incorporated throughout the facility, including management, business, and technical areas. *Figure 1-6* illustrates the "team" concept of integrating these elements.

While this section does discuss tools needed for CE Design, such as cause and effect analysis and problem resolution analysis, the emphasis is on how CE Design can help endow a manufacturing firm with world class status so it can compete effectively in today's marketplace.

1

AN INTRODUCTION TO CONCURRENT ENGINEERING DESIGN

THE BUSINESS IMPERATIVE

For 10 or more years, the globalization of manufacturing has been proceeding at a frantic pace. At the same time, the competitive environment also has reached a serious and sustained high level of intensity.

Every manufacturing organization making more complex products now must:

- constantly reduce product costs;
- substantially shorten time to market and competitive response times, and
- constantly improve product quality.

Quality has become the critical issue. Quality, in this context, is everything about the product and its associated manufacturing organization. Some suggest that, traditionally, quality is available but for a higher price. High quality has become a pre-existing condition for acceptable product performance. Any quality initiative must begin with the product and that product's design.

For example, as shown in *Figure 1-1*, traditional design and engineering processes produced the first acceptable design, or the first design that met all of the design criteria. *Figure 1-1* depicts this first acceptable range in two of the many dimensions of design cost and tolerances. Other dimensions also must be considered but for illustration purposes are not shown here.

Today, the first acceptable design must be close to optimum and rapidly made with little or no need for quality induced modifications. *Figure 1-1* depicts this much higher expectation of design as falling within the optimum range of parameters. The CE Design process provides a stable, repeatable process which increased accuracy is achieved in a shorter time with less variation.

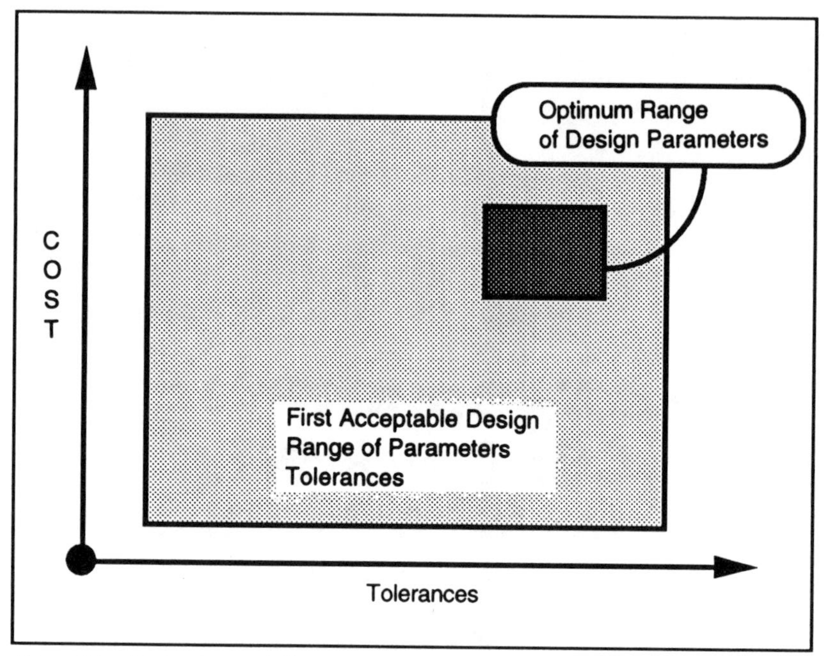

Figure 1-1. *Design Acceptability.*

More importantly, in the manufacturing environment of the '90s, there is no maximum threshold (level of quality) of design performance, after which the organization can "settle down" and continue to operate. The competition is ceaselessly pursuing additional quality characteristics; this means that the quest for improved quality is *constant* and *endless*.

Production manufacturing, or that part of the organization that produces the product, is not capable, by itself, of delivering on these requirements in the traditional manufacturing environment. The entire organization must reorganize itself to meet higher quality requirements. Many organizations are pursuing quality standards of less than four errors in any part of the organization's activities in every one million opportunities. Some firms have called this program their Six Sigma Program, alluding to the statistical figure of a sixth level standard deviation measurement. This level of quality requires more than just a few quality engineers. It requires a re-engineering of the organization and how it operates.

This re-engineering must start with the product design process, the rest of the engineering activities of the organization, and engineering business management processes. This book is about Concurrent Engineering Design, the absolutely necessary business process and "paradigm" shift that is capable of delivering against today's requirements, and is the precursor to a world class manufacturing environment, capable of responding to high quality, speed, and cost requirements.

THE CONCEPT

Concurrent Engineering Design (CE Design) is a term formally describing a set of technical, business, manufacturing planning, and design processes that are concurrently performed by elements of the manufacturing organization prior to the commitment to actually produce something. The CE Design process, in its simplest form, is the integrated execution of four business/technical processes at the same time. These processes are *Process Management, Design, Manufacturability*, and *Automated Infrastructure Support*.

Many technical, business, environmental, and internal manufacturing process needs have created the need to "re-think" the traditional manufacturing process. This re-thinking is focused on several key initiatives, which include, business process re-engineering, high-quality products, and reducing time to market. CE Design is also one of the key initiatives, and is the enabler of the others. *Design* is included in the concept because the general process of design became an increasingly powerful element of manufacturing. Yet, the design process is not well-described, and must be re-engineered if it is to respond to the challenge of world class manufacturing. An engineered *Design* process is integral to the Concurrent Engineering process actually evolving.

These general concepts—when applied in an integrated manner—are very powerful. They could be applied in any type of organization. CE Design has the most immediate impact within the context of manufacturing organizations with the following distinct characteristics:

- manufacturing operations with high rates of product and process definition change (*change rate*); or
- manufacturing operations with high rates and short cycles for new product introductions (*speed*); or
- manufacturing operations with complex configurations that can vary by customer and which could be produced in a single, individual production line (*complexity*); or
- manufacturing operations that require multiple teams for a single product, and typically include business partners with design responsibility for product components as well as suppliers with no design responsibility (*multiple design teams*).

BASIC BENEFITS

Concurrent Engineering (CE) Design is being introduced into a number of manufacturing organizations. Early experiences have produced several emerging benefits. Design incurs, or actually costs, a fraction of the product development total product life cycle cost. Design, however, creates the obligation of expenditures over the product life cycle, for the spending that will be incurred. For example, as reported by Computer-Aided Manufacturing International, Inc., the ratio of incurred cost for design (5% to 8%) and its committed costs

Design is a tiny piece of the development pie, but it locks in the bulk of later spending

	Percent of total costs*	
	Incurred	Committed
CONCEPTION	3% - 5%	40% - 60%
DESIGN ENGINEERING	5 - 8	60 - 80
TESTING	8 - 10	80 - 90
PROCESS PLANNING	10 - 15	90 - 95
PRODUCTION	15 - 100	95 - 100

*Cumulative

DATA: COMPUTER-AIDED MANUFACTURING-INTERNATIONAL, INC.

Figure 1-2. *With Design Decisions So Critical…*

(60% to 80%) reflects the difference between the actual incurred cost and the product life cycle cost of design. This ratio is illustrated in *Figure 1-2*.

The impact of design is not limited to the design's initial release. The typical change made during the development of a major electronics product has impacts as described in *Figure 1-3*. Notice how rapidly the cost of the change accelerates if it is not made during design.

Concurrent Engineering Design can have big impacts on the overall product life cycle. Data provided by the National Institute of Standards and Technology (NIST) describes the impacts of CE Design as shown in *Figure 1-4*.

Because of these substantial benefits, the "prime" or end product manufacturer will adopt CE Design early. In addition, if the organization is, for example, a supplier to an automobile, aerospace, defense, shipbuilder, computer hardware and software, or other complex manufacturer, their prime customer's transition to CE Design will bring the supplier into CE Design as well. The supplier's situation may be complicated by design-related standards imposed by the "prime's" requirements.

Another example of organizations that need to understand and/or adopt at least portions of CE Design to take advantage of its power as applied to complex manufacturers includes buyers of complex products like airlines, governments, or other large organizations with diverse elements. Dealing with these complex product manufacturers will bring them into this approach to speed, and provide

The typical cost for each change made during the development of a major electronics product	
When design changes are made	Cost
DURING DESIGN	$1,000
DURING DESIGN TESTING	10,000
DURING PROCESS PLANNING	100,000
DURING TEST PRODUCTION	1,000,000
DURING FINAL PRODUCTION	10,000,000
DATA: DATAQUEST, INC.	

Figure 1-3. *...And Traditional Engineering So Expensive...*

order to communications, as well as speed and improve the quality of acquisition and product support.

In addition to the positive effects of CE Design, there are negative pressures in the marketplace. CE Design has emerged as a response to several inter-related events.

1. *Processes in Peril.* The lack of success of classical "MRP" extensibility for complex product manufacturing beyond its original material management and scheduling scope;
2. *Process Creating Confusion.* Attempts to turn the "alphabet soup" of acronyms representing ideas and concepts to improve manufacturing into a consistent framework or an architecture;
3. *Technology Push.* The emergence of the powerful technical workstation, networked communications and highly capable software and systems prevalent in the computer industry;
4. *Business Infrastructure Evolution.* The rapid disappearance or demise of industrial corporations and the emergence of "global competition" and

Benefits from designing manufacturability, quality, and ease of maintenance into the product at the start	Percent
DEVELOPMENT TIME	30% - 70% less
ENGINEERING CHANGES	65 - 90% fewer
TIME TO MARKET	20 - 90% less
OVERALL QUALITY	200 - 600% higher
WHITE-COLLAR PRODUCTIVITY	20 - 110% higher
DOLLAR SALES	5 - 50% higher
RETURN ON ASSETS	20 - 120% higher

DATA: National Institute of Standards & Technology, Thomas Group Inc., Institute for Defense Analysis

Figure 1-4. *...Concurrent Engineering Pays Big Dividends.*

the reasons behind the rapid business changes; these changes include the need to:

- constantly reduce product costs;
- substantially shorten time to market and competitive response times, and
- constantly improve product quality.

5. *Business Management under Competitive Pressures*. The recognition that the successful 21st century organization will look substantially different from that of the post-World War II hierarchically structured one. The management methods of "leadership" by itself, and of "division of labor, management and technical specialists," are being replaced by a flatter organization focused on teamwork, coordination, cooperation, and communication.

Figure 1-5 summarizes these inter-related events into the three broad issues discussed throughout the book. The "competitive pressures" being created by the globalization of manufacturing and its subsequent intense quality, speed and cost pressures are creating the need for a competitive response. Processes are "pulling" CE Design because of the wide-spread acceptance of the surveys depicted in *Figure 1-2*, *Figure 1-3*, and *Figure 1-4* and production manufacturing's need for quality design information. The CE Design process's basic product is information. Computing technology is "pushing" on the design process with its powerful, inexpensive workstations, and powerful software. The combined effect of these three forces on design creates a powerful incentive for the design process to change. In changing to CE Design, it becomes the starting point from which to "re-engineer" the entire manufacturing organization into a world class competitor. CE Design responds to negative forces, and yet it also creates a positive environment in which to compete and excel.

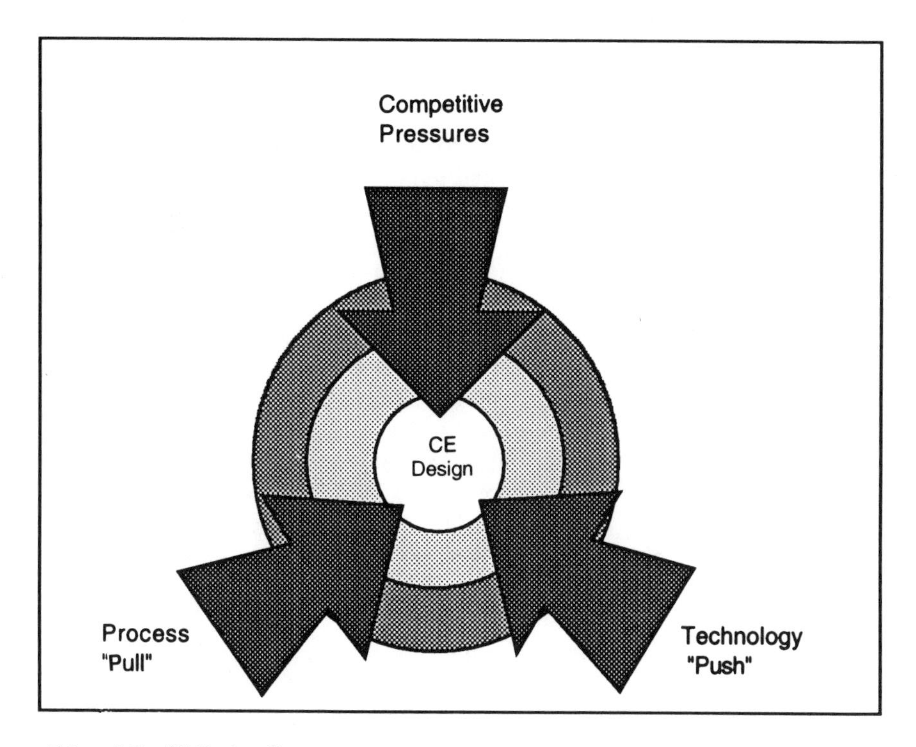

Figure 1-5. *CE Design Pressures.*

There are technical, business, and managerial perspectives to any discussion of CE Design. This book describes the integrated management, business, and technical process framework necessary for its successful understanding, implementation, and execution. It also deals with management technology and process re-engineering or "best practices" dynamic evolution occurring as the organization begins adopting CE Design.

AN EXECUTIVE SUMMARY DESCRIPTION

The CE Design process is made up of four integrated business/technical processes operating at the same time. These processes are *Process Management*, *Design*, *Manufacturability*, and *Automated Infrastructure Support*. *CE Design draws its power from its integrated cross-process operation.*

Process Management may be the most important of the processes. It facilitates and coordinates CE Design activities across many, traditionally independently operated and managed disciplines, functions, and organizational elements within the manufacturing organization.

Design is the orderly execution of a design process for the development of product design and its manufacturing process by many persons in multiple teams at the same time in the interest of speed, diversity, and design robustness exploration.

During *Manufacturability*, representatives of manufacturing, customer support, and other "downstream" activities of the organization interact with designers, as the product is being designed for developing the "as planned," "as manufactured," and "as supported" views of the product instead of just the "as designed" view. This interaction simplifies the resulting manufacturing and support processes.

CE Design occurs when the first three of these processes are executed concurrently with the assistance of the fourth process, *Automated Infrastructure Support*. The *Automated Infrastructure Support* process provides computing capabilities, permitting the other processes to be effective because the CE Design process *manufactures information*. The computing environment has as its main focus the gathering, storage, and retrieval of information. This accounts for the high correlation between computing success and CE Design success.

Interpersonal activity is a key element of CE Design. At the individual level, multifunction design subteam members are part of the overall design team. Subteam representatives are part of the overall product design team, and subteam representatives are part of one or several design for *manufacturability* design or review subteams. At the subteam level of organization, subteams must interact with each other constantly to produce a "signed off design." A heavy emphasis on interpersonal and intrateam communications and coordination results from this arrangement. Multiteam interaction and cooperation at the total product level is also required. This interaction requires electronic proximity because physical proximity across so many different subteams and teams is impossible.

Figure 1-6 depicts representative compositions of the various subteams in a representative complex product environment. In these teaming arrangements, multifunction subteams (performing *Design*) have the design responsibility for the product. Product design engineers, with assistance from systems engineers, coordinate and manage the total product's design and performance characteristics against specifications, using Quality Function Deployment (QFD) or another product requirement development methodology. Product design engineering subteams are responsible for coordinating work across multifunction teams for the CE team chair.

Notice in *Figure 1-6* how the subteams that execute CE Design are matrixed by function and by discipline. For example, a team chair might have subteams of designer engineers, and other subteams of production engineers, and various production manufacturing representatives. These teams also interact with each other as coordinated by the team chair. In the aircraft environment, for example, the multifunction design subteam's focuses include interference analysis or area management, e.g., looking for designed parts that will physically interfere with each other when the product is assembled as currently designed. Other drivers for these subteams are shown in *Figure 1-6*.

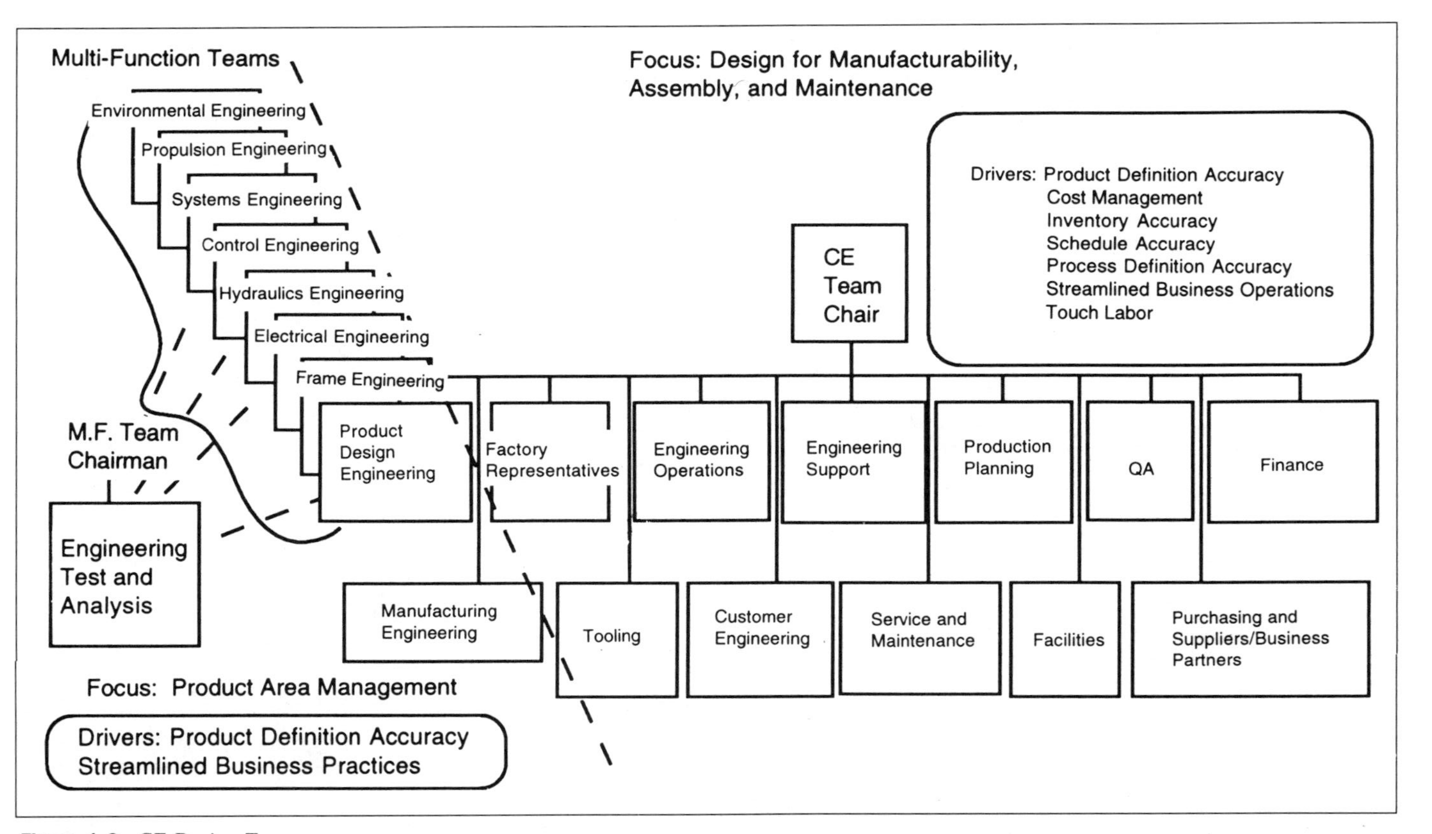

Figure 1-6. *CE Design Teams.*

Along the other dimension of the matrix, the production manufacturing subteam's focuses include manufacturability and cost. These subteams must not only communicate with each other, but with each of the design subteams, and many other CE Design teams and associated subteams across the product design domain.

This coordination and communication quickly overwhelm the teams without a framework for managing these highly interrelated activities. The framework is provided by the three interrelated business, technical, and managerial processes of CE Design (*Process Management*, *Design*, and *Manufacturability*). *Automated Infrastructure Support* provides the distributed, yet integrated information systems to support the rapid, successful execution of these supporting processes. Thus, the need for CE Design with its integrated operations, and this book describing them.

To facilitate the understanding of CE Design, traditional and CE Design should be contrasted. An overview of a traditional manufacturing process is shown in *Figure 1-7*. The process depicted in this figure shows the five major steps through which a product's definition and resulting expression typically pass. These steps are: (1) agreeing to a contract for the product with its associated specifications, (2) design, (3) planning, (4) building, and (5) supporting and maintaining the product. At the intersection of each of these steps, communication must occur.

For complex products, so much product and process planning, and subsequent plan execution information passes across the boundaries between organizational elements, discipline and functions that separate suborganizations have evolved. These suborganizations have developed their own languages, processes, and procedures and identify with themselves first. This inward focus reflects both self-defense against the onslaught of information and complexity, and attempts to maintain control over "their part of the job."

Due to the sequential or serial nature of the overall process and the different views of the same product generated by the suborganizations, significant, unintended errors occur. The cost of these errors grows geometrically as they are discovered later in the process.

Other significant deficiencies of the traditional manufacturing process include resistance to change through specialization, resistance to increased speed because of serial processes and interface between suborganization requirements, and concerns about configuration management and product and process robustness as a result of complexity. As indicated in *Figure 1-2*, one of the three summary "drivers" toward CE Design is the deficiencies of the traditional manufacturing process. These deficiencies are creating a "process pull" for an improvement.

New manufacturing, computer hardware, software, and a host of other *technological* improvements and innovations, sometimes summarized under Computer-Integrated Manufacturing (CIM), Simultaneous Engineering, Concurrent Engineering, Product Data Management (PDM), and design framework or CAD/CAM Data Management labels are all creating the "technology push"

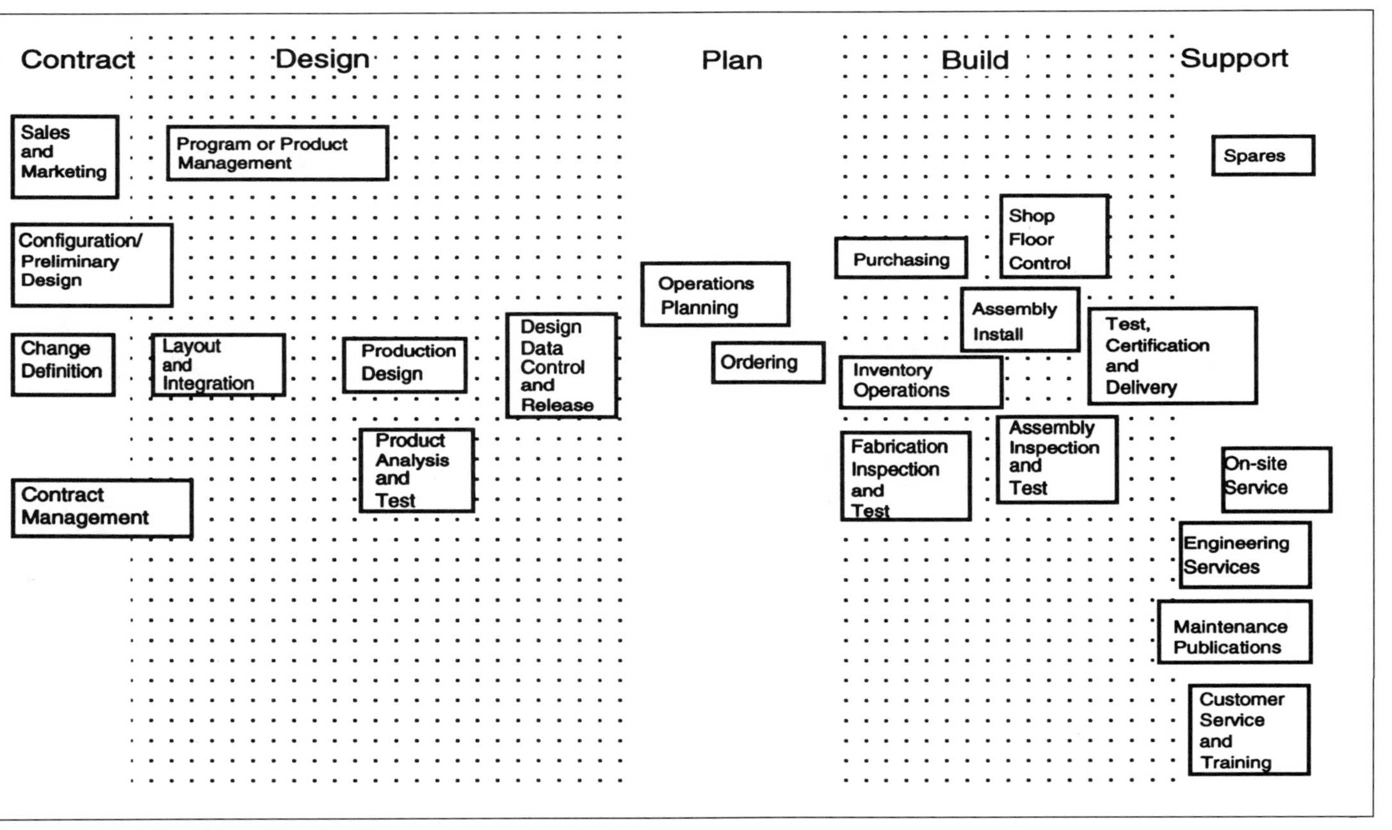

Figure 1-7. *Traditional Manufacturing Process.*

for the adoption of new manufacturing processes to take advantage of these technological improvements.

The best approach in dealing with process pull, technology push, and competitive pressures in combination appears to be adopting "world class manufacturing." This approach may create a permanent competitive advantage for the early adopters. It now appears there are six key initiatives to becoming a "world class manufacturer" and thus capable of dealing with these issues. One of these six, and *the enabler for all of them,* is (1) Concurrent Engineering Design. The other five key initiatives, not in any particular order, are: (2) Quality, (3) Cost Management, (4) Time-based competition, (5) Technology, and (6) Variety and Complexity.

CE Design driven manufacturing reduces the entire manufacturing process to three stages. These stages are *prebuild*, *build*, and *support*. *Figure 1-8A* shows the *prebuild* stage. The CE Design process is part of a simplified and reordered engineering-driven *prebuild* process that retains and encourages specialization,

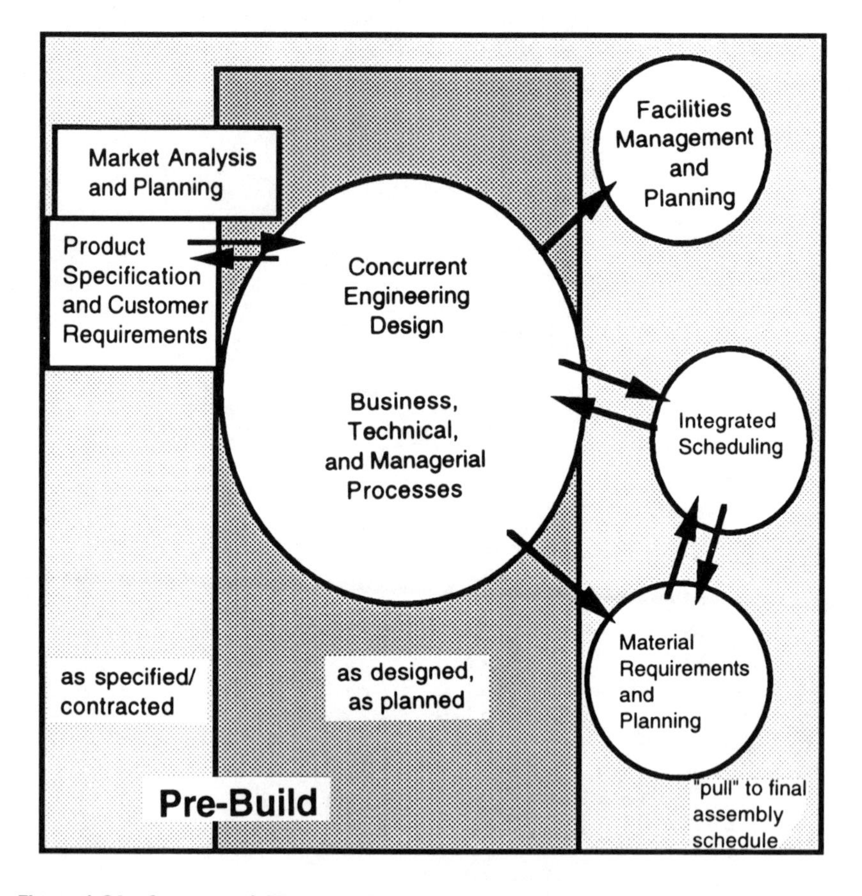

Figure 1-8A. *Overview of CE Design Driven Manufacturing.*

but within an inherent, communications-based environment promoting constant interaction between disciplines. The interfaces between *prebuild* and the *build* stages are reduced but still include a direct ''hand-off'' from CE Design to manufacturing's planning and factory management's shop floor control.

As CE Design feeds its information products to the *build* process, additional activities occur. These activities are depicted in *Figure 1-8B*. The *build* activities are now focused more on the actual build process, and less on getting ready for the *build* process as CE Design and *prebuild* do a more complete job of preparing for manufacturing. Notice that MRP, or Material Requirements Planning, is focused on the proper staging of components and assemblies, not on more ''enterprise'' oriented integration. Also, notice how important facilities and tooling have become. These are high capital and perhaps ''pacing items'' for the production and manufacturing environment.

When the post-manufacturing processes of *support* becomes involved, the final overall manufacturing process built around CE Design is complete. *Figure 1-8C* depicts the fully involved overall manufacturing process. During the *support* process, distribution, after delivery service, and parts management occur.

This shortened, overall manufacturing process can result in an order of magnitude reduction in end-to-end product production cycle time. This is very impressive. An order of magnitude reduction in the automobile business would reduce new car programs from an average of 60 months to six months. In the electronics business, this would reduce new product development from three years to four months.

The CE Design process is composed of the four CE Design component processes previously described. Each of these four processes are themselves composed of key business, managerial and technical processes. These key subprocesses, and other major subordinate elements, can be gathered into the four integrated processes as follows: CE Process Management, CE Design Business Process, Manufacturability, and Automated Infrastructure Support.

Process Management

This top level business process has four integrated, interrelated management processes. They are *Configuration Management, Routing and Queuing, Resource Management*; and *Release and Distribution. Figure 1-9* shows the CE Process management portion of CE Design Functional Processes.

1. *Configuration Management* continuously manages change control processes and multitudes of product configurations, (as differentiated by effectivity), as they move through CE Design, and other overall manufacturing processes. Configuration Management's engineering business operations may include change boards, engineering, manufacturing engineering, quality control programs, tooling, material, manufacturing, and industrial engineering.

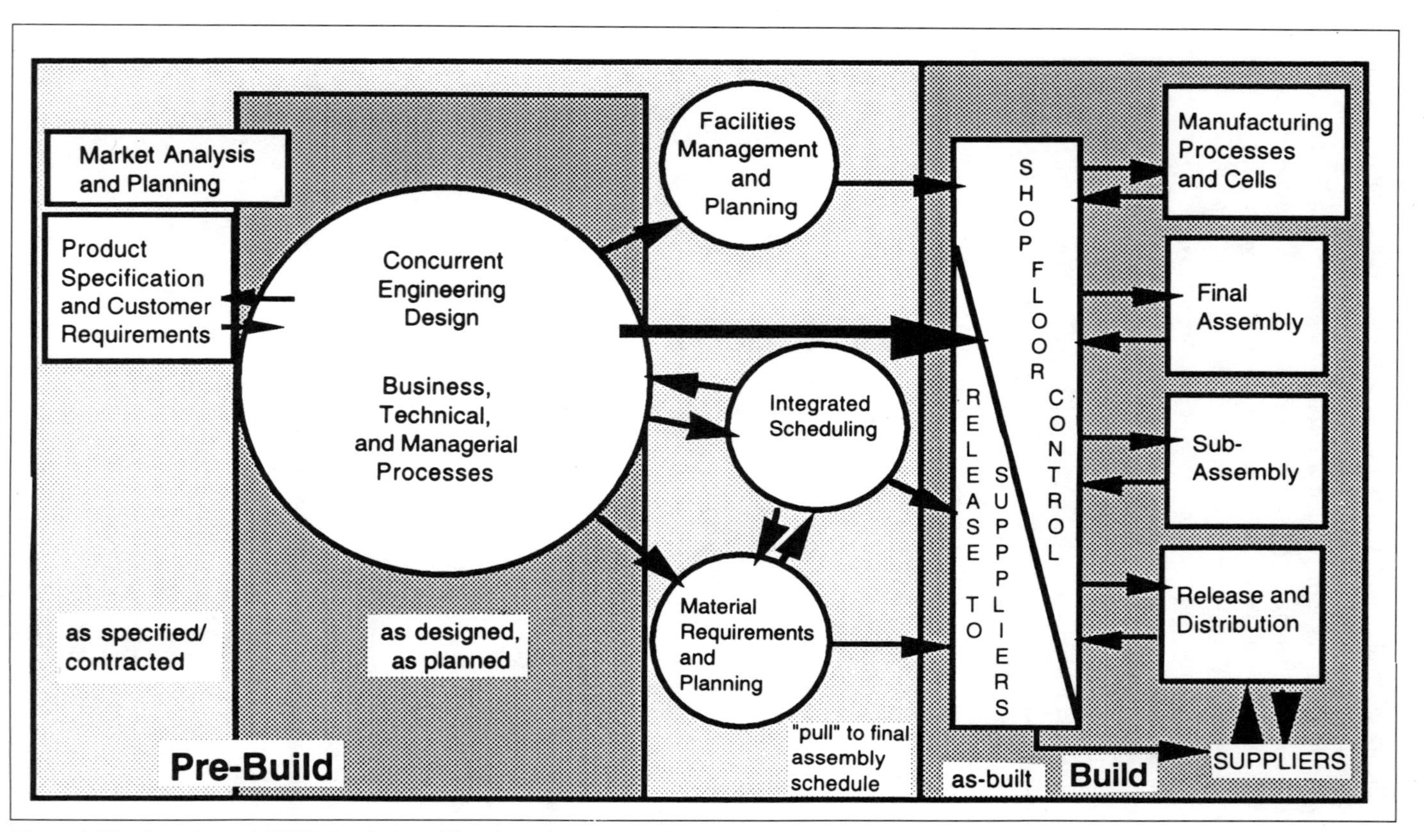

Figure 1-8B. *Overview of CE Design Driven Manufacturing.*

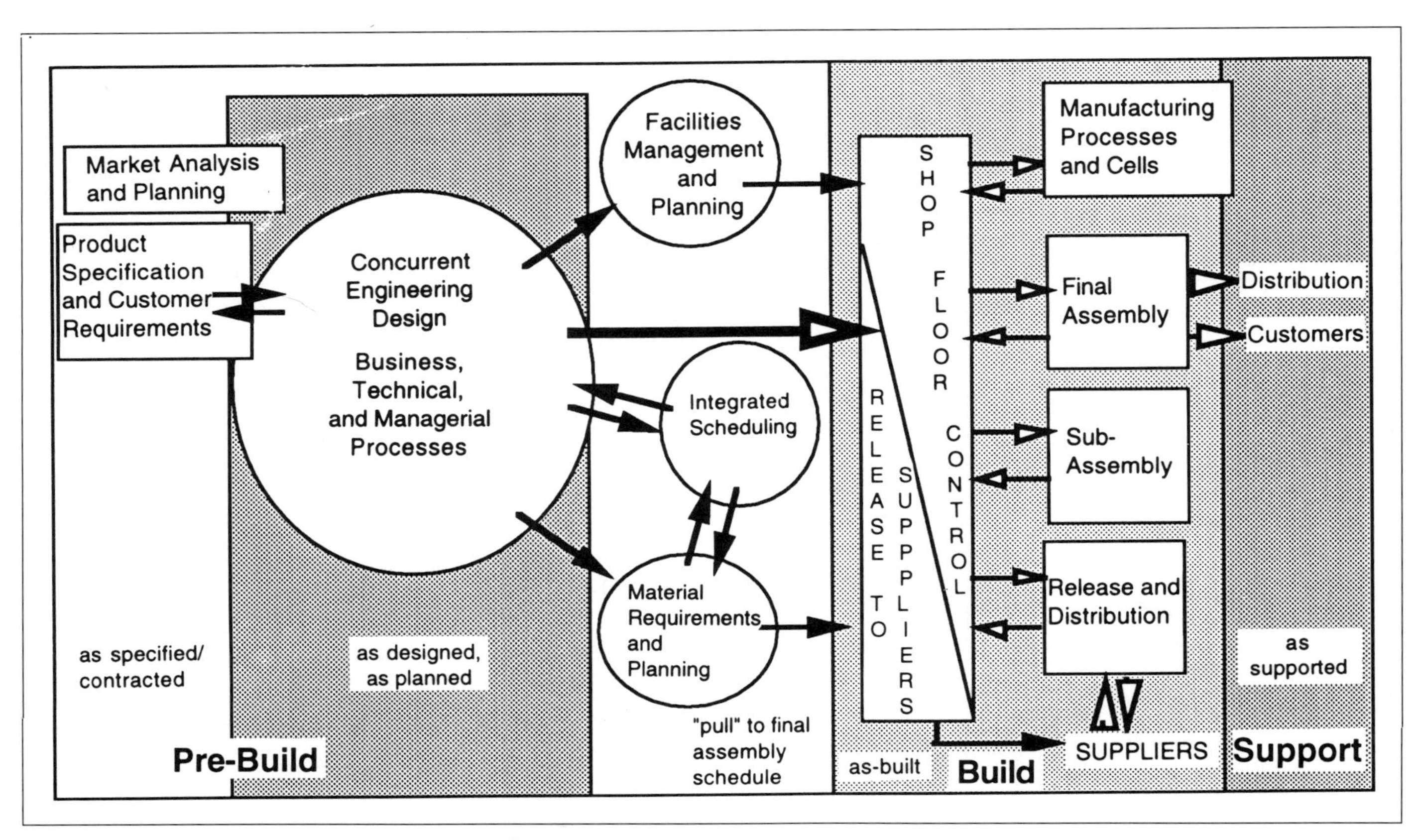

Figure 1-8C. *Overview of CE Design Driven Manufacturing.*

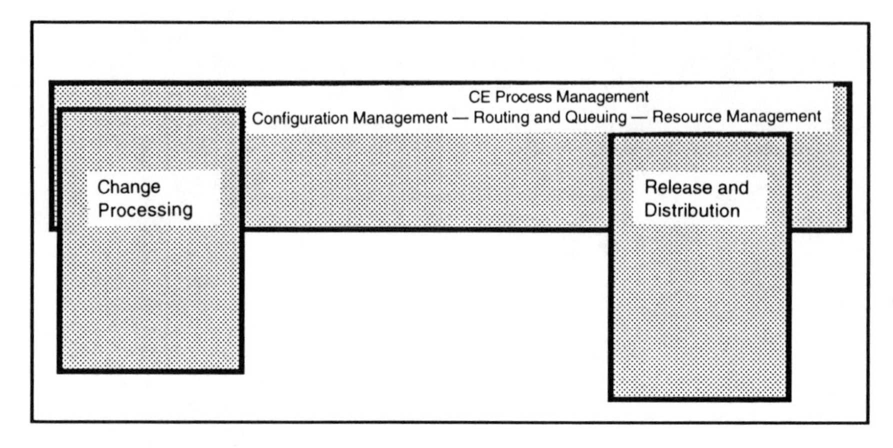

Figure 1-9. *CE Design Functional Processes.*

2. *Routing and Queuing* permits structuring many discipline and management reviews necessary to quickly accomplish the engineering, design, manufacturability, assembly, and maintenance objectives of the CE Design process. This element reflects the dynamic, unstructured nature of the work activities within the CE Design environment. Status Analysis, Authority Management, and apparent electronic proximity between team members are also included or facilitated.
3. *Resource Management* includes traditional program management functions like personnel and infrastructures (plant and equipment) management, cost management, pricing, and budgeting. By coordinating with Routing and Queuing, status reporting and program management functions are also enabled.
4. *Release and Distribution* controls the assembly and appropriate dissemination of the information captured during the CE Design process. This information includes the individual CE Design process itself, (stored as a process model, as executed to support each product design), or design changes. This element permits the orderly accumulation of knowledge and the capitalization of the organization's intellectual property.

Design Process. This second Business Process contains three elements supporting the development of product definition. Each of these three elements is driven and managed within Process Management. These elements include: *Work Statement, Drawing and Geometry Management,* and *Test and Analysis.* The CE Design Business Management category is shown interfacing with the CE Process Management category in *Figure 1-10.*

1. *Work Statements* support the documentation of the goals, objectives, and functional requirements of products and processes. Directly tied to the CE Process Management category; it includes customer engineering, project

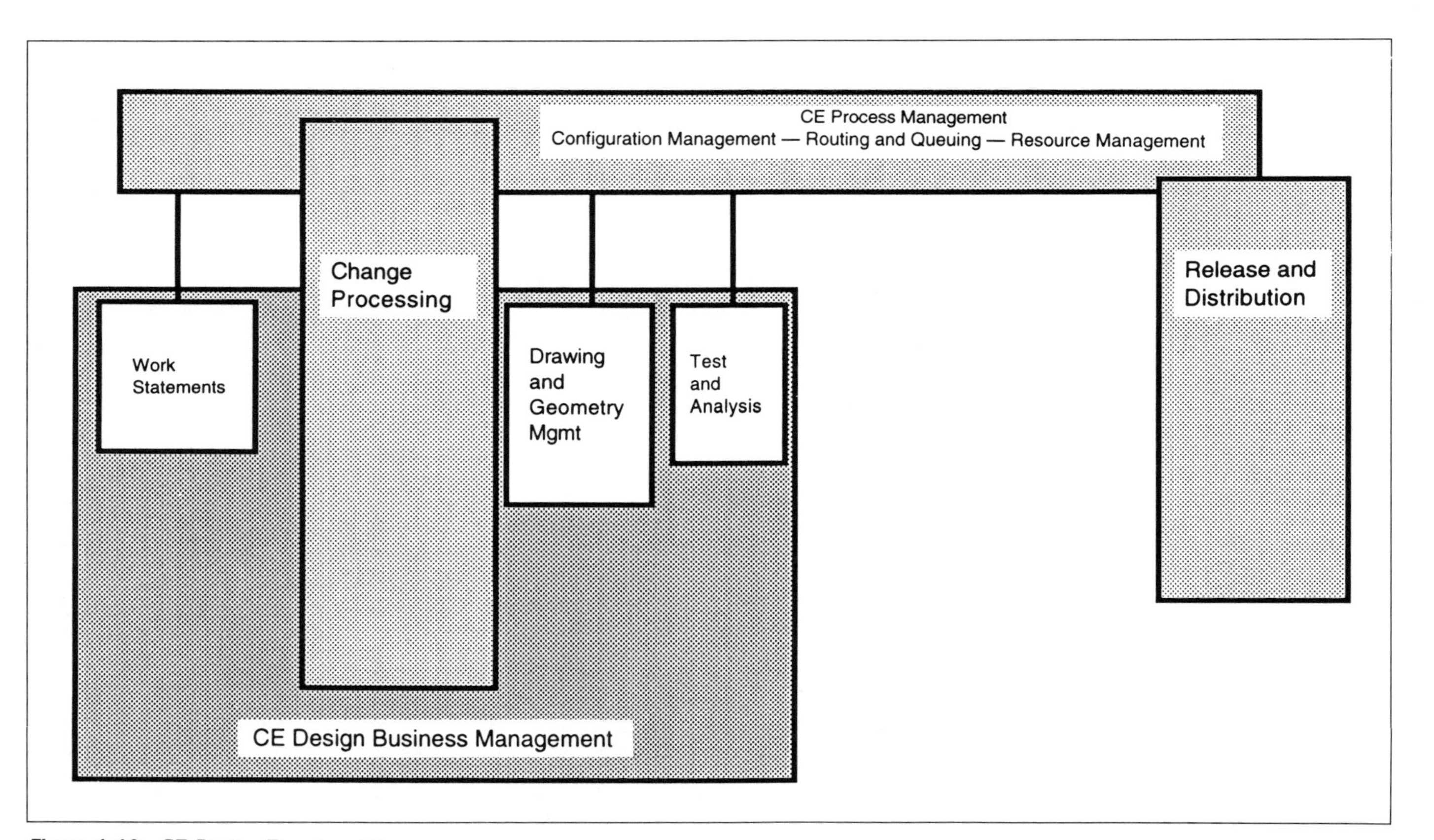

Figure 1-10. *CE Design Functional Processes.*

engineering, industrial engineering, finance, and product management.

2. *Drawing and Geometry Management* includes the business and computer management systems and processes that support and facilitate development and integration of drawings. These drawings may be two-dimensional, three-dimensional, or CAD pictorials and models. This element is also called Product Definition Management (PDM). This element includes product definition, manufacturing engineering, tooling, design, engineers, numerical control programmers, quality control, and finance.
3. *Test and Analysis* includes engineering analysis, simulation, physical and electronic paste-up, mock-up analysis, and other testing, analysis and engineering support activities. Certification and quality control would be a part of this particular element.

Manufacturability. This third business process has two elements which produce the information artifacts of the process, in addition to the Geometry and Drawings, needed by the rest of the organization to produce the product. These elements include: *Bill of Material and Product Specifications and Process, Facilities*, and *Tooling Specifications*. The two elements of Manufacturability are introduced into the CE Design Functional Process diagram in *Figure 1-11*.

1. *BOM and Product Specifications* supports the capture and management of product-related information which includes the Bills of Material (BOM) and the balance of product structure specifications. These specifications include material performance, key features that support product design robustness, standard parts, and other notes and specifications.
2. *Process, Facilities, and Tooling Specifications* formally includes the often overlooked other aspects of product and process planning. In the complex manufacturing environment, the cost, complexity, and change rate in facilities (buildings and fixtures), the process itself (chemicals, equipment, manufacturing cells, and the like), in tooling (jigs, measurement equipment, and the like), and in environmental considerations (waste, co-products, byproducts, consumables, emissions, etc.) can be greater than the manufacturing product.

Automated Infrastructure Support. The previously discussed three categories are supported by Automated Infrastructure Support. This fourth category includes important subelement capabilities such as: *Communication Management; Storage and Retrieval; Digital Mockup and Modeling; Analysis Tools*, and *Statistical Process Control (SPC)*. The addition of these final elements are shown in *Figure 1-12*.

1. *Communication Management* links the various resources. These resources include company staff, computers, equipment, and suppliers.
2. *Storage and Retrieval* manages the archiving, storage, and retrieval of the product and process information, and other work products developed during the CE Design process.
3. *Digital Mock-up and Modeling* supports the pre- and post-digital electronic assembly of the geometric aspects of the product and process

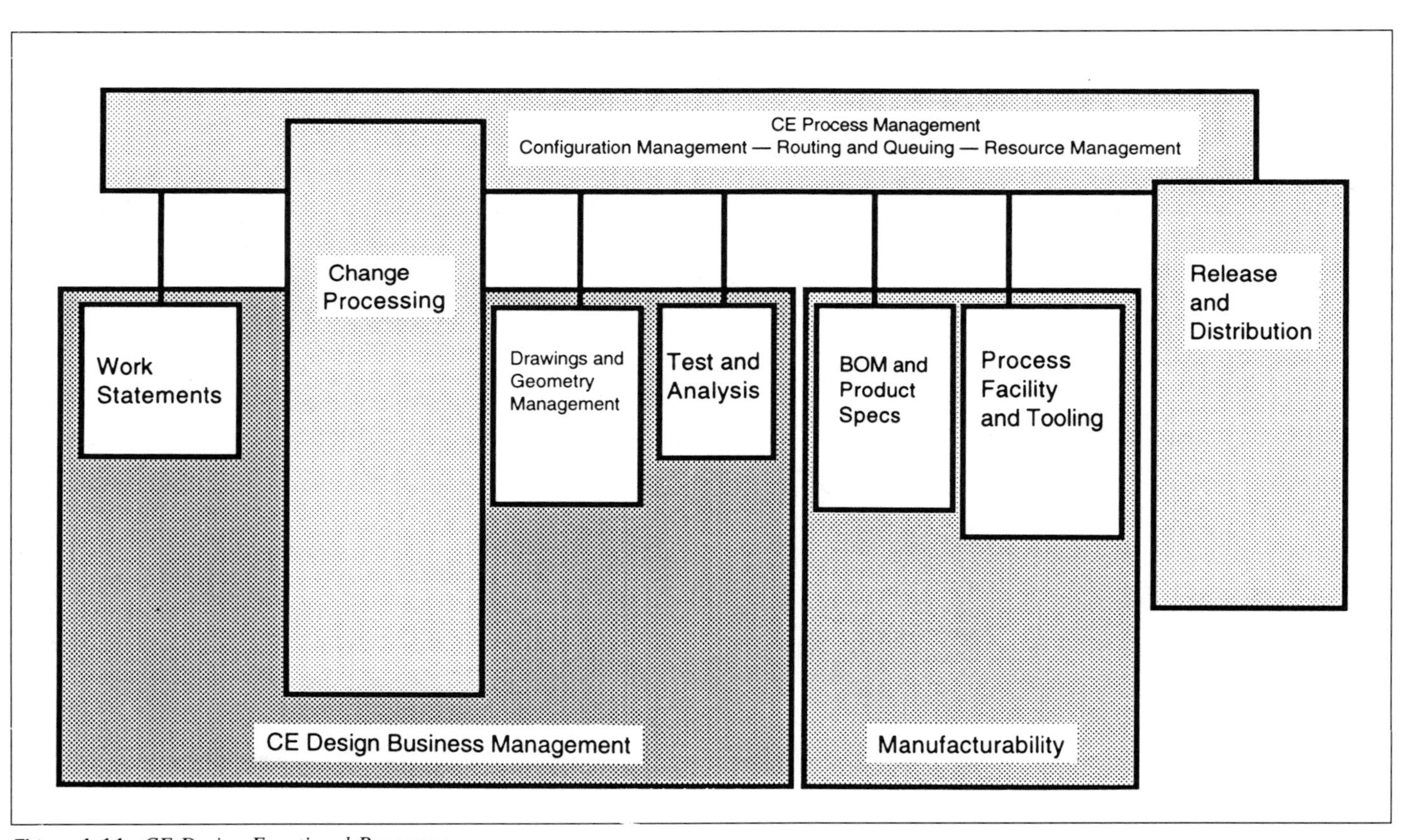

Figure 1-11. *CE Design Functional Processes.*

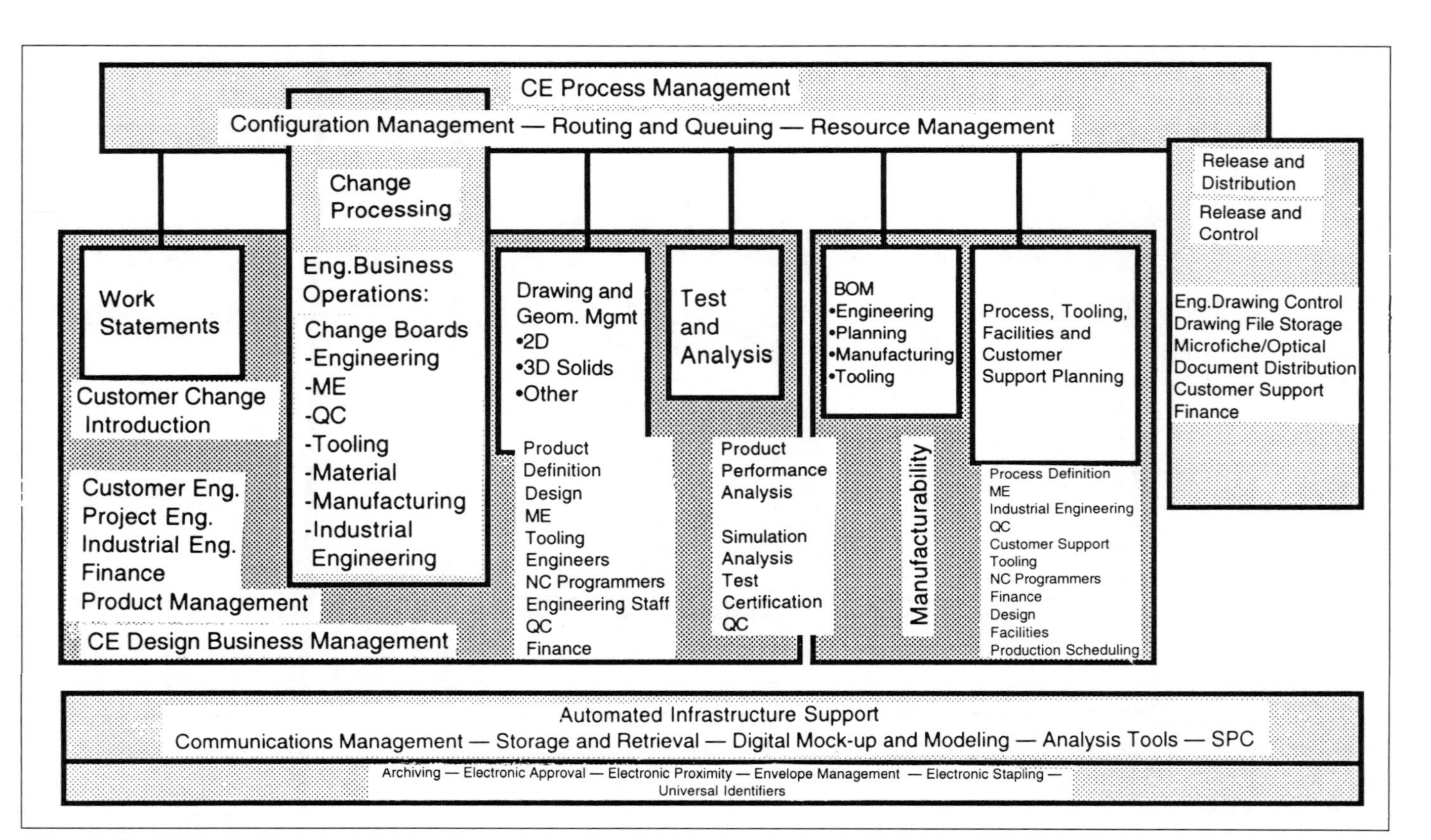

Figure 1-12. *CE Design Functional Processes.*

definitions, including fixtures, tools, jigs, product and process simulations, and other product support items (test kits, support equipment, and probes, for example) as well.

4. *Analysis Tools* manages the computerized and other engineering and manufacturing analysis tools used in the CE Design process. These tools manage the linkages between analysis and resulting product process, and product support item specifications and designs.
5. *Statistical Process Control (SPC)* has proven of great assistance in managing the support product processes and manufacturing support design robustness, the basis for quality.

All the elements and processes in this book are inseparable. While they will be discussed separately in this book, to build understanding, their integrated operating nature also will be discussed because CE Design's power comes from its *integrated* operation.

Because each organization will implement CE Design uniquely, no attempt is made to describe the one and only "how" of CE Design. Since CE Design is at the heart of an organization's future competitive position, differentiation will be important. Also, the Automated Information Support technology continues to change rapidly, and very specific architectures will need to be evaluated over time.

In general, the implementation of CE Design should be forward planning based. That is, it should be a constant, planned evolution toward a final vision, not a statement of an idealized "to be" view with a series of backward steps from it, because this "to be" view will continue to change as well. By evolving toward the "vision" and functionality encompassed by this book, and by evolving in a process "pull" manner, CE Design can be advanced quickly with less disruption and greater chance of success. Expect constant organizational stress as the process of implementation proceeds. The suggested implementation plan contained therein considers the forward planning perspective, and addresses the impact of change.

2

WHY CONCURRENT ENGINEERING DESIGN

The book *Kaizen*, by Imano, describes one way the Japanese adopted quality management and process management techniques introduced to them by Juran and Deming. Included in the book is a guide to process problem analysis and reconciliation. The general steps of that problem analysis process include:

1. Situation description and Cause and Effect Analysis (the "fishbone" charts),
2. Problem resolution analysis (QFD),
3. Near and long-term initiatives, and
4. Cost, benefit and privatization ("Pareto" charting).

These general steps will guide the discussion of "Why CE Design?"

SITUATION DESCRIPTION AND CAUSE AND EFFECT ANALYSIS

A variety of factors when analyzed, support the perception that the current engineering and design process is unacceptable. *Figure 2-1* displays a cause and effect of the "fishbone chart" which depicts the relationships of these factors. These factors are considered throughout CE Design's definition and approach. Chapter 1 provided an introduction to "what CE Design is." Chapters 5-8 describe CE Design's four major subprocesses in detail. The Cause and Effect Analysis description in *Figure 2-1* describes the environment that generated the requirements for CE Design, or "what CE Design is not." The "root causes" associated with the "problems" of the current design process include the larger typeface issues in *Figure 2-1*. A careful review of this diagram is important because management's response to these issues should be CE Design.

Function focused work patterns traditionally divided the complex product manufacturing enterprise and drive the current management framework. *Figure 2-2* depicts an example of the well-known "stovepipe," function oriented, management hierarchy. Engineering typically thinks "top-down" as it takes a complex product and "decomposes" it into smaller, more understandable

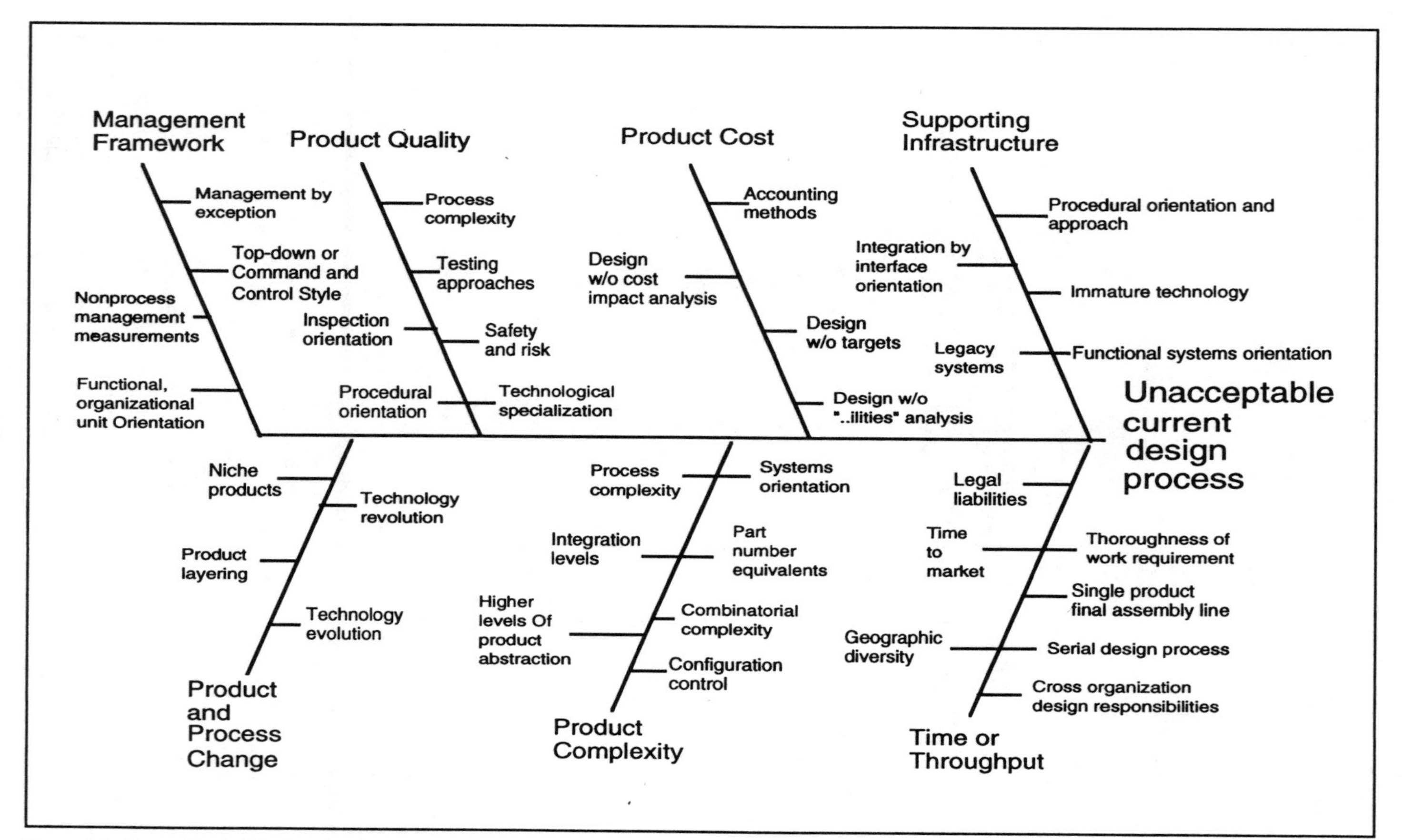

Figure 2-1. *Current Design Engineering Process Analysis.*

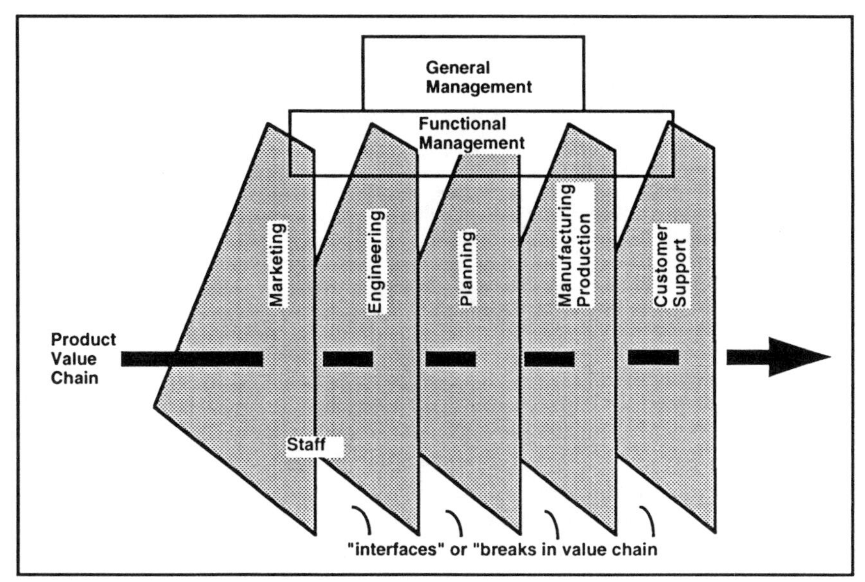

Figure 2-2. *Functional Organizations.*

sections, and finally into individual details. Engineering's work patterns have been dominated recently by the demanding technical requirements of executing product design using various functionally oriented tools. The CAD functions and computer-based "product performance characteristics testing" represent such demanding functions for engineering. Manufacturing Production has been dominated by business issues such as material ordering, staging, supplies coordination, and production planning. The very high discipline requirements of manufacturing resource planning functions represent such demanding functions for manufacturing production. This high discipline requirement reflects the "bottom-up" or detail to assembly to final product thought process of Manufacturing Production.

Engineering and Manufacturing Production also are different in their dominating characteristics. Engineering is concerned with *unstructured* work and must focus on *work flow coordination*. Manufacturing Production needs *structure*, *precision*, and *certainty*. It needs all three as it deals with producing the final *physical* product to a schedule of customers using various *flow control* techniques.

Management activities are what have held the organization together. These management activities are usually carried out at the middle management level. These management activities were largely devoted to negotiation, reconciliation, error detection and correction, using an exception management-oriented technique.

Leadership and decision making are still important requirements of management. But the top-down style of implementing leadership and decision making has become part of the problem. The products are so complex and change so rapidly that managers can become ''dangerous'' if they use knowledge obtained while they were in lower positions to direct new activities. Management's basic ''style'' must change to process-oriented ''management.''

Products have become more complex since the beginning of the industrial revolution, as machines replaced human labor. Since the introduction of the computer in the 1950s, this complexity accelerated as computers provided an additional level of abstraction and the ability to rapidly magnify human capability and potential.

This concept of separating the physical aspects of the product from the management of these aspects, by introducing computing into the product, is a *major* development in manufacturing. By abstracting control, greater flexibility and perhaps two orders of complexity are introduced. This leverage has created ''product improvements'' of many types. The results of these improvements have manifested themselves in many ways.

The results include such diverse perspectives as computer circuits, number of people on Earth, number of chemical compounds used, and amount and number of infrastructures (roads, pipelines, telephones, circuits, satellites, etc.). These reflections of increasing complexity are reaching the point where no one person or group of people can understand the details. The accumulation of knowledge (organized and useful information) has become the main competitive weapon in business. Many authors, such as Alvin Toffler in *Third Wave*, and now *Power Shift*, have written and commented on these issues and their increasing impact.

As manufactured products become more complex in theory requirements, the discipline to manufacture in reasonable time and with good quality also increases. This problem is affecting all manufacturing, but is especially acute in areas of *complex products*. Complex products have the characteristics of:

1. manufacturing operations with high rates of product and process definition change (*change rate*); or
2. manufacturing operations with high rates and short cycles for new product introductions (*speed and/or cycle time*); or
3. manufacturing operations with complex configurations which can vary by customer, and could be produced in a single, individual production line (*complexity*), or
4. manufacturing operations that require multiple design teams for a single product, and typically include business partners with design responsibility for product parts and suppliers with no design responsibility (*multiple teams*).

Some examples of complex products include communication systems (location, voice, data, image, text, and multimedia, for example), commercial airplanes, large software systems, the human genome (the entire human genetic sequence) database, and power generation and transmission systems. It is the effect of the combination of these characteristics, along with the need for the re-emphasis on manufacturability, that has created these trends. Some of the side effects of these trends have been *substituting process for results, designing systems instead of physical products*, and *complexity*.

There has been a typical response to the need for maintaining discipline and configuration control in the face of increasing complexity, size, interoperability, integration within its environment, and the other requirements of these complex products. This response has been to add many different types of steps to the design process. Quality and control were the objectives. Each step's completion was checked carefully. Product changes of even a very simple and overlapping nature were done serially, and with controls, to assure control.

The various responses to complexity created task specialists. These task specialists (engineers and designers, for example) perform a small step in the overall design process. They do not have responsibility for the products of the design process. Of course, they are not required by their job description or sense of responsibility to have any interest in or responsibility for the individual end-product itself. They must *substitute process for results*.

While this seems an indictment of manufacturing organization people, it is not. Most are constantly concerned about this issue. These are not uncaring people or people who have no interest in their product or the organization. Rather, the organizational concept prevents people from having their job description, their procedures, or their organizational unit related to results.

The second trend is designing systems that have physical products instead of products with embedded systems. This trend is a reflection of the continuing evolution of the computer. With its increasing power in smaller packages, the computer and the associated product's systems control various elements of a wide variety of product characteristics and features. These computer-based control devices are "embedded" into other physical products to provide control processes. They are also integrated in new types of products called *systems*. These systems are a new type of product. They are unique because they often have little or no physical manifestation (accounts receivable software package). Sometimes they provide very elaborate control and management functions for physical products (an on-board fuel injection computer for an automobile). Sometimes, in combination with physical elements, they provide the capabilities needed to create other new classes of products that could not operate without these systems (e.g. cruise missiles). These systems evolved to the point where we are designing product systems that utilize physical carriers to accomplish system objectives (genetically engineered instructions; monoclonal antibodies; satellites for communicating and location determining, or advanced attack aircraft). We are now *designing systems with physical manifestations and characteristics instead of physical products with ancillary or embedded controls*.

The third trend, *complexity*, is the result of the impact of the other two trends. Product and Process have become very complex and complexity is increasing. Complexity has become the limiting factor in the continuing development of more capable, flexible, and usefully performing products (systems) and the processes that produce them.

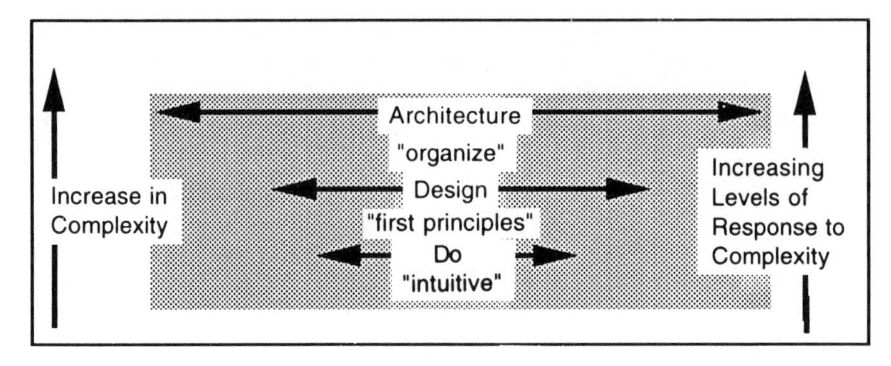

Figure 2-3. *Evolutionary Response to Product, Process and Organizational Complexity.*

As shown in *Figure 2-3*, when the product was simple, or when trial and error was an acceptable approach to the manufacture of a product, one could simply "DO." For many of the products we are now designing, such as airplanes, automobiles, and medical equipment, the "trial and error" design and manufacture process is unacceptable. One or two design-induced crashes, and/or the loss of human life probably would "kill" the program or product because the success of our engineering efforts has created a very high expectation market. This high expectation market now expects close to 100% product success. Any product environment perceived as having serious problems and/or lack of complete success from its beginning, can quickly be deemed a "total failure."

As the situation became more complex, we added a formal "Design" step. This step has evolved into the complex and lengthy process we now experience. During this design process lengthening, the individual had to focus on the *process*, since responsibility for *results* was removed from most individual's job description. As a result, the response of the organization has become part of the problem. This "change to the design process" becoming part of the problem has been recognized. Unfortunately, the current response to the engineering and design "problem" has become part of the problem.

Examples of current response that have created problems include increased emphasis on the thoroughness of work and the transference of procedural control orientation. As a response to the concerns about complex products, the time to do a complete and thorough job at each step in the serial, step-wise refinement, design process, has created a rapidly *increasing* product time-to-market rate. The reason why this thoroughness issue is considered is due to the work of engineers. This work has become highly proceduralized, in part, due to using computing to perform highly specialized computing based functions, such as finite element analysis.

Accounting systems were the first highly successful business management systems using computers. Accounting must be a highly proceduralized activity because the process handles money. The basic intent of the procedures deliberately built into these systems is to prevent error. Computing professionals then transferred this same highly proceduralized approach, with the best of intentions, to the approach to be used *throughout* the business because it was so successful initially.

Unfortunately, engineering and design, especially in a change environment, *are not* highly proceduralized. Every execution of the design process is somewhat different, so the serial process must include a step for every contingency. Thus, much work is potentially unnecessary. Legal liabilities, process orientation, thoroughness requirements, with established procedures, however, seem to force a pattern of procedural compliance. Changing management's attitudes about these issues, their responsibilities, and their "risk" is an important element of CE Design's response.

PROBLEM RESOLUTION ANALYSIS

Most manufacturing organizations recognize the need for change. The problem is what to do, and how to go about changing. An almost universal response to problems is to make gradual change. This is perceived as low risk, and it does not create organizational disruptions. Unfortunately, gradual change alone cannot resolve the highly interrelated problems represented in *Figure 2-1*.

A set of changes that will reset the manufacturing organization's competitive position to resolve, or ease perceived problems is needed. Almost immediately, the experienced person reading this material will think about risk: risk of initiative failure, risk of business failure, and personal, job related risk. Historically, organizations which achieve such basic change, and make it last through the "crises" which enabled change to take place are small in number. Any strategy or near and long-term initiative must directly address risk for the initiative to be even seriously considered.

Figure 2-4 describes the relative benefits of change, and depicts the two basic strategies for CE Design as described. Improvements at the activity or function level have very small returns. Of course, many such improvements, continually pursued, can add up to big impacts. In the complex product manufacturing organization, changes at the entire process level can have better returns. Cross-process and organization-wide changes can have awesome impacts. The CE Design near and long strategies to achieve organizational or paradigm shift benefits combine low risk and moderate risk by employing simultaneous change. These strategies are Continuous Quality Improvement (CQI) and Process Restructuring. CQI is the low risk approach. CE Design is the initial starting point for process restructuring. Its risk is substantially reduced to a controllable level if CQI is already underway.

CQI is known by many labels, such as Total Quality Management and World Class Manufacturing, to name two. There are many articles and books on the subject, of which *Kaizen* is one. CQI is an important counterbalance to CE Design. It is also a paradigm shift strategy. It creates a whole new style of management. A manager who "directs" is now considered a "problem" manager. It re-involves the entire work force in product quality. It enables the organization to institutionalize the concept of constant change under the title of

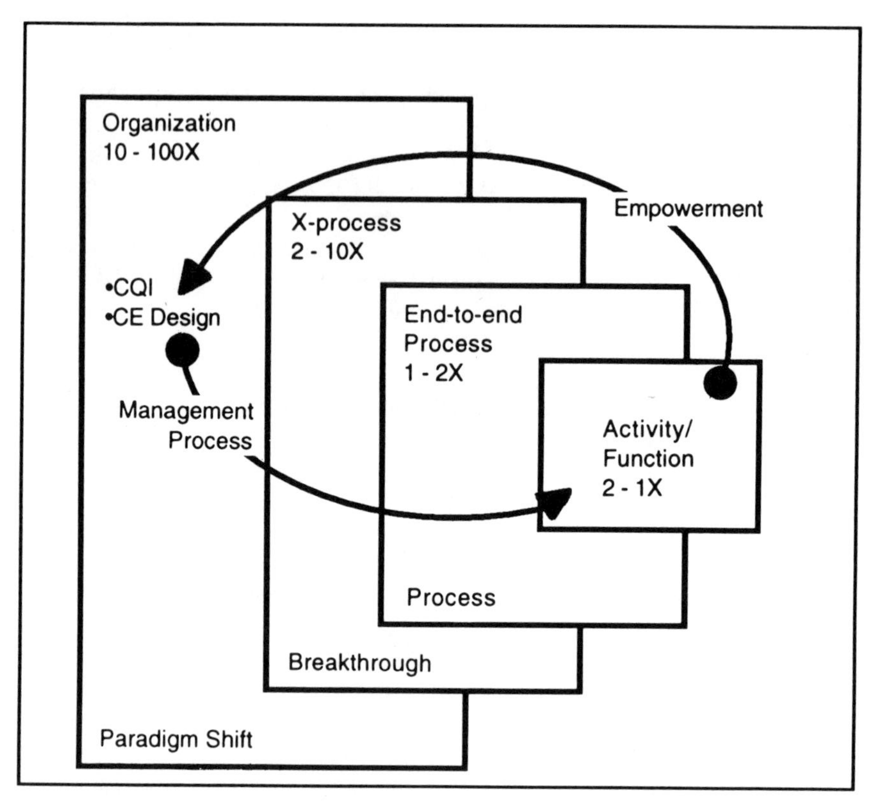

Figure 2-4. *Opportunity Reach.*

improvement. The rallying cry of CQI is "If it ain't broke, then improve it."

CQI creates an overall organizational environment in which re-involving production manufacturing in design seems "natural," and changing away from highly proceduralized design seems "empowering." It is also a low risk, gradualist approach to change. Yet big impacts are possible. Organizations who have accepted CQI and applied it constantly over several years can show startling results. There are many well-known Japanese and American CQI success stories, including Ford and Motorola.

However, these same organizations always find that CQI is not enough. The CQI strategy, when used by itself, appears to work best in a comparatively mature industry, such as conventional automobiles. The product is somewhat complex, but still basically the same, and the strategies of mass manufacturing are still dominant. The competitive strategies as relating to this issue are discussed in Chapter 3. It is important to note from an analysis perspective, that this process improvement strategy, focusing on improving what is already in place, is not enough to *keep* complex product market share.

The processes of the organization also must change and be restructured. *This process restructuring is management's responsibility*. CQI is all about management ''trusting'' its work force. Process restructuring is all about the work force entrusting management with the responsibility of positioning the organization for competitiveness. In the restructured, re-engineered organization, management has two principle responsibilities:

1. *removing obstacles to success*-management is responsible for positioning the organization to succeed via the right people, processes, tools, and training; and
2. *changing how management is conducted*-undertaking a more direct involvement in business and technical process management; i.e., measuring and managing the processes and their cross-process activities for which they are responsible, while supporting cross-process initiatives and other senior management decisions and policies.

CE Design is a cross-process management initiative, done in conjunction with a CQI implementation that precedes it by a period of time appropriate for each organization. CE Design is implemented within a CQI type strategy, and with a CQI style and approach.

CE DESIGN'S ROOT CAUSES PARETO

This book focuses on CE Design as a management process restructuring initiative. This initiative must be organized to deal with the root causes, and in the proper sequence. This initiative's priorities must be based on the frequency and impact of the root causes. *Figure 2-5* is an adaptation of the Pareto analyses technique to the relative ranking of the ''causes'' first depicted in *Figure 2-1*. Notice that the Current Management framework ''cause'' is most important. The most powerful benefits of CE Design are management oriented. Notice that product quality, throughput, and cost are ranked lower, not because they are unimportant, but because the implementation of the proper process framework results in these issues being mitigated already. But what are the characteristics or dimensions of CE Design that address these root causes?

Concurrent Engineering Design, because it is a cross-process restructuring initiative, takes an *integrated* view of the entire manufacturing process. This new design process includes activities throughout the organization. This integrated view is very powerful.

Because CE Design is a *management* response to a set of underlying issues, it also resets the organization's competitive posture. Understanding why CE Design is necessary, and how it resets the organization's role from a competitive repositioning perspective is the subject of Chapter 3.

Intellectual Capitalization

The first of these ''dimensions'' is *intellectual capitalization*. In the manufacturing organizations characterized by the four characteristics enumerated in Chapter 1 and earlier in Chapter 2 (change rates, speed, complexity, and multiple design teams), one could argue that the products produced by the manufacturing organization with the most value are not those sold and delivered

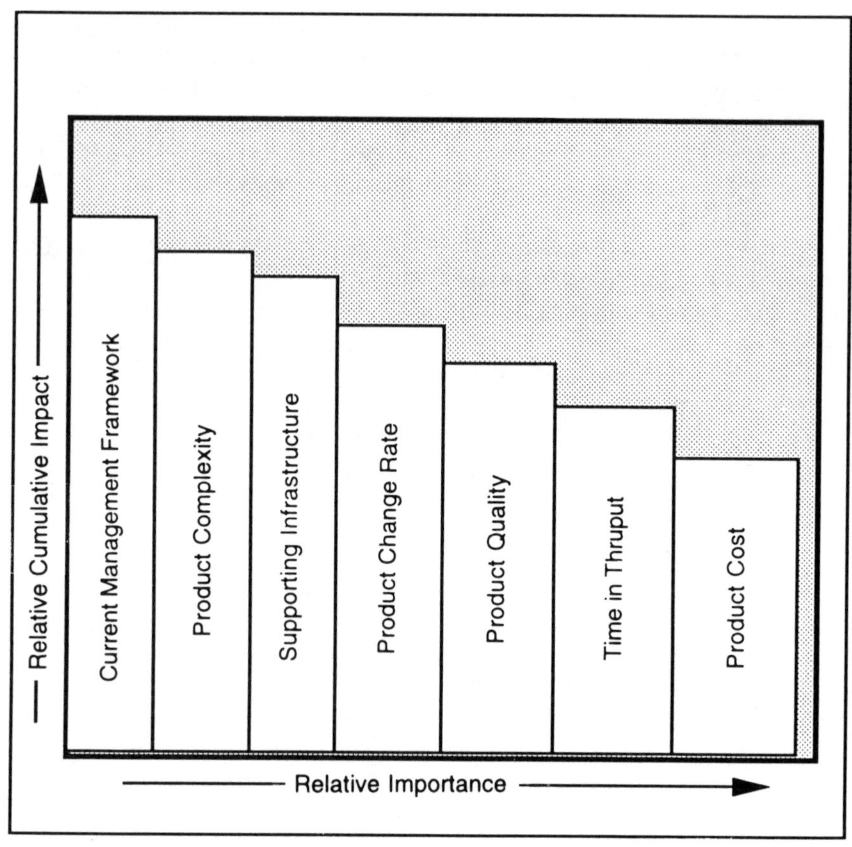

Figure 2-5. *CE Design's Root Cause Pareto.*

to the customer. Instead *it is the product and process definitions that describe the delivered product, and the organization itself (people, processes, and capitalized knowledge) which have the most value*. They create a significant barrier to market entry for competitors. They can produce similar products, and many others, with high quality, reliability of delivery, and reasonable cost.

The intent of Concurrent Engineering, as an integral element of its enabling systems and processes, is to capture, store, protect, and manage the information that includes the product, process and organization. It is important to note that this is information and *not* data. Data itself is not as useful. Information is highly reusable. More will be discussed about information and data in Chapter 8. CE Design's emphasis on information results in a rapid accumulation and capitalization of information and the many decisions and the rationale associated with the production of the product, process, and organization if stored and used correctly. This collected storehouse of intellectual capitalized knowledge is valuable, reusable, and leverageable intellectual property.

Product Cost Reduction

The second of these dimensions is *Product Cost Reduction*. As shown in *Figure 2-6*, many studies and projects have explored this issue. All of these studies seem to have reached the conclusion that most of a product's life cycle cost is fixed early in its life cycle, before the original design cycle is complete.

Why does this occur? Apparently, some of the most basic product design and function concepts embodied in the earliest product descriptions begin this "fixing" process. For example, a ball point pen. A pen that must flow ink to the small rolling ball sufficiently to enable it to write "upside down" probably is more costly to produce than one that operates by gravity feed, and flows ink to the ball only in an "upright" manner. This means higher cost, a different market, a different manufacturing process, and other differences. In more complex products, there are many function choices, shapes, different manufacturing processes, material choices, tolerance, and other design choices being made as the product's definition proceeds. The design engineer has many opportunities to make improvements in the product, or make unfortunate suboptimal product design decisions.

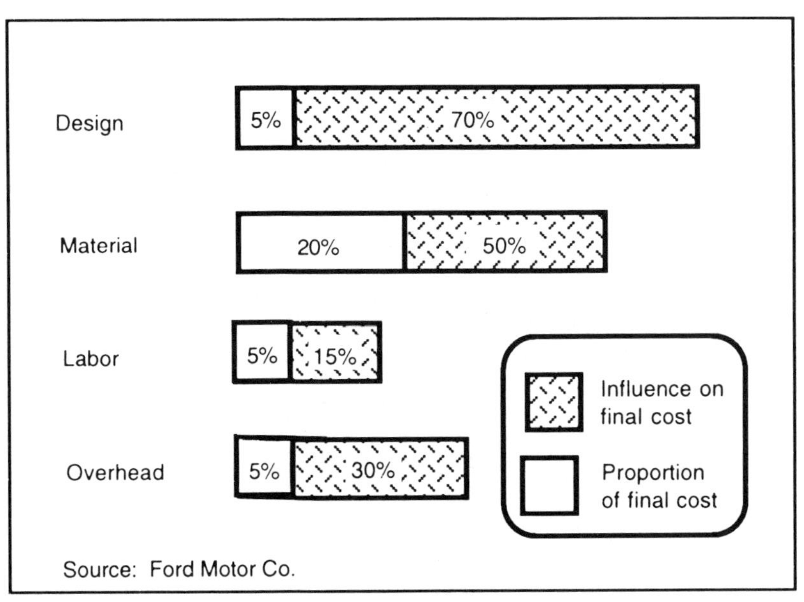

Figure 2-6. *Life Cycle Cost Determination.*

In many of today's manufacturing organizations, a product's configuration and its evolutionary cycle might be depicted in the *Figure 2-7C*, and may involve many parts of the organization (other than just engineers). This early communication and the orderly evaluation of alternatives can have significant cost implications. In this simple example, *Figure 2-7A* shows conceptual design being completed. It is followed by a functional breakdown of the product and more detailed designs as shown in *Figure 2-7B*. Notice that in the CE Design

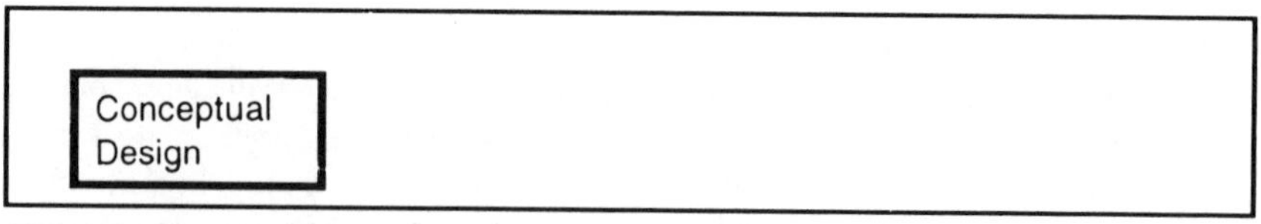

Conceptual
Design

Figure 2-7A. *Simplified CE Design Evolutionary Product Definition Process.*

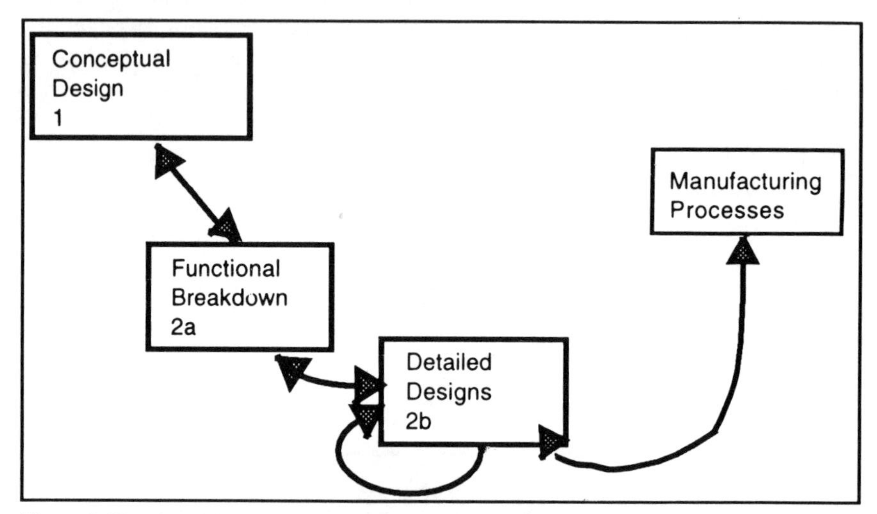

Figure 2-7B. *Simplified CE Design Evolutionary Product Definition Process.*

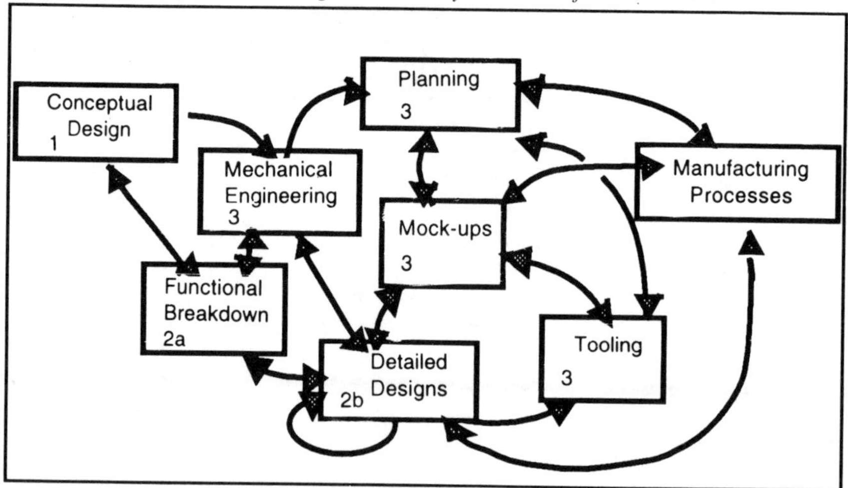

Figure 2-7C. *Simplified CE Design Evolutionary Product Definition Process.*

process (simplified), shown in *Figure 2-7C*, there is instant feedback from all aspects of the prior-to-production manufacturing process (PreBuild). This constant feedback before the "fixing" of the product design counters the too early "fixing." This feedback produces a design that is producible and less costly to volume manufacture.

As shown in *Figure 2-8*, many studies have determined that a mistake discovered and corrected during a design process is comparatively inexpensive to correct. By the time it is encountered during actual manufacturing, it can cost as much as 100 times more to correct, because of the material, planning time, design time, re-manufacturing time, and lost time-to-market implied by a correction process. Preventing the early "fixing" of the design through concurrent review of design corrects overlooked issues. These early corrections and completions are inexpensive.

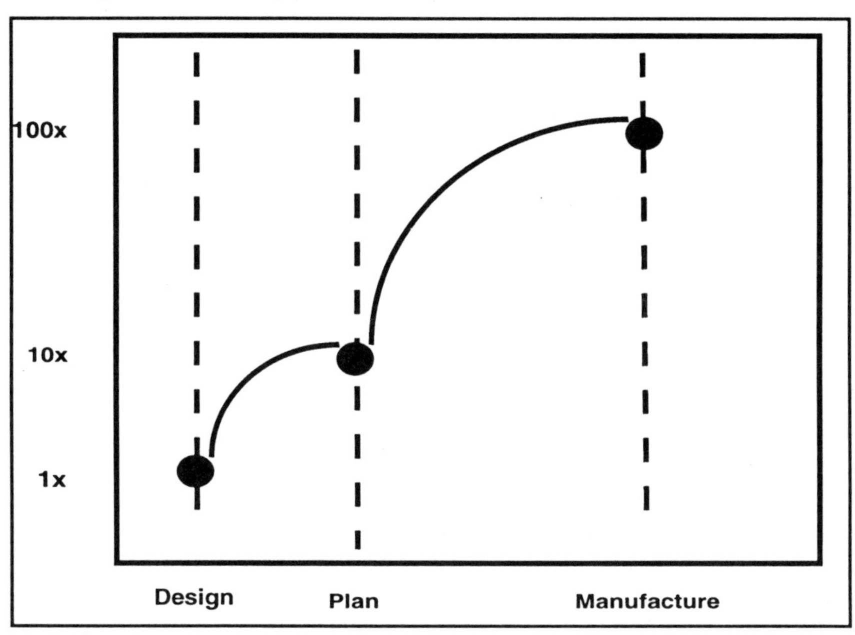

Figure 2-8. *Cost to Fix a Design Mistake.*

Accuracy

A third dimension to the power of CE Design is *accuracy*. The most effectively designed product can still be involved in delays, cost overruns, loss of market share, quality problems, etc. These results occur when product descriptions and manufacturing processes are not accurately developed and accurately communicated to manufacturing. If all aspects of the design are accurately described, *design intent* can be preserved.

An accurate bill of materials is a key element of product description. The presence of an accurate bill of materials always has been an important factor in successful MRP implementation. If the organization does not accurately know what the parts of each of the product's configurations are, how can it produce the product in a cost effective manner? The same can be said of the process description. How could manufacturing produce the product if the manufacturing process is not clearly understood? Finally, how could the product be supported in the field if accurate information is not available to the post-manufacturing or customer service and support portion of the organization?

Systems and processes required to execute CE Design help the production of highly accurate product and process descriptions as a natural byproduct of their control over the many parallel processes intrinsic to CE Design. Once these underlying enabling systems and processes are carried out for their own purposes and associated benefits, higher accuracy comes at little or no additional cost.

In *Figure 2-9*, many traditional data processing projects have yielded the following general error rates or statistical "rules of thumb."

1. An error rate of 7% is typically associated with a standard manual or forms driven system.
2. An error rate of 3% has been associated with a standard computerized batch, update, and error suspense file type system used in business processes, and
3. Error rates of less than 1% are associated with on-line, database, single entry at the source type systems. Many complex products, such as air traffic control systems, must be *much more* accurate. Many organizations are seeking six Sigma quality levels. Higher accuracy rates are possible within each of these systems models, but at much higher costs in traditional design environments.

In the complex product manufacturing environments described, a 1% product and process error rate is several orders of magnitude too high. With high complexity, high change rates, and the overall speed required, a 1% error rate may equate to hundreds of parts, components or process activities in error. CE Design enabling systems and processes support very high accuracy product description and maintenance at low cost. This high accuracy support

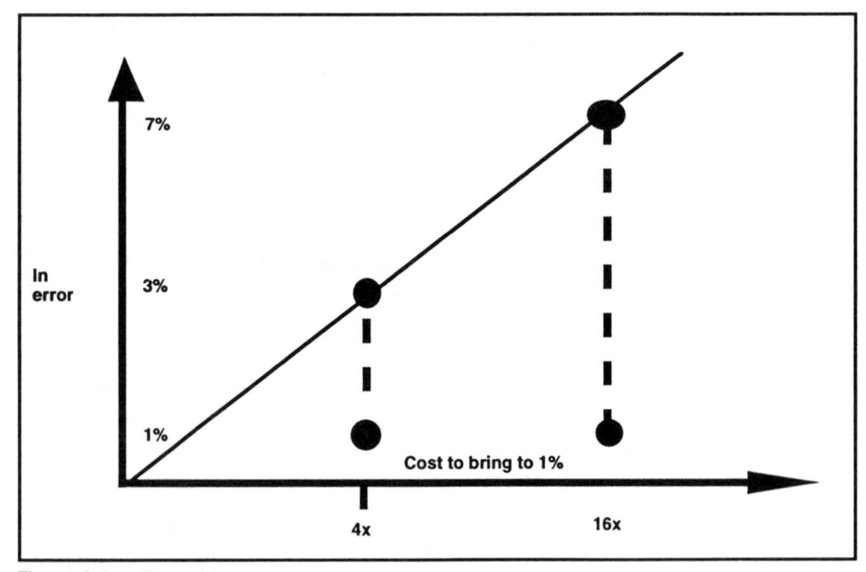

Figure 2-9. *Cost of Accuracy.*

comes from several CE Design process characteristics, including Configuration Management which is also an important dimension of Concurrent Engineering.

Configuration Management

The enabling systems and processes described for CE Design Authority Management, when appropriately applied, provide for configuration management as an integral part of their operation. Integral configuration control is one of the keys to success in CE Design. Configuration management is those business processes which assure that the right part gets to the right product at the right time. Configuration management includes accuracy at all points of the process (e.g., drawing, bill, process, and customer support), and change incorporation control tied to customer specials, basic product change, option incorporation, and product variation. Configuration Management, in the complex product environment, needs several different types of affectivity to operate. Affectivity will be discussed in Chapter 5. Configuration Management improves *all* elements of manufacturing performance.

Time-to-Market

A fourth dimension to Concurrent Engineering, *speed*, may be as or more important than any of the others. There are several aspects to the speed dimension. The first, shown in *Figure 2-10*, describes the various phases of a product and its typical profitability history over time. The major finding, in project after project, is that products first to market enjoy higher margins, and if managed correctly, higher market shares.

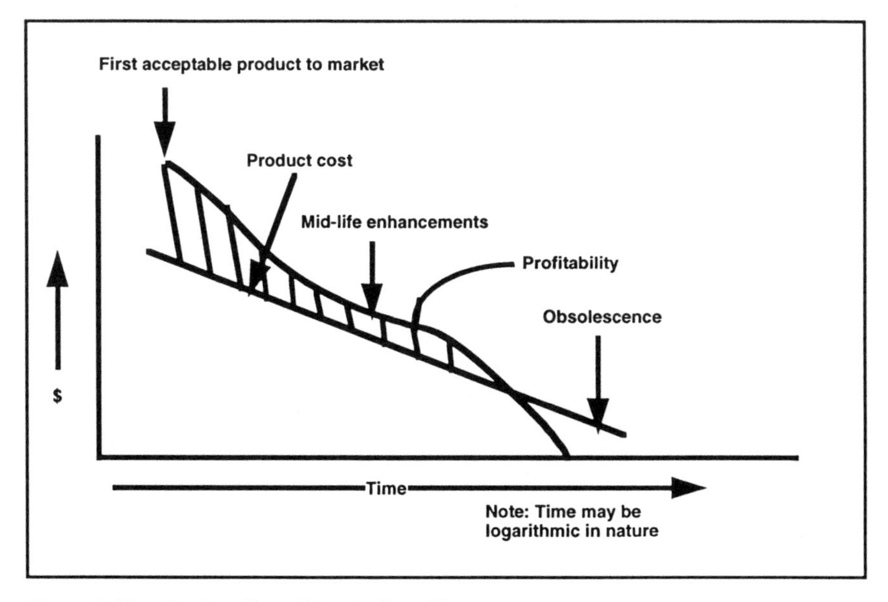

Figure 2-10. *Product Gross Margin Over Time.*

There is no disagreement with this concept of timed-based competition. The first to market with an adequately featured product of good quality and customer satisfaction has the "power position" in that product niche. The first producer can charge a premium for being first. Over time, if volume can be increased, individual profitability may go down. Over time, volume increases and constant improvements in product cost can actually increase profits during that part of the life cycle. Eventually, for the types of manufacturing organizations described, all products become obsolete.

Constant product improvements and the cost to produce the overall product have historically followed a pattern called the experience curve which is based on labor hour reductions. *Figure 2-11* contains two sample experience curves. Curve 1 represents a historical view of manufacturing. The first prototypes of a product actually cost more, sometimes several times more, than the first production run of a product. The earliest production runs set a starting point for product cost, shown as A in the figure. As more production runs are accomplished, incremental improvements and simplifications drive down the cost of the product to point B, and eventually to C. Somewhere along the B-C timeframe, the cost of improvement far exceeds the added profitability and attempts at substantial improvement stop. Usually, this is after the "mid-life improvement" point of *Figure 2-10*. CQI now attacks product and process cost throughout the product's life.

It is possible, with the combination of some of the modern tools now available

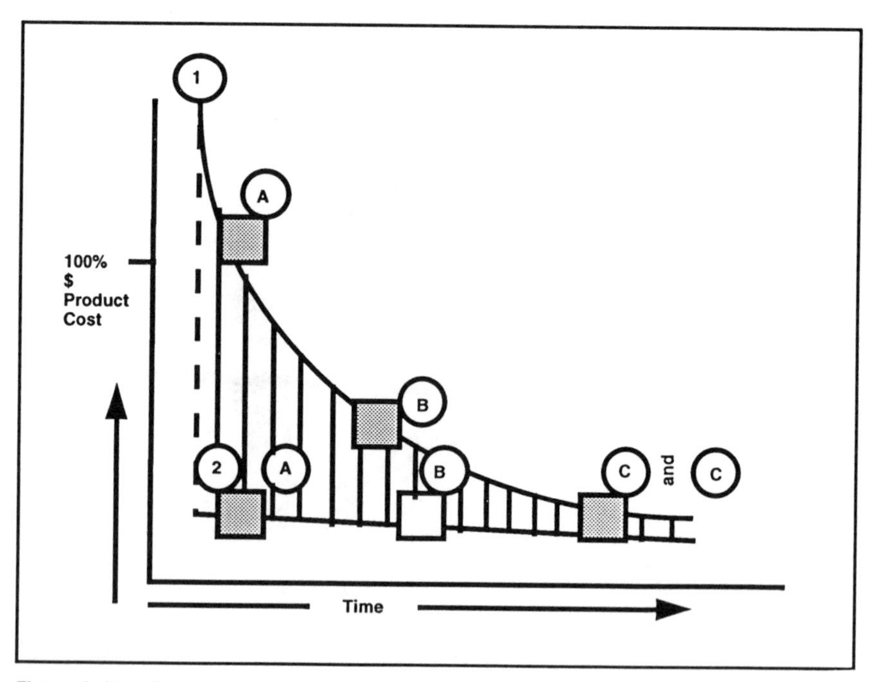

Figure 2-11. *Experience Curves.*

(e.g., 3-D CAD solids, with models, simulation of function, and process simulation) and within the context of the CE Design as described, to bypass these historical experience curves and have a curve much like Curve 2 in *Figure 2-11*. The revenue enhancement benefit created by this revised improvement curve is enormous. The revised cost curve occurs because the accumulation of the various dimensions described here (speed, accuracy, control, etc.) combine to create a high quality product and process definition that can be deployed *quickly* and *accurately* in the first part produced. *It may also be a ''preemptive'' implementation which prevents competition.*

RESOURCE MANAGEMENT

One of the persistent problems in engineering and design is scheduling work and assigning resources. This problem reflects the three basic characteristics of engineering and design:

1. dynamic, unstructured work;
2. interrelated, but only loosely coupled processes, and
3. involved complex information (pictures, words, test data, mail, suppliers specifications, etc.).

Because of these three characteristics, the engineering and design environments have been described as an ''open-inspect-repair'' (''OI&R'') situation. In OI&R, one never quite knows what the task is, or how long, or who will be involved, until it is begun. Thus, the CE Design process must permit scheduling and resource management in spite of this ''unknown.''

The enabling systems for CE Design permit, without added work or intruding on the activities of the staff, the systematic collection and analysis of schedule compliance. The systems also reveal the status of the work of the various disciplines throughout the organization involved in the CE Design process.

This information permits greater schedule accuracy. Sample improvements from the author's personal experiences have included going from 20% to 25% late release to less than 2% late release with completeness improving from 90% + to 99% + . Additionally, this information has resulted in more effective staff size, better training and cross-training, and the best use of time, resources, and equipment. At anywhere from $15-$85 per seat hour for the various pieces of hardware and systems in use, best resource utilization is vital. Only through better resource utilization can the perception of ''computing in engineering viewed as a cost'' be realigned to being viewed as a positive return on the investment (ROI).

A potentially controversial aspect of resource management is a perception of intrusive management control, which could interfere with the conceptualization and invention processes that must occur; that is, ''you can't invent on a schedule.'' This is true perhaps, but most of the engineering done in the manufacturing environments described is spent in the other phases of the intellectual process or in administrative related tasks. The constant interruptions of the surrounding environment and the necessary documentation of progress required in current environments may be a greater interference with invention.

At least from a systems perspective, the engineer's time is spent in three stages. These stages are: invention and conceptualization (stage 1), visualization (stage 2), and realization (stage 3), as depicted in *Figure 2-12*. The enabling systems and processes described for the support of CE Design provide an environment in which the engineer works relatively unencumbered. Many of the business and paper work functions that distract are provided automatically. Rather than exercise more control, these enabling systems provide more freedom. These enabling systems provide less distraction and more time to move quickly and effectively through these stages, and to loop-back as necessary. Productivity goes up, and more products can be engineered and manufactured faster.

AUTHORITY MANAGEMENT

Authority Management is another interesting dimension to the Concurrent Engineering process. Most engineering, performed in the types of manufacturing environments described in this book, is done in a serial manner, that is, one step at a time.

The reason for the serial process is control. In a complex manufacturing world, everything must be carefully reviewed. Appropriate authority sought. Careful preparations made before an engineering release to manufacturing can be permitted. None of these control issues can be "side stepped." The loss of configuration control in these environments can be catastrophic.

The CE Design process uses a combination of routing control and authority review. It is important, in the area of change management, to facilitate very high accuracy authority management as unobtrusively as possible. *Figure 2-13* shows how CE Design is made up of both Horizontal and Vertical Processes.

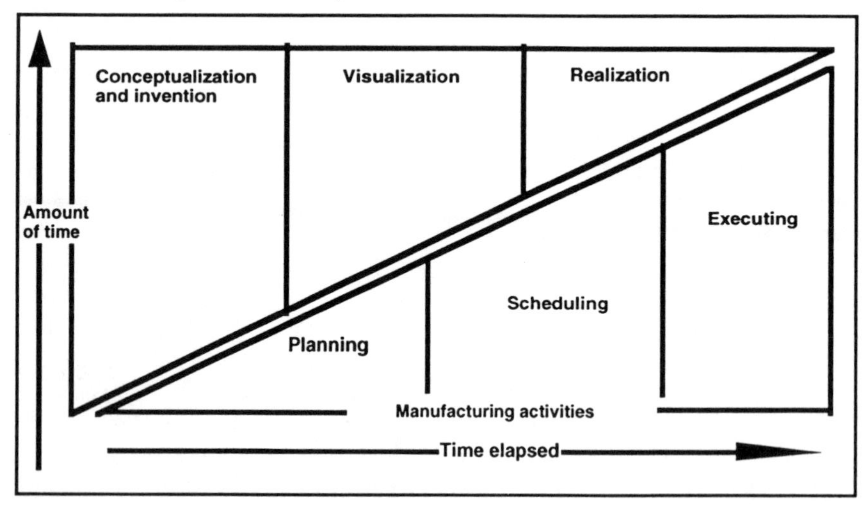

Figure 2-12. *Stages of Intellectual Process.*

Horizontal processes are those which typically make up standard business functional activities, such as design, production planning, manufacturing, and the like. Work moves from person to person using the technique of routing.

Vertical Processes are those processes that involve management and the executive levels of the organization. Vertical Processes are different from the typical Horizontal Processes. Work moves between reviewers and among managers based on impact, scope, and control issues. Authority management uses routing to move the work, but adds an impact analysis dimension for a "running up the organization" movement of work pattern.

These different types of work movements mean that CE Design must include work movement patterns in its processes to be successful, and for control to be effective in the CE Design environment. The cost savings, reductions, and revenue increases from effectively executed authority management have been found by the author to be substantial.

CONCURRENCY AND SIMULTANEITY

Figure 2-14A depicts a typical serial engineering schedule. The simplified CE design schedule shown in *Figure 2-14B* can be contrasted to the serial schedule of *Figure 2-14A*. *Figure 2-14B* shows how the *simultaneity* of CE Design provides for higher speed. It implies higher accuracy because the typical preparation to release step is not present, provides for a mock-up and test

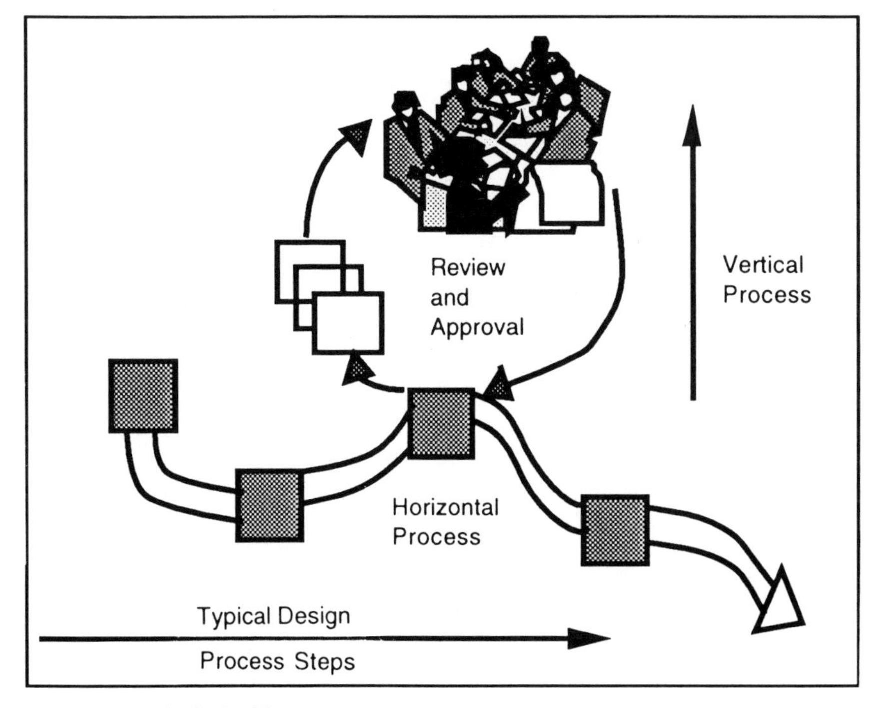

Figure 2-13. *Authority Management.*

opportunity, yet allows for a flatter experience curve. It permits more design options to be explored, while providing for a schedule compression of 40%.

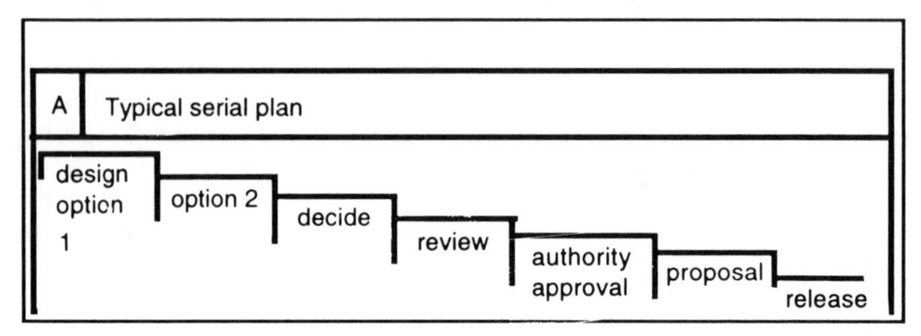

Figure 2-14A. *Serial versus Concurrent Execution.*

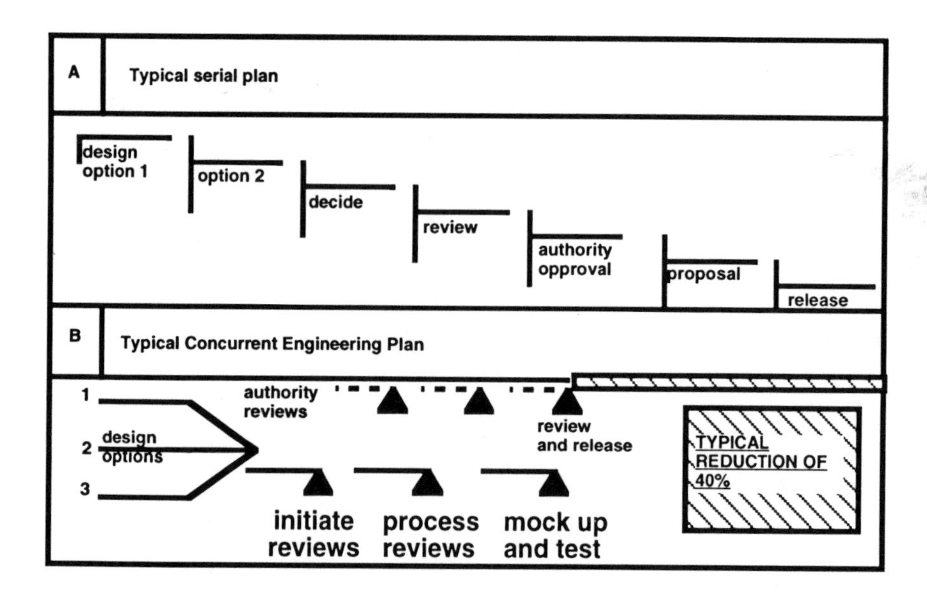

Figure 2-14B. *Serial versus Concurrent Execution.*

ELECTRONIC PROXIMITY

Another significant dimension created by implementing an automated CE Design environment is *Electronic Proximity*. One of the important differences in the CE Design process is the ability to talk constantly with colleagues in the design, manufacturing, and product support functions about the product's design *as the design is taking place*. *Figure 1-2* shows that the design process employs teaming structures that emphasize simultaneity. To accomplish simultaneity, the first approach is always to move these individuals or groups together. This creates physical proximity.

However, as soon as the groups get large, the advantage diminishes because there are just too many people together. Individual disciplines (electronics, mechanical, production, etc.) can find it more difficult to talk about their functional specialty and its issues, problems, and answers. This also diminishes the benefit. Additionally, the complex manufacturer has joint ventures, partners, suppliers, and vendors. These outside groups cannot all be co-located.

An automated CE Design environment provides for *Electronic Proximity*, a compromise between the various advantages and disadvantages. By providing for easy interchange of messages, drawings, sketches, and other communications (data, voice, etc.), a close proximity environment is established which is uniform across all functional disciplines and participants. The CALS (Computer-Aided Logistical Support) initiative from the DoD (Department of Defense), and the PDES (Product Data Exchange Standard) initiative from the NBS (National Bureau of Standards) are two efforts that are trying to standardize the interfaces between automated CE Design environmental support systems to speed the implementation of electronic proximity dimension.

GEOGRAPHIC AND MULTIORGANIZATIONAL DIVERSITY

The complex product, to reach and maintain adequate market share, and sustain itself, requires a global market. This need is reflected in three factors:

1. *Minimum capabilities* – the complex product cycle is very large and long. The product will have several years of useful life; support of the product must be sustained for years. Every aspect of the complex product for manufacturing production, including tooling, the initial product development cycle, and many technical specialists, for example, reflect this complexity;
2. *Minimum size* – The result of the various minimum capabilities is that an organization of a minimum size (i.e., personnel count, capital, physical space, systems, etc.) is required to build even one of the complex products;
3. *Minimum build and sell rate* – There is a cumulative impact of the minimum capabilities and size characteristics of complex products. It is that there must be a certain minimum of the products built and sold to sustain this *necessarily* significantly sized manufacturing organization.

Typically, once the minimum build and sell rate for the product is established, companion market research quickly affirms that no one country can sustain the required minimum output of the complex product for a significant amount of time. Becoming global is an economic necessity. In addition, there will be only a few serious competitors for each type of complex product, even on a global scale.Engaging the competition in its originating markets is always one element of the competitive advantage strategy employed in these circumstances.

Many different aspects of the product, including components, subsystems, and manufacturing production processes are complex. Therefore, many organizations, not just the final assembly and marketing firm are involved in the product. This type of firm is called, for example, the "Prime" in the U.S. Defense industry.

Increasingly, worldwide consortia, joint ventures, or Japanese-style *keiretsu* are being formed to address the complete manufacturing, marketing and support chain for these products. Using this approach, design responsibility is being shared as well. It is the need to provide cross-process and cross-organizational design coordination and communication that is driving these types of organizations to CE Design.

One approach to providing this cross organizational design coordination is to try to "force" organizations in the associated group to use one CAD and other sets of software systems. This approach has had many parallels. In the insurance industry, for example, there has been considerable pressure, for years, for a standard terminal and systems approach for health industry claims entry. Every company understands the issues of the possible competitive advantage if it is their systems of capture that drives the industry. So far, in this segment of the industry, only common paper forms have become accepted. In airline reservations, drug distribution, PC distribution, among others, this issue of capturing markets by common computing systems is being pursued.

The common CAD system approach to shared design responsibilities is likely to be unsuccessful in most instances, because the partner organizations in such complex products consortia are not small suppliers with one or two customers, but are highly capable firms. Many times, these organizations can be competitors in one arena, and partners in another. The CAD system is being pursued not for market capture reasons, but because of another concern, shared design. Most firms do not want to become "indirect sales organizations" for CAD software firms. In fact, the organization was invited into the arrangement in the first place because it was perceived to be a contributor.

A higher level of abstraction, but still common approach for shared design responsibilities, is required. Much will be made in this book of establishing the appropriate management, process, design, computing frameworks, and architectures that enable CE Design as a component of these organizations' implementation strategies for products. Once these underlying infrastructures are in place, CE Design is the vehicle for promoting the careful "sectioning" of the product by organization.

Source Origination and Documentation

A byproduct of configuration control is *Source Origination and Documentation*. It is one of the last dimensions of CE Design highlighted for discussion. In the past, an inventor came up with an idea and turned it into a product. Even today, a simple product can still start with an individual. However, the complex world of the manufacturing firm is complex. An invention of significance is still only a part of a much greater existing production structure and such simplicity of origination and ownership can get lost.

With joint ventures, government sponsored programs, and university con-

tracts with private and government sector sponsorshi
to enjoy the fruits of its expenditures, thus estab
ownership is most important. Enabling systems fo
sort of definitive history of origination and ex
environment. The details of source origination an
in conceptualization, visualization and realizatio
without intrusion, captured and documented.

COUNTER-INTUITIVE IMPLEMENTATION STRATEGIES

Concurrent Engineering Design is a very powerful concept. The highly integrated nature of the underlying processes and systems, however, have impeded the introduction of this concept. The technology is available to build the processes and systems for the realization of benefits arising out of the dimensions described above. Implementing the CE Design concept is, unfortunately, not easy, because the final dimension of Concurrent Engineering Design is that its implementation is counter-intuitive.

CE Design is a high-level business process that is built upon other processes, as shown in *Figure 2-15*. CE Design exhibits emergent properties that do not appear when the individual component processes are themselves examined. The very high ROI high level processes such as CE Design are built upon a moderate level of ROI processes such as Logistics and Design. These are, in turn, built using services such as CAD/CAM. The processes are built on Automated Infrastructure Support, which has no ROI.

The usual approach to a change in an organization is to take a problem or a new idea and break it down into its component parts. This step is followed by the addition of activities to the various functional department's procedures to address the problem or new ideas.

This decomposition approach is not entirely successful for either complex products, or the organizations that produce them. An alternate approach, *differentiation*, is what is used instead. In this counter-intuitive approach, the processes are organized first. The organization is then reformed around them. The manufacturing organization, adopting CE Design, has altered how the various functional disciplines operate and interact. It has also altered how the business systems and processes it uses to accomplish its activities operate. The benefits of this improved operation are enormous when the dimensions described are realized. However, everything about the organization is altered, and a return to prior conditions is not possible.

CE Design requires a distributed, yet integrated, computer-supported system for the various business processes for its success to be achievable from a practicality perspective. CE Design depends on the flow of information, materials, and intermediate end-product-related work materials from many disciplines, functions, departments, and individual systems. Interestingly, the proper computing setting for CE Design doesn't require a large central computer. In fact, highly centralized control over the CE Design environment is

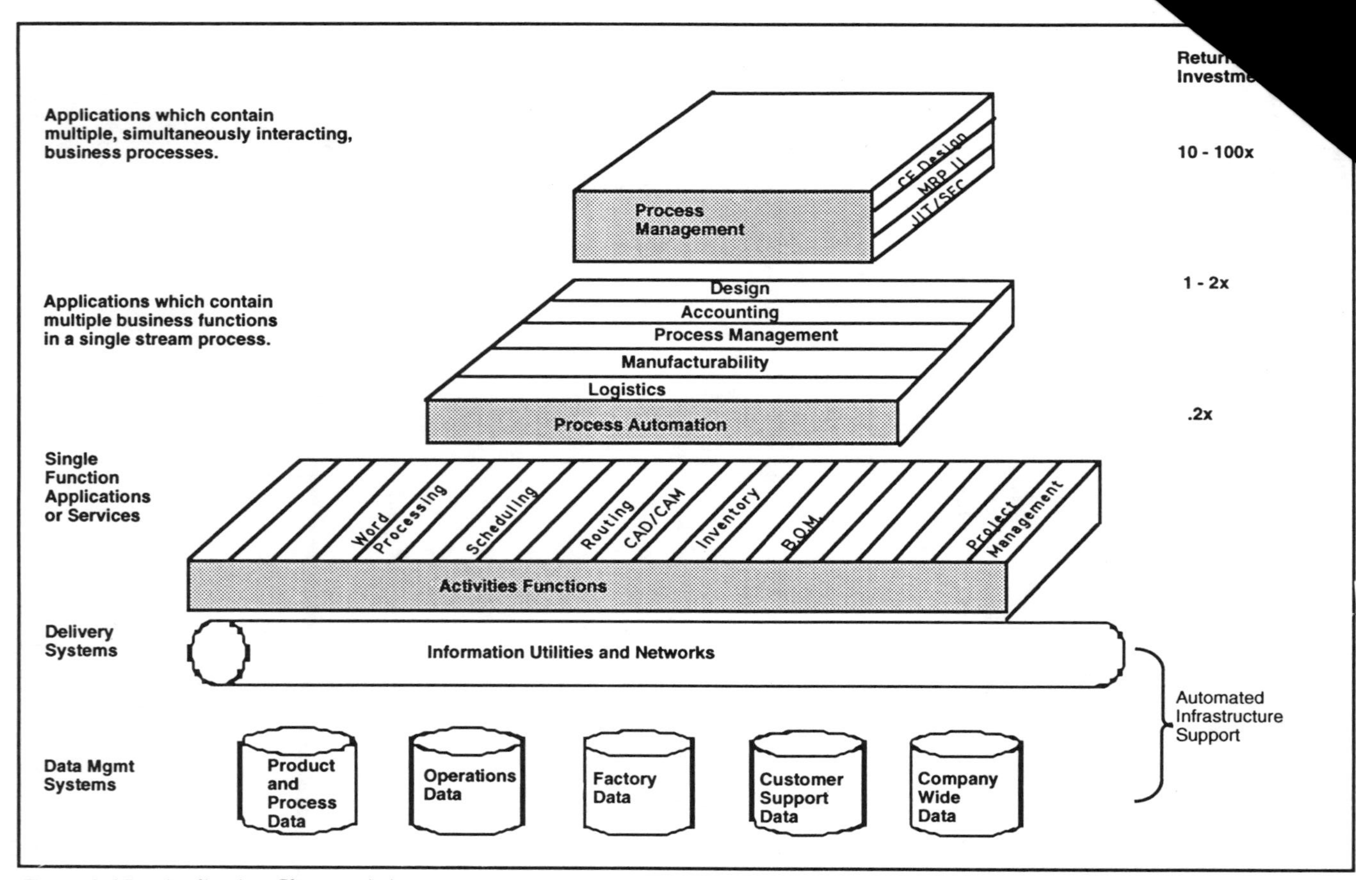

Figure 2-15. *Application Characteristics.*

not effective. CE Design emphasizes local autonomy and flexibility. Yet, centralized control of the coordination of design, design intent, product configuration control, and emergent product characteristics only apparent or designable at the final product level is highly desirable. Another counter-intuitive aspect of CE Design is its computing architecture, focusing on *distributed* computing supporting business processes that require an *integration* element in their execution. Chapter 8 discusses the computing architectural dimension of CE Design.

The key to implementation success in this environment is an integrated planning and implementation process *equal to* the integrated nature of the Concurrent Engineering Design process which provides significant local or group flexibility. Chapter 9 provides an overview of such an implementation strategy. Each organization will need to develop its own unique plan tailored to its own unique environment.

3

CONCURRENT ENGINEERING DESIGN CAN CREATE A COMPETITIVE ADVANTAGE

Figure 3-1 (based on *Figure 1-1*) shows that there appears to be three major pressures forcing a manufacturing organization into changes in "how they do business." These pressures can be focused, and with CE Design, can create a competitive advantage.

Competition worldwide is increasing. Moving product around the world has become easier because of improvements in communications, transportation, and the global financial infrastructure. This has resulted in increased competition in most manufacturing niches. If there are not global competitors in every market, there will be soon. Seeking and gaining a comparative competitive advantage in this environment has become a compelling business management issue.

SEEKING COMPETITIVE ADVANTAGE

There are over 160 countries which in some manner are important markets in the world. Industrialized nations are attempting to continue to grow. Lesser developed nations are seeking further economic development. The recently emerging general pattern is to emphasize economic over military development. The search for competitive advantage in manufacturing in these markets is following a general pattern of evolution.

This chapter explores competitive advantage from a product and process enabling competitive advantage perspective. The discussion begins by discuss-

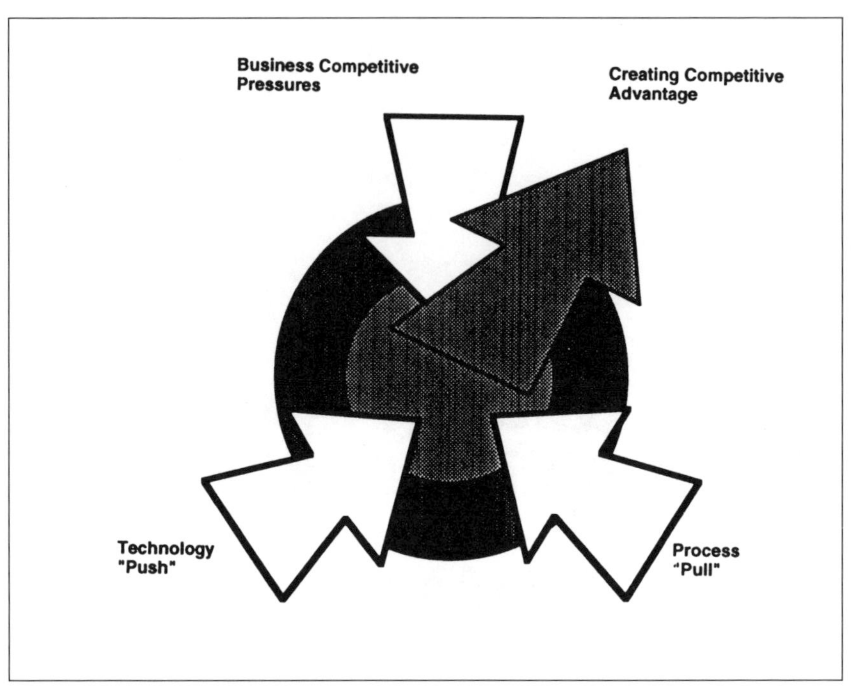

Figure 3-1. *Concurrent engineering design pressures.*

ing the evolving nature of competitive advantage. Most organizations have largely focused on a strategy based on a mass production internal strategy, while competitive advantage was sought external to the organization. New product- and process-centered strategies have been developed; these are largely internal components of the organization's competitive advantage seeking strategies. Product design can be an important part of these strategies. Only recently have competitive strategies based on process excellence become important. CE Design, as a competitive advantaging strategy, is discussed within this context.

THE EVOLUTIONARY PATTERN

The evolutionary pattern, starting in the post WWII era (circa 1950), has followed progressive stages of seeking competitive advantage. These stages are described in the sequence in which they have occurred and/or are still occurring, depending on the development of the industry, country, and markets. The stages include:

- Establishing basic industrial capacity;
- Acquiring the technical knowledge and capability to execute product design and manufacturing process, and
- Employing product design and process excellence as competitive strategies.

Establishing Basic Industrial Capacity

This stage is based on the availability of plant, equipment, and adequately trained people to produce goods. The U.S. was one of only a few major industrialized nations that did not have its industrial capacity destroyed during World War II. A significant percentage of its working population was not killed or displaced by war. Thus, the U.S. had an initial competitive advantage across most industries. A variety of assistance efforts, such as the Marshall Plan, have with time, allowed virtually all industrialized countries to have developed or redeveloped needed basic capacity.

Knowledge and Capability for Product Design and Manufacturing

This stage includes the knowledge necessary to produce, in adequate quantity, higher part count and complexity products such as automobiles, refrigerators, and chemical compounds. Time, education, and experience in manufacturing organizations have led to many nations with at least these skills in low technology, or small part number count products and their associated industries. For process industries, the complexity of the process has been a limiting factor for most nations. Complex products, with part number counts of more than approximately 5000, or complex processes, have historically been dominated by the most industrialized nations.

These highly industrial nations have recently encountered competition in even the higher part number count product or more complex process areas. Other nations have or are interested in developing these capabilities as well. These industries improve the quality and technology of their industrial base. This improvement raises product prices, and the value of their industries. This, in turn, leads to improvements in the overall standard of living of the nation. Once significant manufacturing process capability is established, a set of substrategies, based on a manufacturing strategy of economies of scale, are possible.

Outside the Organization Focused Marketing and Sales Strategies. These substrategies primarily focus on elements *outside* the manufacturing organizations. Typically, they affect product design and internal processes, but are not driven by these internal issues. They include:

- Building knowledge of markets,
- Marketing-based strategies,
- Taking advantage of comparative labor rates,
- Encouraging invention,
- Amassing capital,
- Free market restrictions.

Building knowledge of markets. The issue of market building is especially true in the modern global economy. For many products, knowledge of the local consumer has become very important. This was originally dealt with by moving the company to within adequate geographic proximity. The early availability of goods and capital provided U.S. firms an initial competitive advantage. However, foreign competitors bought U.S. personnel and companies and/or came to the U.S. and learned. Now many different firms from many different nations have gone global and done the same thing.

Marketing-based strategies. These strategies include various technologies—such as packaging and other differentiating but minor product ingredients—to distinguish products which may not be distinctive without some customer education and advertising guidance. This can be provided by brand name establishment and maintenance targeted advertising campaigns, distribution channel control, market share dominance and price management, customer service, superior direct selling, retail display space control, retail location, and discounting. These marketing techniques are clear differentiators in many industrialized nations but their effectiveness may be declining in some markets.

Taking advantage of comparative labor rates. At one time, labor rates offered comparative advantage. With automation and many other factors, "touch" labor or direct labor is now less than 15% of product cost and in many cases well below 10% for complex products.

Originally, labor rates moved factories. This movement was constant and toward the lowest labor rate possible commensurate with adequate labor pool skill. Increasingly, as comparative labor rates came closer, and labor who touches the product goes down, labor rates have become less of a dominant issue.

Using the U.S. as an example, many competing industrialized nations have high labor rates and have moved operations to lower cost areas of the U.S. to take advantage of lower labor rates. This was not the case 10 years ago. The competitive advantage of labor rates has become less important in the complex product environment. Access to market by labor usage may become more important. A basic, underlying principle of CE Design is that the aspect of competitive advantage associated with labor rates and product cost, is now found not in the primary product value in the factories, but instead in those areas of the organization previously described as "overhead."

Encouraging invention. Differentiating inventions have provided opportunity for manufacturing success. It only provides an opportunity, however. In the U.S., invention has been a trademark of the individual-success-oriented U.S. society. This has become an international issue. The ability to invent new things and processes is still a U.S. strength but many global competitors have their own R & D activities. Each is using the other's ideas to compete. In addition, many lesser developed nations now have built focused R & D capability, and are world leaders in some specialty field. The global ability to invent is increasing.

Amassing capital. From a market share or dominance perspective, there should be comparative advantage to manufacturing organizations whose operations are in countries able to provide a long-term cost of capital advantage to the organization.

Much has been made of the strong relationship between government and business in Japan. There also has been much discussion of the strong relationship between industry and government in U.S. defense. One could argue that the Japanese government focused on consumer-oriented collaboration because of their WW II experiences. Their constitution prevented them from concentrating on extendable defense-related activities. Their options were limited to business-related activities.

There is a widely held view that the U.S. Constitution and its attitude toward free markets has, in contrast, limited U.S. government activities to supporting defense-related activities and allied industries. The perception is that this narrow government support has been due to the general success of business and the history of business in the U.S., including its antitrust laws.

This is not completely true.

The U.S. government, as an extension of its interests in public health, became heavily involved in directly funding or subsidizing R & D in health-related industries, including disease detection, genetic engineering, bio-technology, and related health issues. This indirect support through regulations has influenced communications, power generation, environmental control, and other industries. Such indirect funding has led to many entirely "new" industries.

Being the dominant, low-cost producer in a market creates a large barrier if substantial capital is required for market entry. A present example of this strategy might be the Japanese trading companies and their dominance in consumer electronics. These firms have spent billions perfecting their high definition television capability. This includes satellites, studios, special cameras, etc.

Under some circumstances, dominance can be "leapfrogged." "Leapfrogging," in the consumer electronics case, for example, might occur if a new high definition television technology, based on computing, is deployed in the U.S. Such a "leapfrog" could be supported through regulation and might be based on the U.S. strength in complex systems and software. Using a variety of competitive strategies to "leapfrog" is discussed further in succeeding sections of this chapter. This strategy takes advantage, in a competitive environment, of the breakthrough or paradigm type opportunities provided by strategies such as CE Design.

Free Market Restrictions. To permit the growth of domestic industry as it goes through some competitive advantage stages and to allow domestic manufacturing organizations to handle these issues, governments used bans, tariffs, import and domestic business operations laws, taxes, procedural inspection rules, and local or social customs (by encouraging their continuation) to create real and artificial barriers to a true free market.

The world is beginning to move toward an environment where restrictions will become more difficult to sustain. If disparity in the availability of goods and services exists, the nation with this disparity will go to others, and seek their goods and services.

Information, entertainment, and communications, (especially via television), are the first products on a general scale to become global products and a global political issue. The facsimile machine, or FAX, is another excellent example. Originally, the FAX machine was viewed as a minor device used by businesses to communicate, at low speeds, and as a substitute for telex.

Because many oriental languages use complex characters, standard telex was a difficult medium for their communication. When the computer chip made more complex character representation and the process of scanning images easier, FAX machines quickly replaced telex. FAX prices dropped. Individuals could afford them in their homes. Its impact has reached the point where the FAX machine is reputed to be one of the key communications devices assisting the resistance during the failed *coup d'etet* which resulted in the dissolution of the Soviet Union.

Improvements in allied technologies brought about growth in communications in a few years. When this allied technology was adopted into products in this category, it allowed FAX technology to ''leapfrog'' telegraph and telex to the point where they are hardly a market factor.

Internal to the Organization Strategies. This evolution of competitive advantage via *economies of scale* has another dimension. This second dimension is based on the manufacturing firm's basic product characteristics and its associated internal manufacturing *process* strategy.

Economies of scale. The first approach to economy of scale used a basic process competitive strategy based on mass production for reaching adequate economies of scale for mechanical products and process industry facilities. This well-known strategy formed the basis for competitiveness for most of the industrialized nations through most of this century.

In this volumetric-oriented strategy, one attempts to become the low-cost producer. Low-cost producer status comes from producing low change and at least moderately complex product in large volumes through factories changing little over a number of years. The cost of the factory, the product's initial development, and any minor changes can be spread or amortized over many hundreds of thousands of product units over many years.

Being integrated in operations, serving large markets, and introducing change slowly, such a manufacturing organization can dominate markets through adequate quality at a price, or dominate markets through controlled and crowded distribution channels. The customer satisfaction survey for such an organization, ranking the best at five and the worst at one, measured success based on customer satisfaction of at least a 2.5 on the scale.

The classic volumetric strategy is represented by automobiles. Complicated, but not highly complex, the passenger automobile's basic product structure (four wheels, four to eight cylinders, reciprocating gasoline engine in front, round steering wheel, two rows of seats, etc....) has not changed much in 75 years. Many novel designs have been introduced over the years but these had little effect on the basic concept of the automobile.

The automobile became an absolute necessity of personal freedom and personal mobility after World War II. Everyone wanted their own automobile. Change, until the 1970s and 1980s, came slowly. Most plants were single product facilities. The single product with limited options (less than 2000) per facility remains largely true today.

Until recently, much of the basic organizational approach to manufacturing has reflected the fact that competitive advantage has been sought outside the organization. Major emphasis was placed on marketing and finance. Many senior executives came from the marketing and finance groups within the organization. These individuals were paid high salaries. Masters in Business Administration (MBA) degrees were entry points to these important areas of major firms. Small improvements were made in manufacturing, and were done when a labor hour reduction could be demonstrated. Many factories changed little in 10 years.

A variety of factors have, since the late 1970s, forced the manufacturing organization to consider *internal change as a competitive strategy*, or at least a response to others employing such a strategy. These factors include the introduction of abstraction via computing in a variety of manufacturing products and processes, and the widespread implementation of quality-driven processes. First introduced in Chapter 2, *Kaizen*, or the constant improvement of manufacturing processes for producing higher-quality products, resulted in factories which have changes so often in a year or two that they become unrecognizable. New competitive strategies are thus based on factors which are *internal to the product and process*.

Product and Process Simplification Using Abstraction. The basic concept in this internal improvement strategy is to include a higher level of abstraction components as product elements, permitting change in the product to become a competitive, product differentiator. The abstraction and substitution concept will be described using watches as an example.

Before the microprocessor or highly integrated circuits, available since the late 1970s, a chronometer or a watch that lost accuracy consistently was a high value, high cost item. The concept of watches has now changed. A development in the computer field permitted substitution of a higher level of abstraction product component, the microprocessor, for time keeping.

This substitution in the watch industry resulted in the industry being completely reset in several ways. The product has become complex and simple. The process of assembly of a highly accurate, electronically powered, quartz crystal oscillator regulated watch, with consistent accuracy to within seconds a year or less, is highly automatable and can be assembled in a few seconds. These watches are now available for just a few dollars. The old products, with over 100 parts when fully assembled, many with partially hand-crafted components, took many hours.

The high-cost watches still available are sold on a different basis; these watches are sold as jewelry (not timepieces), on the strength of product ruggedness, or on the basis of name brand recognition. Many are sold not as accurate timekeeping devices, but as status symbols. They are no longer as accurate as quartz watches selling for 10 to 1,000 times less.

Now the price of the watch is not based on timekeeping accuracy, the previous measure of customer quality, but on other factors. The cost of the case of the watch, the wristband, and marketing and distribution are now the major cost items of the watch. This means that a highly accurate watch can cost a few dollars, even after all costs are considered. Yet the quartz watch with day and date, an alarm, seconds indicator, as well as AM/PM or military time, is a more complex product in terms of delivery function than the old style chronometer.

The mechanically driven equivalent of the microchip watch, to emulate those electronic functions, would require a design composed of hundreds of wheels, windows, indicators, and gears. When the watch was mechanical, the individual watchmaker had to actually make the components. The skill of the watchmaker was the principle, governing ingredient in timekeeping quality. Labor hour cost was directly linkable to timekeeping quality. Since labor hour costs were the key cost ingredients, these hours and the skill employed in them were directly reflected in the price. In the electronic watch, the key quality ingredients moved to a design-centered strategy. Component capabilities and ‘‘final assembly’’ process quality are now the key determining factors of product cost. Product cost may be reflected in the price, if the watch is sold on an other than standard cost-based mark-up.

In product after product, and industry segment after industry segment, the level of abstraction of the product is being changed. Many times this change is induced by technology, many times the change is process-driven. The abstraction raising depends many times on the industry segment. In oil and gas, it is process-oriented change. What does raising the level of abstraction mean? How does simplifying add complexity?

Referring to *Figure 3-2* and using the watch as an example, notice first that

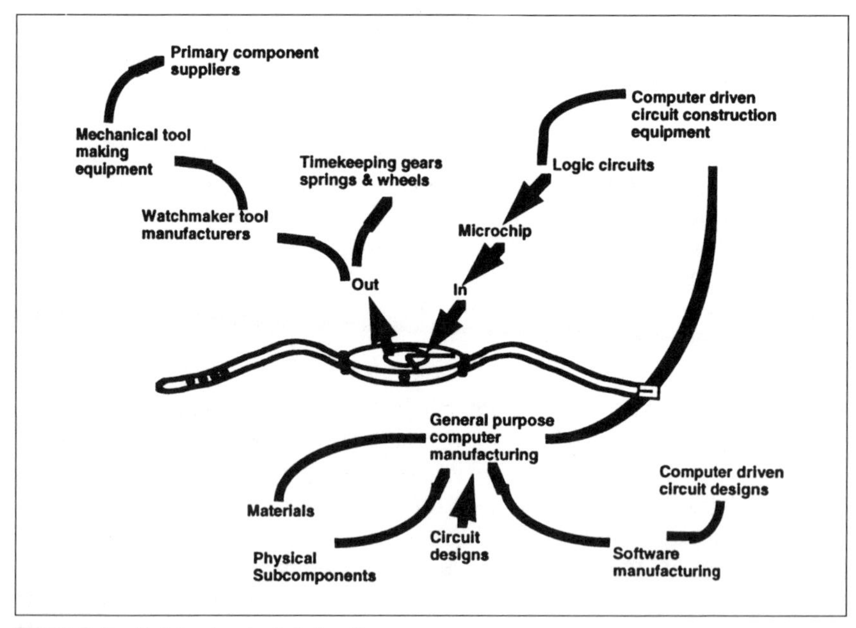

Figure 3-2. *Raising level of abstraction.*

the craftsman actually making the watch has been replaced by an electronic timekeeping designer. The electronic watch designer still is concerned about timekeeping. However, a general purpose "thing," the microchip, has been substituted for the mechanical gears. The chip does not do anything mechanical, of course, but it controlled the indicators of time, and substituted its logic for the gears and wheels, which had embodied the watchmaker's timekeeping logic. Introducing a general purpose computer and preparing a program, or designing the logic of timekeeping using general computer circuit elements introduced into the circuits, is raising the level of abstraction from mechanical design to electronic and mechanical circuit/logic/software design. In the watch example, this raising includes the transfer of physical objects (gears) to concepts (circuits) as the implementation medium, and the change from a mechanism (gears) strategy to a policy strategy (timekeeping).

In this abstraction raising process, the simplified product is also more complex. The industry, to support the traditional watchmaker, had two or three elements to it, all mechanical. By introducing computing, each element of the industry value chain has more levels of abstraction. Each has supporting elements with more levels of abstraction. The process in each element can be more complex, and costly, if greater features and functions are introduced into the timekeeping circuits.

This level of abstraction raising process has moved the designer/engineer into an even more important role. Using the higher level of abstraction concept has permitted change to be introduced rapidly. The introduction of change, even in previously "mature markets," is an important issue. One firm has created an art collectables market for their inexpensive, highly accurate watches as an inducement to buy them. Well-known artists develop watch faces. Watches are changing all the time, but only the faces and their cases. The watch face actually becomes the collectable.

In Chapter 2, substituting process for results was described as a problem to which CE Design is responding. Shifting responsibility for product internal operations to design engineering is also a concern for CE Design. CE Design forms the basis for new strategies which are not as constrained as the mechanical quality/process strategy dominating the first two-thirds of the 20th century. The computer removes many design constraints which mechanical manifestations of the design required. Making change may have meant completely disassembling and reassembling the product, and even then changes might not be significant. Changes, introduced as rapidly as the market will allow, can be made in the embedded computer itself. The computer is not the only change agent, of course, but is an excellent representative.

Product and Process Quality. Product and process quality became the second evolution of the economy of scale or volumetric strategy. For most of the 1980s, manufacturing has attempted to respond to the competitive pressures of the quality/process strategy.

The volumetric strategy, even for "mature" products, or products which have reached the perceived end of their high change engineered lives, has been under attack since the 1970s by a quality/process strategy. The examples of automobiles and watches are representative of these pressures. The quality/process strategy *redefines* the concept of volumetric success. All aspects of the manufacturing operation are subject to change and can form the basis for a competitive advantage. The most important aspect of this new competitive strategy, and the underlying competitive repositioning which is occurring, is a repudiation of the concept that *higher quality means sharply higher prices*. As a result, the watch example has moved into more complex products, such as automobiles.

Sharply increasing product and process quality *lowers* prices, or at least provides for higher margins. The Japanese are the earliest and most aggressive adaptors of this approach originally identified by Juran and Deming. This quality manufacturing process strategy has received much attention. The *Kaizen* book, among many on the subject, has greatly reoriented manufacturing management's strategies.

One of the outgrowths of the need for a product and manufacturing process quality strategy has been the major process entitled *design for manufacturability (DFM)*. DFM is one of the major subprocesses of Concurrent Engineering Design, but not the only one. Chapters 5-8 will discuss each of the the CE Design processes thoroughly.

The importance of this product and manufacturing process quality strategy, when combined with a CQI implementation strategy, cannot be over-emphasized. Undertaking a quality-based manufacturing strategy is a necessary pre-condition for CE Design. However, design, built on a pre-existing product and process quality strategy using CQI, is now becoming *the* competitive advantage strategy.

Product Design as a Competitive Strategy

CE Design can become the basis of new competitive strategies because of the raising of levels of abstraction in products trend. The impact of this "raising" to the design process itself is discussed in Chapter 6, which describes design process strategies and the evolutionary path to a Concurrent Engineering Design Process needed to respond to the levels of abstraction issue. Here the discussion will focus on the impact of design-centered competitive strategies on the manufacturing organization's competitive advantage.

It is important to first discuss the cumulative nature of these strategies. As described in Chapter 2, the use of a design-centered competitive strategy must be proceeded by CQI, process restructuring, product simplification, and complexity introduction via the raising of the product's level of abstraction activities. The initiation of one, several, or all of these preceding activities, or even with CE Design introduction activities, or near simultaneous activities, depends on management's perception of competitive advantage gain versus risk and cost.

Once the precedent strategies are adequately initiated, a product design-centered strategy can be initiated. There are several design-centered product strategies. These product design-centered strategies permit many products or their components to address simultaneously different market needs while creating a ''pull'' or reasons for their customers to stay within one product family. Examples of product design features and functions characteristics that generate competitive advantage via the product design-centered competitive strategy include:

- *Consistent presentation and use*. The customer can move easily within the product family, as well as across product families. After learning to use the first product, use of succeeding products becomes easier. Examples of this product design-centered competitive strategy include the user application software interface of the Apple MacIntosh computer and the cockpits of Boeing 757, 767, 747-400, and 777-200 commercial airplanes.
- *Color and style recognition*. The customer can immediately identify the name or brand of the product. Examples include the Oreo® cookie and its six product derivations, the Ferrari® automobile, with its color and style; and the color identified with John Deere® and Caterpillar® equipment.
- *Interlocking or inter-related features*. In this product design centered competitive strategy, various products are designed in modules or components to physically or functionally interoperate. They can be sold separately or together. Examples include:
  — *Lego® building kits*. These interesting products, having unique pieces for particular designs and assembled structures (windows for houses), but whose other component parts interoperate across many company products, are excellent examples of a product design centered competitive strategy. Some might call this the Lego® system.
  — *Systems of many other types*. Examples of this product design-centered competitive strategy include IBM® PC-type computer software and hardware, and many different offerings of stereo components. By interlocking discrete components, each sold separately, the designer has greater design freedom and can evolve the components more rapidly. The product also can have a combined price and market share that is greater than a pre-combined unit, if sold as a single discrete product. However, the danger of this strategy is that competitors can compete at the component level. An example of this component market opportunity that has been created in the PC market, in which IBM® has a minority share. A competitive strategy response to this danger is addressed as the variety/complexity strategy.
- *Customer endearing qualities*. This product design-centered competitive strategy focuses on developing products which appeal to customers beyond the usual set of product features and functions. Examples include many different types of products designed with ''cleverness'' such as:
  — *Products with extra utility*–Simple products such as the ''Swiss Army Knife'' shown in *Figure 3-3*, are excellent representations of this product design-centered competitive strategy. Even the simplest ones have several pleasant surprises for the buyer. Recent editions have

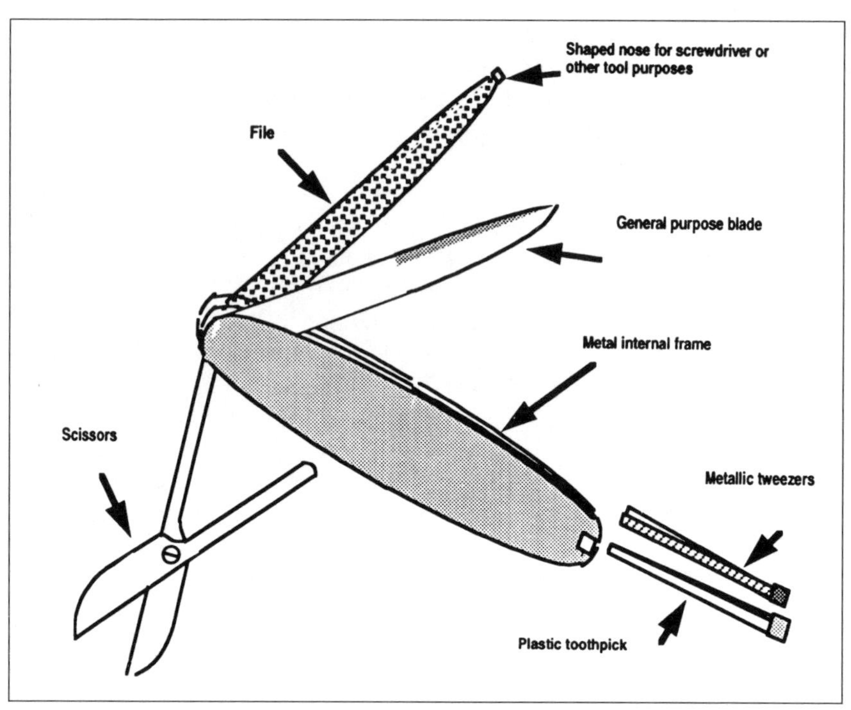

Figure 3-3. *Swiss Army Knife.*

added a screwdriver nose to the file blade small enough to handle Phillips head as well as bladed screws. Yet its original purpose, as a pointed end, has not been seriously affected. These tools also display another key type of product strategy, "unexpected parallel uses" discussed later.

— *Products that assemble and disassemble*–Another design approach that is enduring to customers is how products can be disassembled and then reassembled. The "smoothness" and ease of assembly/reassembly, in repeated use, is an important feature for many products such as cleaning instruments (vacuums, rug shampooers, etc.) and automobile tire-changing equipment. In the case of the Swiss Army® Knife *(Figure 3-3)*, the plastic toothpick snaps into a side slot.

This utility and cleverness can be very important in complex products because the product's enduring features can be further organized and then advertised. For example, the use of a "heads up" display and other related equipment by Alaska Airlines for its pilots allows planes to operate in almost zero visibility. Customers appreciate this adaptation of military technology that projects current flying instrument conditions on the pilot's window. This allows the pilot to never have to look down at instruments and thus to fly safer. It endears itself to customers because they can depend on airline schedules in bad weather.

- *Unexpected parallel uses*. This product design-centered competitive strategy is similar to the design process organized around a standards parts design strategy discussed in Chapter 6. This strategy tries to produce products with extra uses. Standards parts design attempts to re-use component parts on new or different assemblies. In the ''unexpected parallel uses'' strategy, the product itself can find new uses as well. For example, the video cassette recorder (VCR), used to record and playback magnetic videotape, has now been used as a substitute for digital data magnetic tape devices for personal computers.
- *Complexity and variety*. This last product-centered design strategy is becoming increasingly important. Almost a new category of strategy altogether, it builds on the level of abstraction concept, and requires at least many of the capabilities of CE Design to be operative before being completely enabled. In this strategy, the manufacturing organization deliberately introduces a constantly expanding family of products. These products are extensions of existing products, but are directed at new markets, not confined in one market or market segment. For example, an automobile manufacturer extends its engines into other parallel uses to include backup power engines, aircraft engines, boat engines, or snowmobile engines. While this variety of products are being introduced, product complexity (from a manufacturing process perspective) is also deliberately introduced. Products are constantly changed, with new features and functions being added. This constant change creates a difficult target for competitors to match. This change is introduced without confusing or driving away the customer. These different products are produced in the same factory, on the same line. This drives up process quality, adds to flexibility, while reducing costs across all products. This complexity and variety strategy requires CE Design when such a strategy is desired for complex products.
- *Systems of complex products*. This product design-centered competitive strategy integrates a set of complexity and variety strategy products into a higher level of abstraction yet. This strategy will also be further described in succeeding chapters. Preemptive execution of this strategy can create ''permanent'' or very long-lasting competitive advantage. An example might be all digital, direct broadcast high definition television (DHDTV). Such an integrated product, entertainment via computerized television, could be extremely difficult to overcome if established as a major country standard. Another example might be worldwide, all digital cellular phone technology based on satellites. This ''personal communications'' product offers a very appealing concept product using a closed or predefined and engineered set of standards controlled by a single consortia. This system, composed itself of digital systems, creates complexity barriers to potential competitors.

Figure 3-4 summarizes the ''internal'' manufacturing competitive strategies already discussed. These include: basic functionality, volumetric economy of

"Internal" Strategy	CE Design Relationships
(1) Basic Functionality	• Formal design process optimal
(2) Volumemetric Economies of Scale	• CE Design creates 'low cost producer" opportunity
(3) Quality in Product/Process	• CE Design plus product/process architecture create substantial competitive advantage for complex products
(4) Design Product	•CE Design plus product/process architecture can lead to preemptive competitive advantage
(5) Design Process	• Additional aspects to CE Design can be added to further distance competitors; to include: --geographic diversity --cross organizational design integration --Keiretsu --design process quality

Figure 3-4. *Competitive advantage "internal" strategies.*

scale, quality in product manufacturing process, and product design strategies. They are "internal" because the organization itself must execute these strategies. While engineered requirements and design cost targets drive the product's characteristics, there are no "artificial" external influences. *Figure 3-4* also provides for a final "internal" strategy based on the design process itself.

The Design Process as a Competitive Strategy

Once products are produced, they can be copied, or at least their functionality emulated. It is much harder to copy or emulate an internal process which cannot be observed or understood without active participation. Such a competitive advantage process is CE Design. Fully operative Concurrent Engineering Design resolves many product design quality issues, identified as cross-process, during typical CQI efforts and creates the opportunity to establish new design-based strategies.

As manufacturing organizations evolve during their adoption of the quality manufacturing process competitive strategies, the complex product manufacturing organization also should evaluate the organization's *core competencies*. The objective of the core competency evaluation is to identify and establish intellectual property content components, processes, and the processes of the organization itself, which provides its competitive advantage for each product. Many firms, after undergoing such an evaluation, are deciding whether to remain in a market if they are not be the best, or one of the top two or three performers in each potential core competency area. "What is the value added to the product?" and "How much of the product's key characteristics and price can be assigned this added value?'' These are the beginning questions for core competency.

The basic CE Design process for complex products, when deployed on a large scale over many components, products, and teams can become such a core competency. Very high contribution to product value can be attributed to CE Design. Besides core competency and competitive advantage created by basic CE Design, there are two other added design process elements which can add to this competitive advantage. They are: geographic diversity and technology ''push.''

Geographic Diversity

In the Geographic Diversity extension of the basic CE Design strategy, the capabilities of the CE Design infrastructure permit the distribution of design authority across geographically diverse elements of the organization. For example, an organization which has been segmented into a variety of MBUs, or Managed Business Units, can now share not just production management responsibilities, but design responsibilities for the various internally developed components as well.

The use of a MBU organizational strategy reduces the organization's resources (people, facilities, time, communications circuits, money, and management attention) devoted to coping with the size and complexity of the organization itself. Coping with size and complexity does not add value to the product. By establishing MBUs, the extra resources are freed up to perform other needed product enhancing functions.

Many integrated, complex product manufacturing organizations have reached the point of *diseconomies of scale*. Such integrated manufacturing organizations usually include most of the industry value chain's activities within one organization. The market analysis, design, production, distribution, direct or indirect sales, service, and most of the major subassemblies and their components are all provided by a horizontally and vertically integrated operation. This integrated operation is now being replaced by the MBU organization. It seems counter-intuitive, but having more organizational units seem to be less expensive, creates an easier to manage environment, and is more adaptable and responsive to the markets.

The problem in this MBU environment is what to do about product design. Historically, products are designed on an integrated basis; after all, it is a single product. The complex product manufacturing organization can adopt three different product design approaches to provide this integration in an MTU organization. These strategies can be supported by CE Design. The most appropriate can only be performed if CE Design is used as *the* design process.

Maintain Centralized, Integrated Product Design. In this process-centered product design strategy, CE Design is used, but MBUs are relegated to mainly ''focused factories'' roles. The focused factory produces a single portion of the overall product. Final assembly is performed by the same organizational unit that retains central design control. Most components supplied by outside firms are built to central design specifications. The obvious advantage to this strategy is the "tight" control potentially retained by the central design team. This advantage may be illusory when the number of design teams reaches more than a few dozen, because they will be so far out of touch with each other that they might as well be more geographically dispersed. This tightness, if sustained, can lead to cycle time-to-market improvements, a very important benefit in the complexity and variety strategy. The disadvantage is the inevitable "gap" which will exist between the designer and the producer in the probably geographically dispersed focused factory. Design for Manufacturing, a major CE Design process, is hampered by this physical distance. This "gap" can be offset somewhat by technology such as interactive-across, multiple-device 3D CAD, video conferencing, and other "electronic proximity" enablers.

A Fully Dispersed Approach. In this process-centered product design strategy the MBUs each have full design and manufacturing responsibility for a portion of the overall product. High-level design is coordinated between the MBUs. Central engineering acts as a communication coordinator, and resolves design conflicts. It may have full product performance characteristics responsibility, but it has no other direct design authority. In this strategy, the focus of design is on interfaces. For example: a large aircraft consortia in Europe divides its commercial airplanes into major responsibility sections. The front section, or the cockpit, the rest of the main body, the wings, the landing gear, the engines, and the avionics (the electronic guidance and control systems) are all designed and produced separately. Final assembly consists mainly of "mating" the pieces together, performing final tests, and marketing and delivering the end-product. All other aspects of design and manufacturing have been completely dispersed. The obvious advantage to this strategy is that the full benefits of MBU organization are enabled, and the designer-producer "gap" is diminished. A significant potential disadvantage includes potentially poorer integrated product performance. Poorer product performance and maintenance arises because of the lack of design integration resulting from inevitable design communications errors and omissions.

CE Design must be implemented with this approach to significantly mitigate its disadvantages. Concurrent electronic mock-ups can be substitute for co-location, and fully integrated product simulations allow design teams to achieve the same benefits of integrated design without the "diseconomies of scale" encountered by the fully centralized, integrated manufacturing organization. Hence the term "design by simulation." Finally, greater design process direct cost can be gained.

Multiorganizational Design Integration. In this process-centered product design strategy, the MBUs each have full design and manufacturing responsibility for a portion of the overall product. Design is coordinated between the MBUs. Central engineering acts as a communication coordinator, and resolves design conflicts. In addition, design responsibilities are shared by partners, other consortia members, and qualified suppliers. This is the approach most interesting to suppliers, nations, or markets desiring to "move up the design chain, or wanting to participate in the complex product production process but, at the same time, lacking the capital or personnel skill base to be co-partners, or serious competitors. Increasingly, this sort of participation is a precursor to access to these markets. The highly paid jobs, the high value products, and the capital generated by participation are of great interest to every nation.

Using CE Design, this process centered product design strategy can be mixed with the *Fully Dispersed Approach*. MBUs can themselves work with key suppliers and partners on a smaller element of the MBU's major product element. Suppliers should be constantly seeking to participate in product design. This provides them a competitive advantage in their tier of competitors.

Extensions of the CE Design process competitive advantage strategy can further distance the CE Design capable manufacturing organization from its competitors. The *Fully Dispersed Approach*, when combined with the *Multi-organizational Design Integration* approach, yield a pre-emptive competitive advantage if combined with other world class initiatives, as discussed in Chapter 4.

Technology "Push"

The second of these major pressures "pushing" manufacturing organizations into Concurrent Engineering Design is technology "push." This is a heavily discussed topic. A variety of techniques, tools, and technologies are part of "technology push." These technologies can be incorporated into the product, into the manufacturing processes, or into the design processes. Usually, when one thinks of CE Design processes and technology, one is thinking about computing in support of the processes.

The software and systems necessary to provide all the support described as a part of the Automated Infrastructure Support process review may be currently available as needed by each manufacturing organization, in the manner necessary or appropriate. In general, there are several "baseline" or infrastructure support technological capabilities that are necessary for Concurrent Engineering Design to function effectively. These requirements and functional capabilities are described in Chapter 8.

Technology itself cannot create lasting competitive advantage. Any technology available for sale will reach competitors. How technology is deployed into processes, and the improvements made through technology in those processes which cannot be easily duplicated, *can* be part of a competitive advantage. The discussions of technology in this book will focus on aspects of computing that can improve CE Design and its processes, and the manufacturing organization's competitive advantage.

Process "Pull"

Technology alone is not enough to create a successful implementation of Concurrent Engineering Design. It is the identification of improvements that could be made to internal business processes that "pull" improvements such as Concurrent Engineering Design into the organization.

One of the keys to the implementation success of any change is the desire of the organization's people to change. Typically, some sort of "crisis" has been necessary for change to be accepted with little resistance. An organization's staff is more aware today than 10 years ago of the world, its economic and competitive environment, and the effects of the overall situation on their current employer and themselves. Their awareness of competition and its impact, their awareness of technology and its impact, and their general perception of the "intensity" surrounding their individual job environment has created an overall environment more conducive to change and more apprehensive about change.

One of the perceptions associated with "intensity" is speed. Everything has to be done faster. Everyone seems to understand the need for speed, but they are concerned that they will not be provided the tools to go faster. Another of the perceptions associated with "intensity" is complexity.

Figure 3-5 shows the various aspects of complexity and its management that affect the manufacturing organization. Increasingly complex products, higher speed, and a more complex business and organizational environment all are creating a situation in which the need for control and the resulting conflict with speed is apparent. This conflict has many different symptoms. One of these symptoms is shown in *Figure 3-6*. The "easy" solution to this conflict between complexity, speed, and control is not apparent. Yet, competitive advantage can be established if the solution to these issues is found.

The other important aspect of process "pull" is the integrated nature of Concurrent Engineering Design. While it is important to have an environment in which people identify improvements that "pull" change and improvement, these changes are usually focused on limited scope intra-organizational elements, not across many element improvements. Concurrent Engineering Design, being an

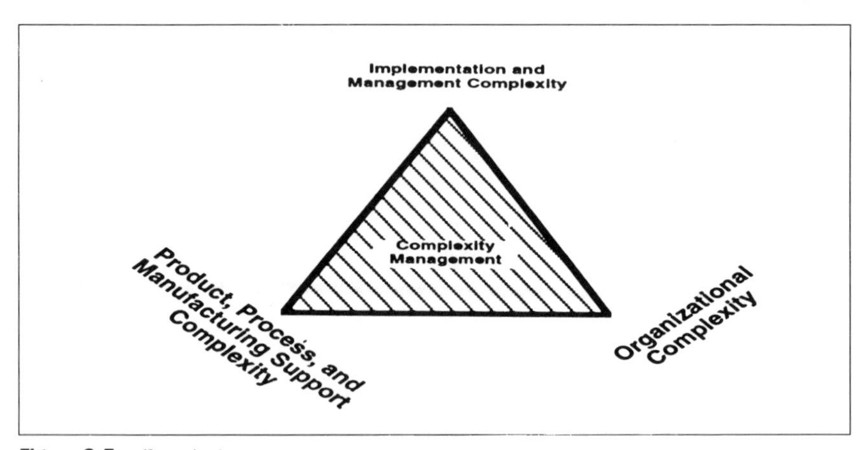

Figure 3-5. *Complexity management.*

integrated process which changes "how things are done," in many organizational elements simultaneously, requires more than just simple improvement projects put together, as valuable as they may be.

As shown in *Figure 3-6*, after extensive analysis, one firm found individual or typical team generated improvement suggestions rarely affected more than 15% of the overall process. Reasons for this include: scope of responsibility even at high levels in the organization; scope of vision, where due to specialization, it is hard to visualize improvements in nonspecialization areas; and reluctance to make recommendations outside their scope of responsibility and vision, because the success or failure of the change is not controllable by the recommender. Cross-functional teams have more success in larger scope recommendations, but they must be empowered by executives and be orchestrated as a part of an integrated vision.

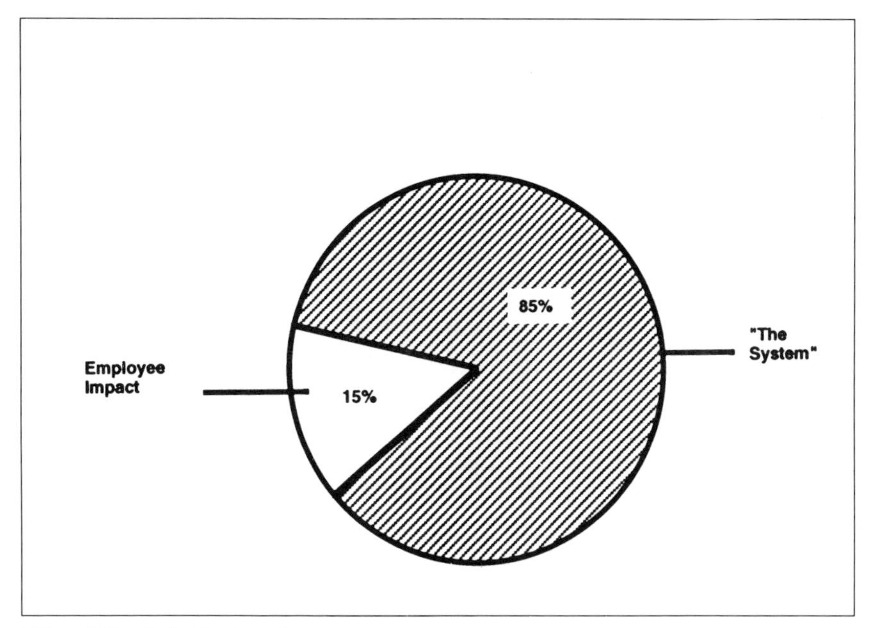

Figure 3-6. *Individual employee improvement impact.*

As shown in *Figure 2-10*, integrated processes have very high rates of return (10-100 times). However, the implementation of an integration level process must be different; the implementation process must be focused on the integrated process view from its inception. The key to Concurrent Engineering Design implementation success is to identify potential processes that will benefit from Concurrent Engineering Design, get cross-functional teams involved in these processes to document their present processes and identify improvements (as described in Chapter 7). While this is proceeding, one must train these cross-functional improvement teams in Concurrent Engineering Design and its integrated nature. This generates ''process pull'' and creates the successful implementation desired. CQI is an excellent vehicle for performing these prior to CE Design identifying activities.

Historically, manufacturing in the U.S. has focused on ''production manufacturing.'' The design and engineering processes were viewed as preparatory to manufacturing; important, but manufacturing actually ''built'' the product and did what they had to do to get product of adequate quality out the door. Unfortunately, initiatives begun with just a manufacturing focus, have shown improvement, but as shown in *Figure 3-7*, improvements made, even with world class manufacturing initiatives, cannot support improvements which can reach the target levels appropriate for world class manufacturing. For example, manufacturing cannot overcome inaccurate bills, designs that need to be corrected, or tolerances that are inappropriate. However, with heavy screening, bill accuracy can be improved, work-arounds to deal with poor designs can be developed, or more expensive equipment can be acquired to meet tolerance requirements. Improvements are possible.

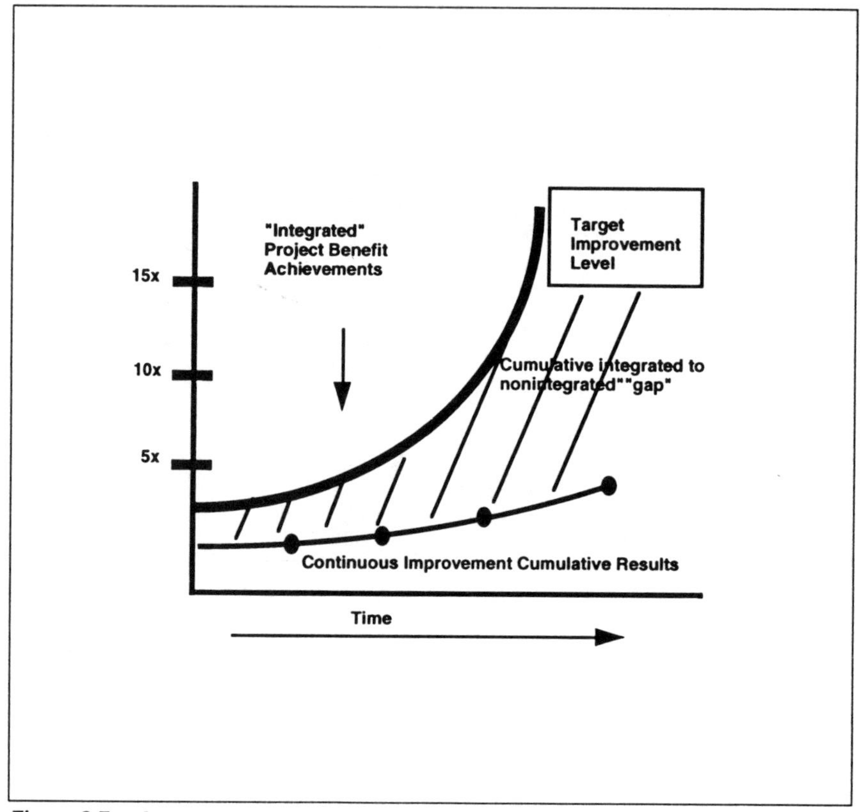

Figure 3-7. *Improvement level achievement.*

The problem with these "manufacturing only" solutions is their sub-optimal goals of defense against someone else's error, and their internally focused performance improvements. Of course, this would be true of any functional area's initiatives, not just manufacturing's; this includes CQI in production manufacturing only. Thus, production manufacturing must integrate itself, with engineering and design, if only in self-defense.

An example of how technology can, inside processes, create a competitive advantage, is shown in *Figure 3-8*. The diagram in this figure uses the results of a project that evaluated the activities which engineers actually performed. The result of the study was that engineers spent approximately 20% of their time actually doing design.

Using the CE Design process and its associated technologies can result in significant improvements in the amount of time actually made available to the engineer. This improvement in time availability, if used productively, can result in, among other developments, a direct reduction in engineering flow time, reductions in product costs, and improvements in product quality. *Figure 3-8* presents an example of benefits. A 5% reduction in other than design activities translates into a 25% improvement in design productivity and a 4% to 6% direct reduction in product cost. Much larger reductions in non-design activities, and resulting benefits can be expected. A significant cost-based competitive advantage in engineering is possible through CE Design, along with time-to-market, quality, and the other competitive advantages possible as described in this chapter. Technology, as a part of the CE Design processes, automated infrastructure support, can indirectly create significant competitive advantage.

EVALUATING AN ORGANIZATION'S CE DESIGN READINESS

Governments, businesses, and academic groups are all actively debating the issue of competitive advantage. In this new environment, what can a nation or society expect? Is it reasonable to expect that a nation can be dominant at everything, or eventually, at anything? For the complex manufacturer, the issue is what, if anything, should be done to change the current organization's approach to competitive strategy?

Appendix A contains two questionnaires. Questionnaire 1, "Complex Manufacturer Determination," determines if the organization could be considered a complex manufacturer; it also contains several interpretation examples. Many more manufacturing organizations can be classified as "complex manufacturing" than usually believed.

Questionnaire 2, "Readiness for Concurrent Engineering Design," evaluates the organization's status relative to the Concurrent Engineering Design process

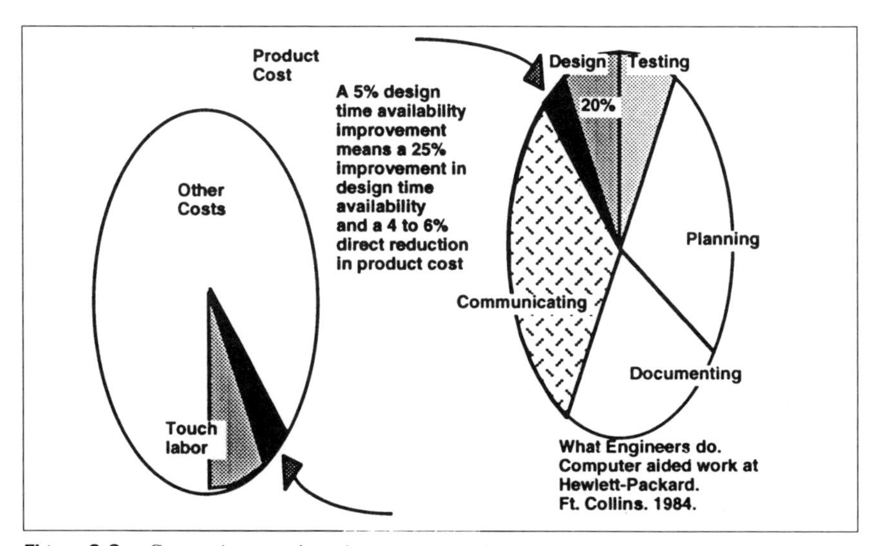

Figure 3-8. *Generating cost-based competitive advantage.*

and its process and systems requirements and capabilities. It assists in establishing a baseline from which implementation plans can be developed. This questionnaire is best used after the book has been read and understood.

A question of significant interest to senior management and/or business owners, is whether any new business strategy is necessary. Their first question must always be "Must I?" and "How much is it going to cost?" If the organization has complex manufacturing characteristics and has or anticipates having competition, then world class manufacturing capabilities and their primary enabler, Concurrent Engineering Design, will be implemented by someone. The resulting implementations within a direct competitive niche will have to meet at roughly the same time, at least, or permanent competitive disadvantage for some may result. Recovery will be more costly, the recovery time substantial, and the profits lost significant and unrecoverable. For most manufacturers, implementing Concurrent Engineering Design as an element of a world class manufacturing competitive advantage effort is a necessity now.

If the organization does not have complex manufacturing characteristics, management should determine if competitive position would advance through some of the dimensions of power of Concurrent Engineering Design, and/or the business characteristics of world class manufacturing. If so, implementation of the Concurrent Engineering Design process as a precursive enabler will be important, and worth the effort in this circumstance as well.

Finally, if an organization supplies another, it is important to determine if the firm supplied (or whoever is the final assembler and marketer of the end-product) is a complex manufacturer.

To keep this business, the supplier organization must adopt the Concurrent Engineering Design enabler for the following reasons:

1. To maintain the level of quality and timeliness demanded;
2. To manage the increased "risk participation" shown by longer term contracts with more responsibilities borne by the supplier which will come with this new environment;
3. To prepare a significant capability from which adopting the other aspects of world class manufacturing is reasonable, and
4. To supply whoever the "winner" is in this environment.

If the end-product manufacturer currently supplied doesn't adopt Concurrent Engineering Design quickly, a competitor will. The supplier organization's ability to survive in that market will then depend on it being an attractive supplier to someone who is not now a current customer of the organization's component products who has adopted Concurrent Engineering Design.

From a manufacturing perspective, responding to this competitive pressure with a shift to competition on the basis of world class manufacturing built around CE Design is beginning to occur. The premise of this book is that a long term, permanent competitive advantage will result of an organization's adoption of "world class manufacturing." The CE Design concept is the absolutely necessary component and enabler of world class manufacturing.

4

CONCURRENT ENGINEERING DESIGN AS A WORLD-CLASS MANUFACTURING ENABLER

As depicted in *Figure 3-1*, there are three major pressures driving manufacturing organizations to adopt Concurrent Engineering Design. The most important of these is business *competitive pressures*. Responding to *competitive pressures* can take many directions. The best way to seize the initiative is to develop and to carry out a business strategy that provides an advantage over its competitors. Seeking a competitive advantage, especially a permanent competitive advantage, requires a change to world-class manufacturing status. This status reflects very high performance in five critical areas simultaneously. These areas are:

- Quality,
- Cost management,
- Time-based competition,
- Technology, and
- Variety and complexity.

Competitive advantage can come from the prior stages described in Chapter 3. These advantages, however, are transitory. They do not create insurmountable barriers for potential competitors, except for invention. When there is direct competition, it now appears that a lasting advantage may only come from doing better business through basic changes to the organization's operational processes and procedures.

This competitive advantage is not artificial as are the preceding stages. The artificial barriers to market *entry* are only partially effective. This type of competitive advantage is enduring. The competitive advantage is in market *sustainability*. Competition through Continuous Process Improvement in many industries has goals and objectives such as:

- Quality improvements in both product and process used to produce the product that yield cumulative defects (in both) of less than 10 per million, (as championed by Motorola in their Six Sigma programs).
- Cost reduction of five to 20% compounded per year in constant money value.
- Cycle time reductions for new product or change incorporation schedules of 30 to 200% over a five-year period.
- Time or speed to market with such goals, for example, as the ablility to order a customized car and take delivery 96 hours later.
- Using "best practices" and core capabilities to jump industry categories with instant success.
- Quickly incorporating a new technology across all product lines, and
- Introducing an increasing variety of products, many of them new, with more options and customer-featured changes during every product life cycle for an ever-increasing number of product niches.

Some are trying to address these goals by doing the same activity faster. Just speeding incorrect processes will not work and will more likely decrease quality. Others are only making minor process changes.

The major improvements are not usually achievable by small measures. Some firms in every industry are going to accomplish these goals. This achievement will result in a paradigm shift because the organizations will have reorganized and redesigned themselves to achieve these goals. They also will have built a continuous improvement environment that is self-sustaining. This will allow their competitive position to continue to improve for an extended period or time. Thus, they will have established a long-term competitive advantage through *business* process redesign that continues to improve itself. This business process redesign is the next stage of competitive advantage.

When dramatic improvements in these competitive process attributes occur, the net effects appear as in *Figure 4-1*. Costs drop dramatically, soon, and permanently, and profitability rises quickly and is sustained at a higher level for a long period. Such increased profitability feeds R&D, provides increased

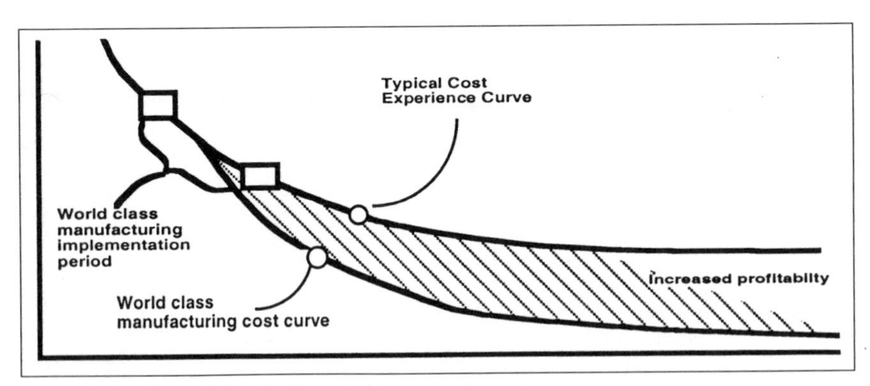

Figure 4-1. *Competitive Attribute Achievement Impact.*

capital, and promotes the ability to keep a competitive advantage. Unfortunately, firms can squander away such an advantage.

There has been a tendency to follow a concept embodied in the phrase: "If it ain't broke, don't fix it." Unfortunately, it is the success of the past that impedes the success of the future. To gain this next competitive advantage, organizations must constantly look for examples of the "best of the best" business processes, not just in their own industry but throughout the world. Increasingly, competition, especially the most devastating type, comes from unexpected sources. Many times the organization that is the source of devastating competition is a newcomer to the industry. New entrants with new technology, new concepts, and much higher quality can create an imbalance and destroy their competition. These organizations employ best-of-the-best practices, and then look for new niches in which to employ these practices. *Increasingly, the concept is "If it ain't broke, improve it, and constantly."*

WORLD-CLASS MANUFACTURING QUALITY

Moving to the next stage in manufacturing competitive advantage includes permanent changes in how a firm addresses its general business activities. Correct reimplementation can transform the firm into a world-class manufacturer. But how does Concurrent Engineering Design relate to world-class manufacturing?

It is probably most important to start exploring how Concurrent Engineering Design enables world-class manufacturing by discussing quality. There has been an explosion in discussions about quality. It turns out that product quality is more than free. In fact, quality generates revenues and competitive advantage.

Reviews of the "cost of quality" (e.g., all the various elements of the organization's attempts to improve and control quality) show that improvements in product quality can have a substantial and pervasive impact on competitive advantage. Process quality, however, can be very expensive; some surveys indicate that up to 25% of the total cost of a product can be attributed to process quality.

Figure 4-2A describes the various stages through which both product and process quality is pursued. The typical product QC program uses inspection and sampling techniques to detect failure. Testing catches internal failure. Sometimes the product fails in the hands of the external customer. Preventing a quality problem costs 10-100 times less than fixing a failure with a customer. The objective of advanced quality programs is to move quality from "external failure," (in the hands of the customer), progressively back in the overall manufacturing process to "prevention." The basic premise is that quality cannot be "inspected in," but must be "designed in." As shown in *Figure 4-2B*, the basic approach of advanced product quality programs is to catch errors during design and to make in-line corrections. This is a critical step before pre-delivery and post-delivery.

It is through the quality of design, where products have design robustness, and exhibit utility and reusability, that the greatest impact of quality on overall

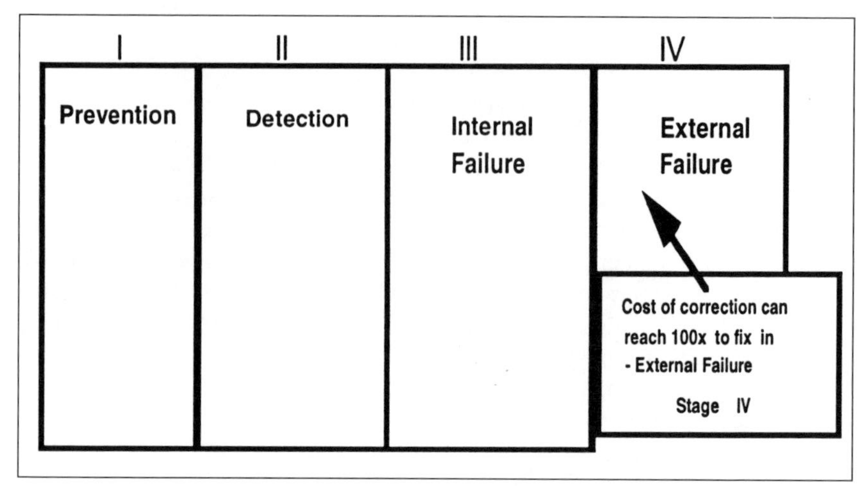

Figure 4-2A. *Quality Stages.*

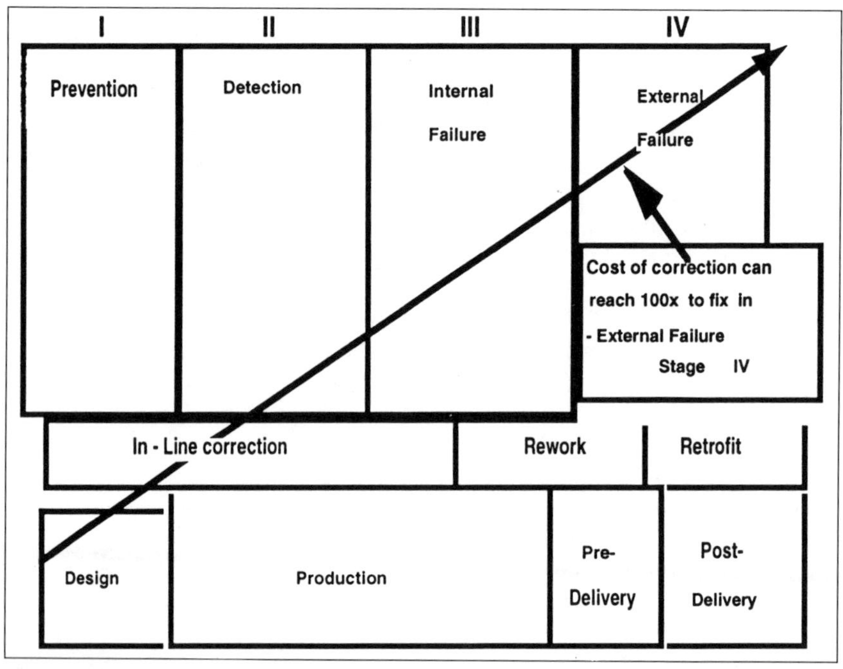

Figure 4-2B. *Quality Stages.*

product and organizational profitability can be achieved. Aspects of the design process which generate robustness and reusability include design for manufacturing (DFM), where simplification, and ease of manufacturability and reproducibility are realized.

An additional aspect of design, that of design for utility, occurs when design focuses on ease of assembly and ease of disassembly and maintenance. There are literally hundreds of tools and techniques which have emerged in recent years as the focus on design for quality has been increased by world-class manufacturers.

The focus on quality in design permits cost savings through a reduction in the overall cost of quality by preventing quality problems through designing problems out of the product. Costs are avoided by reducing the size and cost of (1) inspections and other internal quality control processes, and of the (2) field service operation. Revenue is generated by reducing the cost of manufacturing the product and getting to market more quickly, thus realizing additional profits. In addition to internal improvements generated by quality through design, improvements through quality can generate revenue because as customers become more sophisticated, superior product design and product engineering generate greater sales and revenue. Quality allows an organization to move from "customer satisfaction," at best a neutral perception which does not imply any customer loyalty, to "customer delight," which brings customers back.

Concurrent Engineering Design permits and promotes an emphasis on quality by providing several added capabilities, including:

- Additional time,
- Additional comments,
- Inputs from production manufacturing and a number of other disciplines,
- Application of statistical process control to the design process.

Most of all, it is the integrated environment of Concurrent Engineering Design which promotes quality by stressing all of the aspects of product design simultaneously. This is Concurrent Engineering Design's most important capability; that is its focus on quality in the design process itself.

Concurrent Engineering Design turns out to be so pervasive, so able to cross over, that it generates competitive advantage across the product lines of an organization which adopts this best-of-the-best process. Some highly successful firms are already deploying the cross-over impact of this best-of-the-best practice, and produce consumer electronics, computers, cameras, copiers, automobile components, and the like in the same facility with many of the same people.

WORLD-CLASS MANUFACTURING COST ACCOUNTING

Addressing how world-class manufacturing, enabled by Concurrent Engineering Design, is advanced by costing may seem far afield.

The prior section on quality's cost reductions, savings, and revenue generation describes how cost is improved through Concurrent Engineering Design. However, the cost management "best practices" element of Concurrent Engineering Design can, in and of itself, also lead to cost reduction, savings, and/or revenue enhancement.

Erroneous cost accounting is currently one of the major reasons for a loss of competitive advantage. Traditional cost accounting theory has focused on "burdened labor hours." A burdened labor hour is a "touch" labor hour with all other costs allocated to it. This concept has been used without much change since the 1920s, and is based on the principle of direct or touch labor consuming 70% or so of the product's cost. In today's complex manufacturing environment, touch labor is more likely 10% or less.

The problem with today's approach using burdened labor hours is that in complex manufacturing, the allocation for overhead rules used have become dangerously distorting. The touch portion of the cost has gone from 70% to a range of 10% or lower. Thus, allocations now comprise the overwhelming part of the product cost, and if applied via traditional measures, such as burdened labor hours, cause misunderstood improvement objectives or allocation distortions.

In the area of *misunderstood improvement objectives*, many cost systems continue to focus on headcount reduction or labor hour reduction, based on information captured 30 to 45 days after the fact. Analyses show that more labor is sometimes put back into the product in quality inspection and other allocated costs, than originally in initial product manufacturing. Changing the focus to what are the "best practices" and overall process improvements can cause touch costs to go up for a while—at the same time—overall product cost is substantially reduced.

In the area of *allocation distortions*, many allocation pool approaches have been found to substantially distort product cost because products do not consume all overhead and indirect costs evenly. "Make-buy" decisions in particular often are found to be backward, in that a company is buying what it should be making, and vice versa.

These distortions focus senior management away from the nontouch portion (or overhead, indirect and the like) of the company, where the cost-containment opportunities come from, and from where the competitive advantages also emanate.

Most companies desire to provide line managers and operators with total cost per-completed-unit statistics. This allows those same managers to vary controlling inputs (i.e., costs such as direct or indirect labor or capital) in any combination, without micro-managed cost variance reporting traditionally computed from standard product cost oriented accounting systems. Manufacturing cells and inventory-bufferless material flows should make this form of accounting easier.

The advance in cost accounting necessary to achieve world-class manufacturing is to move to activity based costing (ABC). ABC reflects how complex manufacturing is actually conducted today, and will be in the future. Most non-touch or direct labor and materials actions are just that, activities.

Activity Accounting serves two purposes. First, it produces substantially more accurate product costs by detaching the costs of activities which do not vary in direct proportion with unit volume (e.g., labor or machine hours, good units produced). Second, ABC data provides managers visibility on activity costs for a variety of management initiated programs, including continuous improvement initiatives. Costs and cost driver rates can be trended, or compared to target costs or best practices. As shown in *Figure 4-3*, the establishment of the

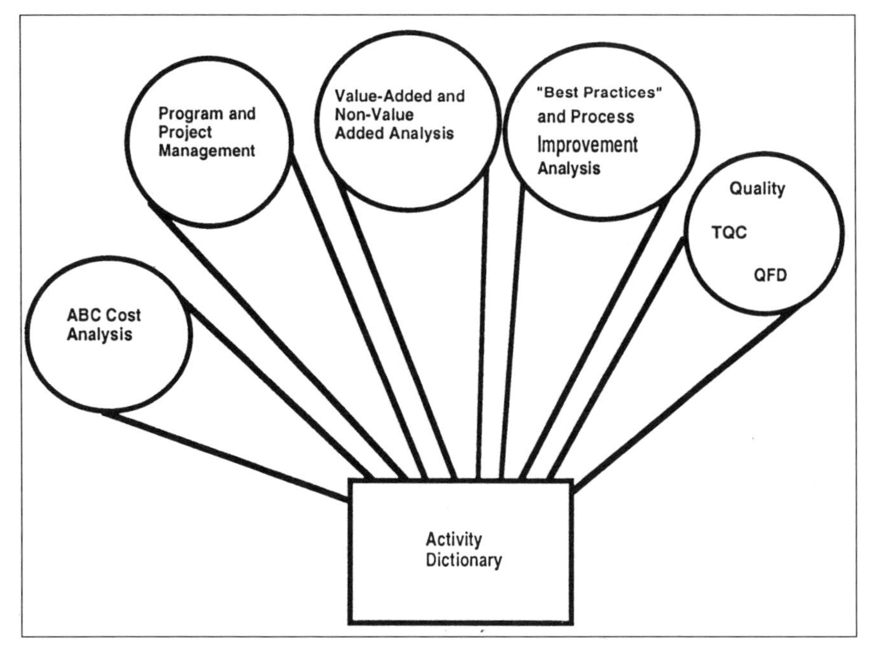

Figure 4-3. *Activity Dictionary as a Basic Element.*

Activity Accounting dictionary for activities actually allows many other parallel improvements in measurement to occur.

A simplified approach to ABC is possible under certain circumstances. For example, in simple subassembly manufacturing, tracking shop labor and material is no longer necessary. These costs are so small that they are immaterial to the product's overall cost. Alternatively, if the environment is a focused factory, which makes a few noncomplex products which do not change, and which has little nondirect cost, then its total cost can be blended. In both these cases, costs can be aggregated and spread over end products as they are produced, or "backflushed." This technique can be called "process or process center" accounting, and is seen frequently in Just-In-Time or Kanban environments. In a complex manufacturing environment, both techniques may be used in organizational areas where applicable.

Process Center Accounting's appeal to companies comes from its simplification and support of the trend toward empowering work teams. It serves a different purpose than Activity Accounting; consequently it is acceptable to include some overhead costs traced to Process Centers despite their variable cost behavior (by batch or by product) which may not be in synchronization with output volume (in completed units or product design features, like number of holes drilled).

The most probable approach to this coexistence in complex manufacturing is to evolve the cost accounting approach to that shown in *Figure 4-4*. *Touch labor* and *direct materials* are those actually used to manufacture the end product.

Indirect labor includes those costs which can be tied to products or products lines, such as marketing, purchasing, etc., where a majority of activities are product-related. *Production overhead* includes people, engineering, plant and equipment, and other elements used directly to execute the manufacturing process, but not attributable to an individual end product. *Other indirect overhead* includes those functions which are company-wide focused, including personnel and certain executive functions. Product-related activities include indirect labor and touch labor and materials. These costs are usually captured during production manufacturing data capture. The processes which occur during production overhead are linked to product processes and their costs are captured. Other indirect overhead is allocated if process linkage is not possible.

As shown in the *Figure 4-4*, the ''Future'' approach is really a return to a basic cost accounting philosophy, allocating small elements of the cost and directly measuring and managing its large elements. Using this philosophy, the touch category can usually be distributed via an output-oriented allocation methodology, such as good units produced as used within Process Center Accounting. If, however, change and variety are intense, ABC will still have to be utilized at this level. If Process Center Accounting can be used, this approach brings this portion of costing into concurrence with JIT and other throughput-oriented and velocity-based manufacturing techniques. The large amount of the remaining cost is directly measured and attributed to products via activity collection.

ABC, or activity accounting, is most useful for the majority of costs because almost all of these nontouch costs are accumulated in activities or actions which affect the products or processes. They are traditionally managed at the level of individual activities. For example, engineers keep time cards by project or product activity; computing equipment is used on a metered or time/activity basis, etc.

Activity Accounting also is useful in this complex environment because it removes most of the distortion in allocations. As shown in *Figure 4-5*, different products consume different amounts of overhead and indirect costs, just as products used to consume varying amounts of direct touch labor and materials. There will be gaps in this approach which will still need to be allocated because, like direct costs, they are small and will not distort the overall cost picture. Thus, this approach returns a company to its cost focus of the past, restructured with a forward-looking orientation to reflect the changing circumstances and management needs of today.

By collecting costs as described in the Complex Manufacturing Cost Accounting Approach Model *Figure 4-4*, predictive accounting becomes possible. Models of processes, built during value chain or valued activity analysis, now can be cost accounted. Individual changes can be measured and/or predicted. Only with a technique such as ABC are we not doomed to constantly study for improvements and then have the data used to make the decisions obsolete because it cannot be updated on a routine basis.

In a world-class manufacturing environment, continuous improvement and

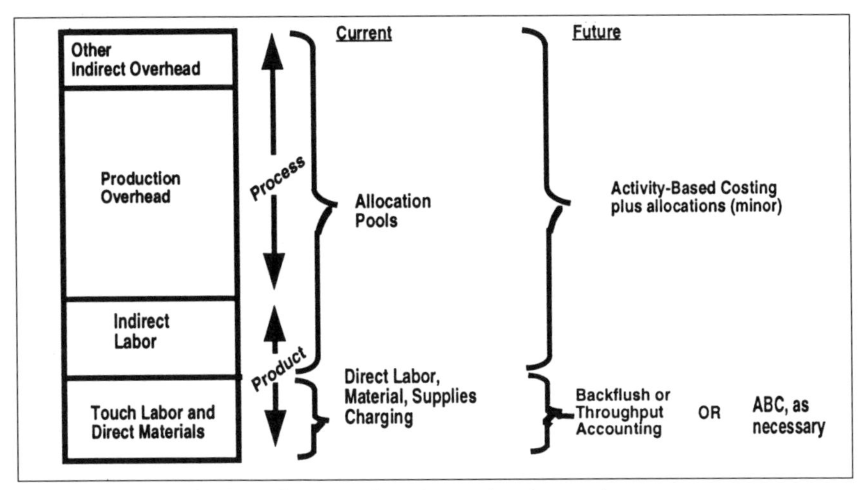

Figure 4-4. *Complex Manufacturing Cost Accounting Approach.*

business process redesign are keys to success. Cost improvements are important, since they correct decision-making and priority setting. Among other benefits, adopting the Concurrent Engineering Design process creates substantially lower engineering process costs by raising engineering productivity through a variety of means as measured by activity dictionary-based costing.

Concurrent Engineering Design and Cost Accounting have three additional principal interest points from a world-class manufacturing perspective:

1. Product cost versus target cost;
2. Engineering process cost, and
3. Cost of quality.

The CE Design environment supports the development of engineering process cost directly by capturing, in an unobtrusive manner, most engineering activity. CE Design does this by using its built-in ABC processes to develop process cost. Modeling, budgeting, predicting, and managing engineering process costs are thus supported. Since engineering process cost and resource balancing are interrelated, additional discussions on those points follow.

Product cost versus target cost. This is a key issue today. The market determines target cost; a variety of techniques assist in this determination. Techniques include using features and functions targeting in focus groups, surveys, competitive product analyses, and direct customer contacts by both marketing and engineering personnel. The key is to establish target costs early in the design process, and then to design and build so the first of the new product out of the manufacturing process is produced at that cost. Techniques, such as Quality Function Deployment (QFD) from the American Supplier Institute, help structure product specifications.

The Concurrent Engineering Design process is so important because target cost establishment techniques become institutionalized and embedded in the systems supporting Concurrent Engineering Design. The Concurrent Engineering Design process is important in the product cost area because the whole intent of the process is to drive down engineering and manufacturing process errors and drive up the product's design robustness. Both of these factors drive down product and process costs. ABC permits the proper cost picture to be understood.

The cost of quality. This concept reflects the perception, as reviewed in *Figure 4-4*, that design is the best place to ensure product quality (design robustness) and prevent errors. Error detection and correction is expensive. In a traditional engineering environment, time and engineering process cost (the project budget and schedule) are usually the enemy of robustness and error reduction. The traditional serial process reduces the number of potential design alternatives which can be analyzed, compared to the rapid design conceptualization-realization iteration cycle times available with Concurrent Engineering Design. The traditional serial process also is more likely to permit error because it is essentially a positive feedback loop. People must take the initiative to examine and find errors. Procedural errors can be spotted easily, but with time pressures and a lack of design intent understanding by reviewers, design and manufacturability errors are much more likely to be caught after manufacturing has begun or, unfortunately, even in actual product uses.

The Concurrent Engineering environment is focused on correcting this positive feedback loop feature of the traditional serial process. By building in

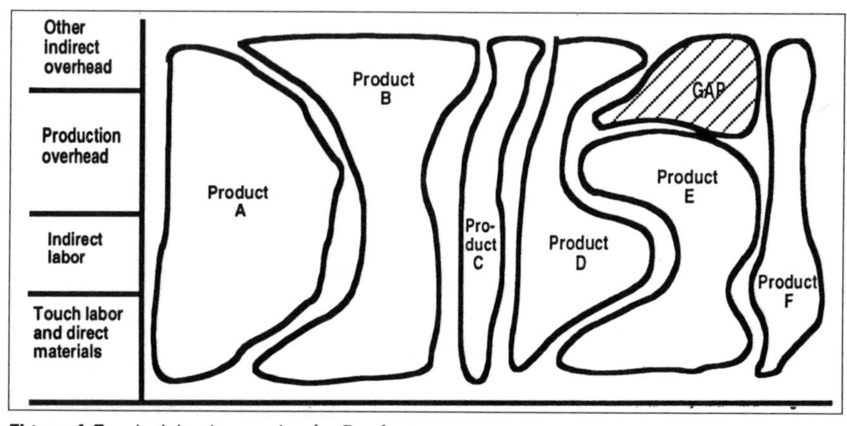

Figure 4-5. *Activity Accounting by Product.*

automated routing and queuing, high-speed automated support for design alternatives and by providing for simultaneity of understanding of design intent, errors are more likely to be caught during the design process.

Design robustness also is supported by this simultaneity characteristic. Understanding design intent permits a more thorough and faster evaluation of design. But robustness has other dimensions to it; these include manufacturability, assembly and disassembly, reliability, and maintainability. All aspects of design robustness have a positive impact on product cost (lowers cost of production and maintenance) and engineering process cost (lowers cost of engineering by reducing engineering rework). And design robustness permits target cost achievement to occur.

All of the costing analysis performed to support design robustness revolves around activity analysis. For example, analyzing various process alternatives includes deciding which activities to perform, in what sequence, and at what cost. Correct costing in a world-class manufacturing environment is a key to many of its processes; Concurrent Engineering Design and cost accounting are intertwined elements which lead to world-class manufacturing.

WORLD-CLASS MANUFACTURING AND SPEED TO MARKET

As described in preceding sections, if a product can be delivered first to the market, marketed adequately, and priced aggressively with high quality, then desired success will surely follow. If this faster cycle between product need identification and product delivery in quantity can be repeated consistently, a lasting competitive advantage is attainable. The Concurrent Engineering Design process is the enabler for this world-class manufacturing characteristic of time-based competition, or faster speed to market.

In *Figure 4-6A*, a diagnostic representation of the serial process of design, including suppliers, is shown. As represented in *Figure 4-6A*, the flow time between initial idea and product in volume is the aggregate result of numerous processes, each with its own cycle time or elapsed time from initial contact. The traditional total flow time or throughput reflects the serial aggregate of these cycle times. This total flow time can be aggravated by intra- (inside) and inter- (between) process errors. Especially damaging are errors which are inter-process in nature. For example, a moving arm which should have been at a two-degree greater angle might pass all cycles until testing or even field use before the error is discovered.

Each manufacturing function, from engineering to planning to distribution and sales has its own systems and procedures with barriers to keep errors out of their processes from upstream processes. These barriers might include checkers, inspectors, planners, etc.

As shown in *Figure 4-6B*, the problem of cycle times gets larger when production manufacturing and post sale support is included. The interfaces between these islands of process and operation are formidable barriers to flow time reduction, because they are perceived as necessary, yet are time consuming and expensive. They are necessary because existing systems and processes are optimized for each group or function's internal purposes. Many times, such internal optimization inadvertently presents downstream functions with information and instruction, material and processes, or other matters which seem correct but are ambiguous, contain errors of omission, or specification misunderstandings. These generally occur because not all of the elements of the communication (such as design intent) are written and/or communicated correctly. When the manufacturing organization is small, or the product is simply from a final assembly perspective, or change is slow, these miscommunications can be kept to a minimum by personal contact.

Using aggressive JIT, Kanban, and other flow minimization techniques, organizations have substantially reduced inventories (90%), reduced internal cycle times (95%), and improved quality in a period of six months to a year. Making the cost accounting changes (ABC and/or backflush or process center accounting) and utilizing statistical process control (SPC) techniques for quality control facilitate these dramatic improvements. In some cases, automated systems, such as MRP II can be impediments to success in this environment. MRP is a "push" operation. JIT and Kanban are "pull" oriented manufacturing environments. MRP actually pulls to the master schedule, but is seldom used in this fashion. MRP remains important and useful for that portionof the product's structure which exhibits complexity and high change rates.

The organizations which are the main targets of this book and of CE Design do not have the simplified environment described. Instead, they have a complex interplay of suppliers, constant change, build cycles which are longer than the final assembly cycle, and complex products. In this type of organization, the final product's content might include elements manufactured in process-type manufacturing, and in discrete, electronic, and composite-type manufacturing as well.

The problem in the complex organization is that a local optimization of cycle time does not improve overall throughput. The objective of world-class manufacturing is to increase overall throughput. The key to resolving this in the complex product environment is Concurrent Engineering Design.

Concurrent Engineering Design addresses overall product throughput in several ways. These include:

1. Reducing the number of processes and their internal business cycles which are operating independently, thus reducing the number of interfaces and inter-cycle barriers while providing a more integrated set of design and design to manufacturing processes. This integration must provide for both physical and electronic proximity, while facilitating the rapid and essentially simultaneous communications among the many functional disciplines which previously had considered their element of the design in a largely serial fashion.
2. Designing quality into the product from the beginning, reducing error cycle times, and thereby increasing total throughput. As shown in *Figure 4-2B*, errors cause several interrelated, throughput-slowing problems. These include diversion of personnel time. Engineering managers actually must plan for less than a full day of work for each day's work. Typical allocations reflect the large amount of nondesign and engineering time actually spent by the professional. Notice in *Figure 4-7* how a small reduction in the amount of time required to communicate means a substantial percentage increase in design time, which creates significant product cost leverage.
3. Eliminating proces errors reduces the total throughput cycle. Concurrent Engineering Design's intent is to move product and process problem

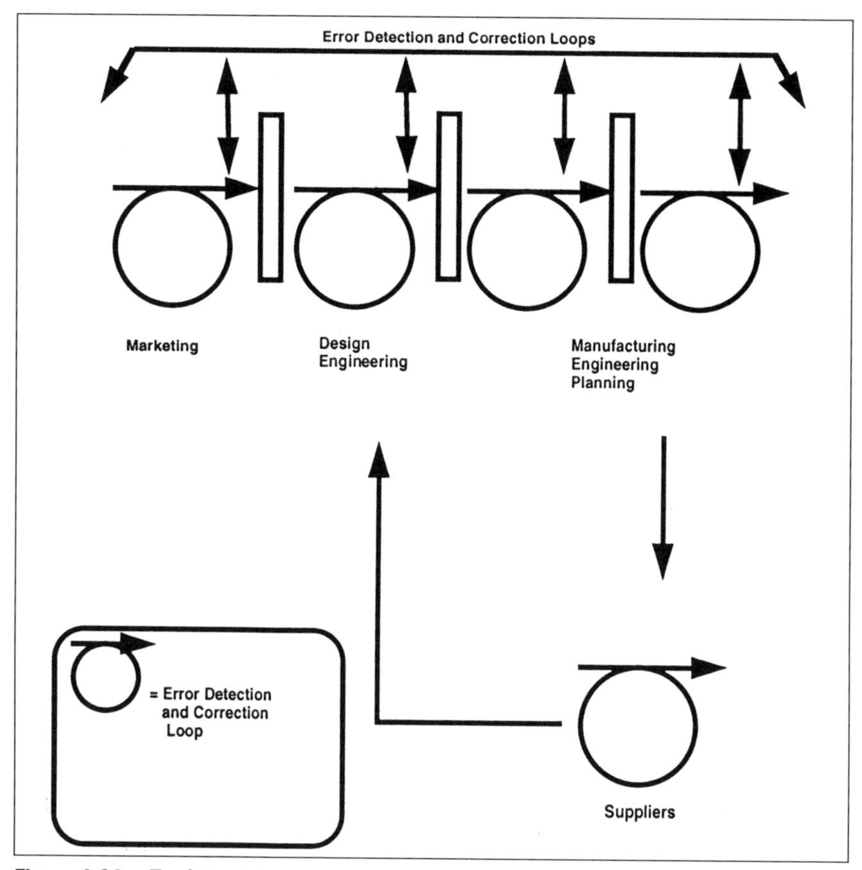

Figure 4-6A. *Traditional Total Flow Time.*

identification and repair back through the total throughput cycle into the design process. By using the simultaneous nature of Concurrent Engineering Design, more design options can be explored and taken through manufacturability analysis quickly. This process reduces errors significantly. Error reduction means smaller, shorter, and less complex error loops, more time to devote to productive activity which further reduces errors, and so on. Concurrent Engineering Design can lead to JIT-type throughput reductions even in complex product circumstances.

4. Additionally, Concurrent Engineering Design provides the designer more time to concentrate on quality. Within the context of this book, quality's definition includes product attributes, (ease of use, sense of rightness, utility, etc.) as well as elemental attributes (manufacturability, ease of assembly and disassembly, and durability). Concurrent Engineering Design makes contributions in both areas. Throughput, however, is affected by elemental attributes.

Design robustness is an important elemental attribute. Certain critical or key

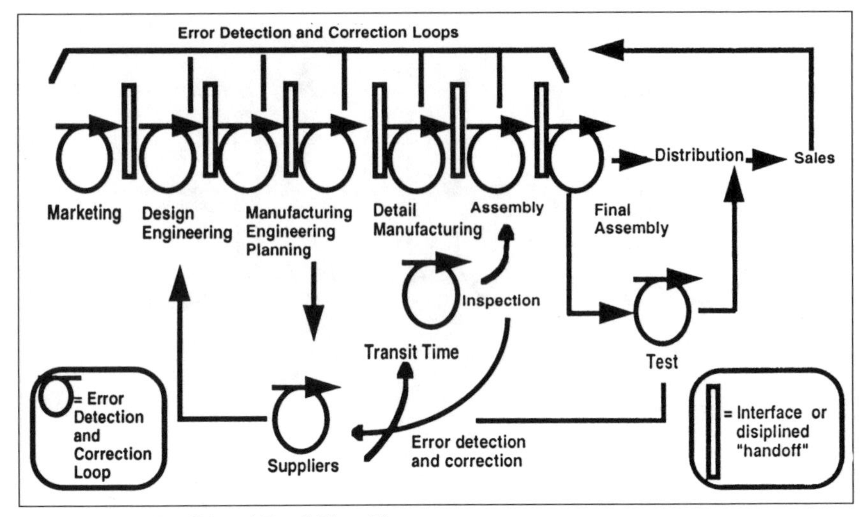

Figure 4-6B. *Traditional Total Flow Time.*

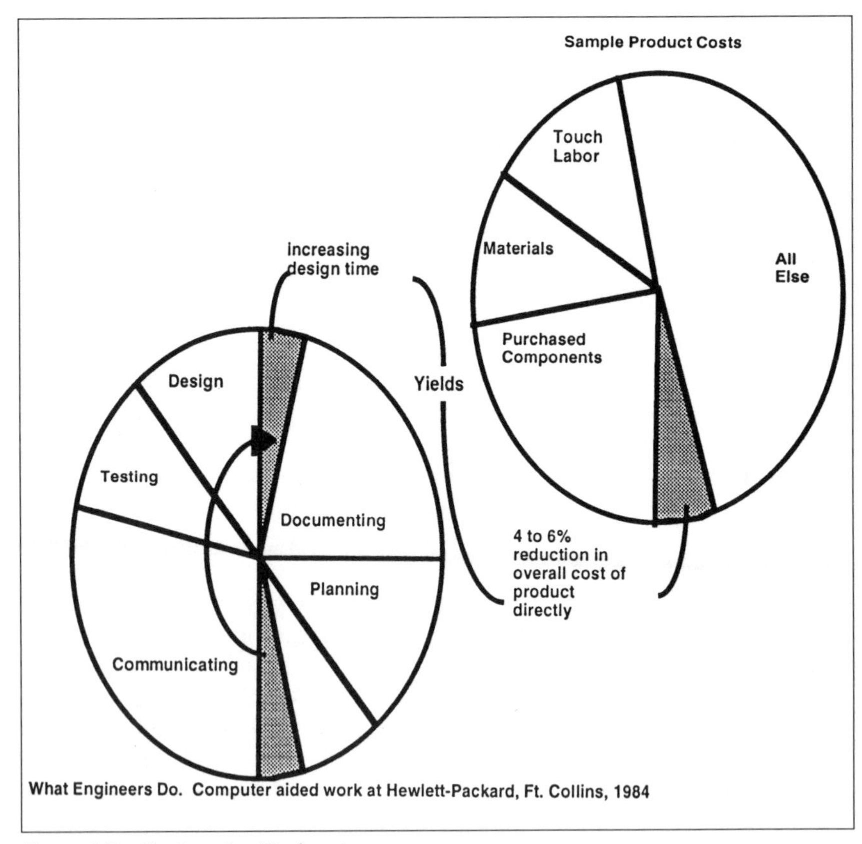

Figure 4-7. *Engineering Workstation.*

features of each part or element of the design are noted by the designer/engineer. These features are critical because they determine component form, fit, and function. For example, as shown in *Figure 4-8*, a discreet moving arm in an

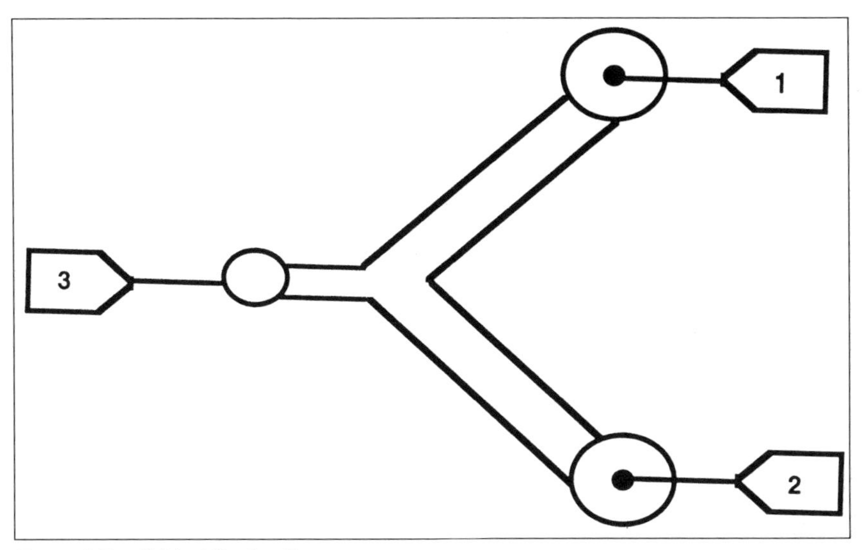

Figure 4-8. *Critical Design Features.*

assembly may have three contact points with other parts. These may have holes for shafts in two points and a ball joint as the third contact point. The three points have specifications which must be very tightly controlled. If these three points are tightly controlled, the assembly goes together more easily, is easier to disassemble for maintenance, and operates through the entire range of motion more smoothly providing for better operation and a longer useful life. The balance of the part's specifications can vary somewhat without hurting performance.

It is through design iterations, simultaneous revisions, manufacturing tooling, and assembly/disassembly digital modeling, all in a Concurrent Engineering Design environment, that such features are more easily discovered, tested, and identified. The manufacturing tooling and processes to produce to those critical or key features are facilitated. Design robustness reduces errors, but more importantly, it simplifies manufacturing and thus reduces flow time and increases throughput. Thus, design robustness and design quality can facilitate substantial improvements in supplier throughput and quality as well.

There are several related activities which impact suppliers in a Concurrent Engineering Design environment. Better designs and the callout of critical or key features aid the suppliers. But if the supplier is a business partner as well, that is, if the supplier has design responsibility and/or provides a completed product which is a subsystem in the final complex end product (for example, a radio tape player in an auto), then other activities become important.

Three of those other activities are EDI, CALS, and PDES. EDI, or Electronic Data Interchange, provides for the computer-to-computer communication of purchase orders, shipping advices, receipts, claims, returns, and other trading transactions under a contract with preset terms and conditions. This speeds orders, shipments, and payments.

CALS, or Computer-aided Acquisition and Logistic Support, is a DoD (Department of Defense) initiative which is evolving into an industry-wide effort to standardize many elements of the ordering, manufacturing, and supply chain which serve the DoD. Concurrent Engineering Design has become central to its efforts. These standards will permit easier interchange of digital information used to execute the manufacturing business.

PDES, or Product Definition Exchange using STEP, is a standard for the interactive, interchange of product specification and geometric information intended to facilitate the shared design responsibility of a partner situation as well as the supplier receipt digital function. In all of these situations, the intent of these initiatives is higher design quality and robustness, as well as throughput improvements.

Quality, robustness, error reduction and overall throughput translate into *Time or Speed to Market*. When combined with Time Phased Procurement and Scheduling, as discussed in Chapter 7, manufacturability of a 96-hour custom automobile is possible.

WORLD-CLASS MANUFACTURING TECHNOLOGY INCORPORATION

Incorporating technology in a world-class manufacturing environment is probably the most, as well as the least understood, of the various aspects of world-class manufacturing and Concurrent Engineering Design. It is the most understood because many of the technologies to be incorporated or utilized are widely communicated via marketing activities, industry oriented magazines, seminars, etc. It is least understood because technology is constantly changing, and its impact often can be misunderstood in product and process. Training, education, innovative thinking, and new management thinking are necessary and usually on a constant basis, if technology is a major issue.

There are two aspects to technology: "push" and "pull." "Push" is where new technology creates an opportunity for improvement; "pull" is where technology, whether it be new technology or technology which has been in use for some time, is incorporated into product or process, because the *need* for the technology was identified first. *Figure 4-9* describes some of the various ways in which technology can impact a manufacturing organization. New or existing technology can be used as a component or as a multiple component substitution in a new or existing product or can be the basis for a new product. New or existing technology can be incorporated into the business and management processes themselves, the machines which are utilized in the manufacturing process, or the infrastructure, such as buildings, communications, transportation, and product packaging areas. It can be incorporated into the production manufacturing process itself, as with Automated Storage and Retrieval Systems (ASRS) or a new chemical bath for treating a component.

The balance between "push" and "pull" is an important issue in world-class manufacturing and Concurrent Engineering Design. There are many books and articles, many large, complex organizations, and whole industries devoted to technology. The emphasis in this book, from a technological perspective, is on what technology creates ("push") an opportunity to implement world-class manufacturing and Concurrent Engineering Design, and what necessary activities within the Concurrent Engineering Design processes can only be realistically accomplished ("pull") with new technology. A Concurrent Engineering Design environment provides support for more rapid alternative modeling of designs, and thus product-related technology incorporation is facilitated. A more detailed discussion of the elements of technology discussed in *Figure 4-9* but not directly related to CE Design (such as how to incorporate technology into products) will be left to others.

One of the most significant dangers to the implementors of Concurrent Engineering Design is the belief that if the technological components of it are put into the organization, then the rest of the Concurrent Engineering Design process will *evolve* to meet the technology and achieve the technology's potential. As we observed earlier, the key to implementation success in this environment is an integrated planning and implementation process equal to the integrated nature of the Concurrent Engineering Design process itself.

The discussions in this chapter on world-class manufacturing are also integrated. To achieve quality, one needs costing, technology, a revised design process, a revised cycle and throughput philosophy, etc. For each area of world-class manufacturing, the other elements are needed for the program to be successful.

And technology alone is not the answer either.

For world-class manufacturing and Concurrent Engineering Design to be implemented and successfully sustained, there are some technology requirements ("pulls"). These include:

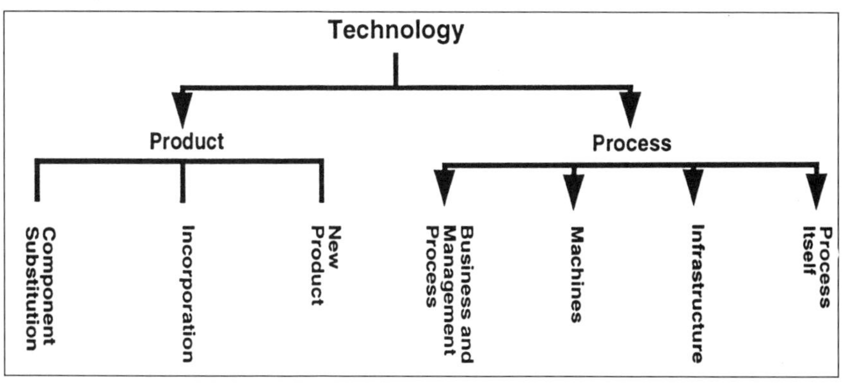

Figure 4-9. *Technology Options.*

1. Computer-Aided Design (CAD);
2. Communications systems between workstations and computers both inside and outside the firm, and
3. Various application program systems which provide for various capabilities.

Computer-Aided Design (CAD) is a widely discussed topic. There are many software products which provide CAD support. What is needed from a Concurrent Engineering Design perspective in CAD are three characteristics:

1. Adequate dimensional accuracy;
2. Solid modeling capability, and
3. Application-specific support.

Currently prominent CAD products are capable in general terms. They have relatively few bugs, have most of the needed commands, run on acceptable equipment, etc. However, they must support the ability to portray and print design objects with adequate dimensional accuracy because design robustness requires the key features to be produced with little variance.

Solid modeling is necessary to resolve interfaces and/or interferences between various design elements during mock-ups, or model/prototype development and/or product performance simulation.

Application-specific support means that for product design areas such as:

- Electronic circuitry,
- Hydraulic piping,
- Hard parts,
- Packaging,
- Tooling and fixtures, and
- Other elements of the complex product.

There are CAD features which are specific to the area's requirements (yet which hopefully act similarly in these environments) are available. These application-specific support areas are also different enough to need specific additional commands and capabilities which facilitate their individual area design needs. A single CAD environment for every element of the product is highly desirable but not absolutely necessary. Representations of various subassemblies with interfaces can be built in higher level models and synchronized with their detailed CAD representations.

While this may seem duplicative in nature, key features can be determined in the higher level models and efficiencies can be achieved by using the technology most effective for each component category. For example, high level designs may be done on workstations, along with most elements, with several different types of workstations using several different application-specific CAD software environments. Solid modeling might occur on workstations, minicomputers, or even mainframes, as necessary.

This varied CAD environment is the reason why communications systems are so important. For the purposes of enabling world-class manufacturing, the Concurrent Engineering Design communication environment must also exhibit certain characteristics:

1. Support for interoperability;
2. Support for the movement of CAD information, and
3. Standardization of functions of network and communications-based application systems of certain types.

Interoperability, as shown in *Figure 4-10*, means the ability of workstations

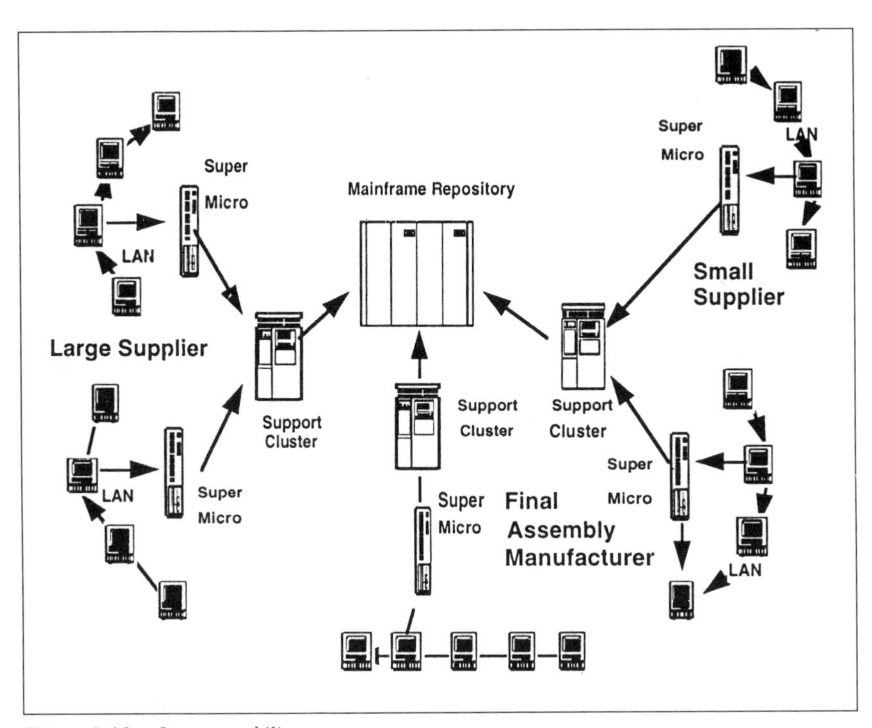

Figure 4-10. *Interoperability.*

and terminals to reach each other's computers with applications and data, and to reach through these systems to other computers both inside and outside the organization. This is a requirement because the near-simultaneity aspect of Concurrent Engineering Design is based on electronic proximity, a key dimension of Concurrent Engineering Design.

The movement of CAD information in the Concurrent Engineering Design environment is also very important because the base concept of the environment is to communicate visually via electronic means. Words, written and spoken, and specifications alone do not suffice. The movement of CAD is currently accomplished with file transfers using simple ownership-oriented configuration management techniques. This is a temporary situation because it is a high interest area of technology for many governmental, international, and business organizations. These organizations have participated in various initiatives, such as Product Definition Exchange using STEP (PDES), the U.S. contribution to the international STEP standards effort for the exchange of CAD and other product-related data. PDES is the translation standard which is intended to replace IGES, the current U.S. file translation standard.

In PDES, the CAD image is translated into a neutral database from which any commercial CAD system can then regenerate the image and its associated data. A pictorial representation of this movement is shown in *Figure 4-11A* and *Figure 4-11B*. *Figure 4-11A* depicts the current method of file exchange. *Figure 4-11B* depicts the PDES/STEP approach of translating the CAD image into a neutral format. Other Electronic Data Interchange (EDI) standards are being established as a part of the DoD CALS initiative. The CALS office of the DoD is emphasizing Concurrent Engineering Design because of its perceived substantial positive impact on DoD costs.

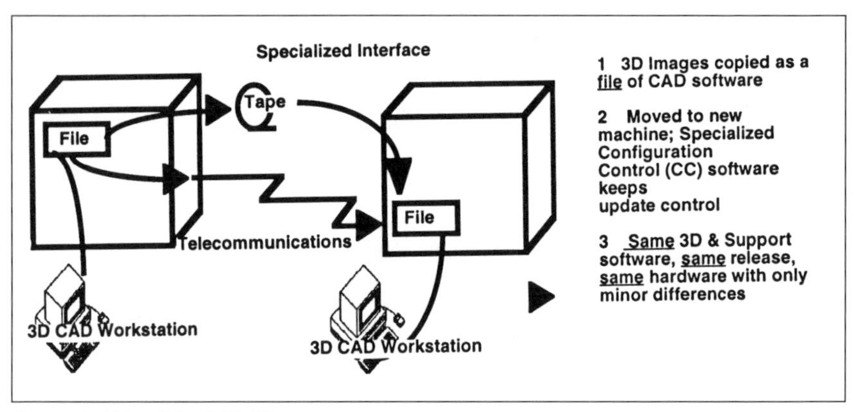

Figure 4-11A. *3-D CAD Movement.*

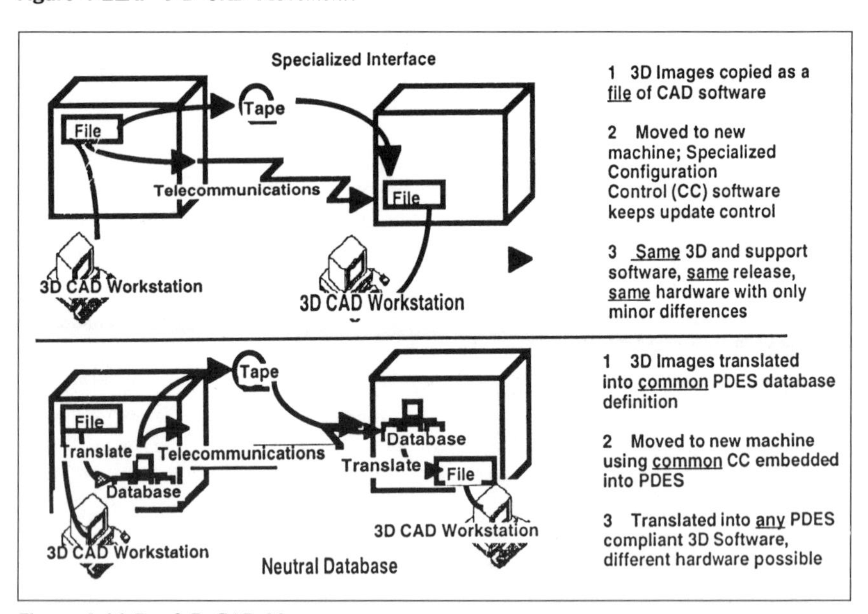

Figure 4-11-B. *3-D CAD Movement.*

The standardization of network and communications functions based application systems is important for intraorganization reasons. As will be discussed later, a number of Concurrent Engineering Design activities depend on these application systems functions, which include scheduling, routing and queuing, configuration management, and resource management.

VARIETY AND COMPLEXITY IN WORLD-CLASS MANUFACTURING

Increasing product variety is a key competitive strategy element in manufacturing. As shown in *Figure 4-12*, the intent of a world-class manufacturer is to take each product or concept and quickly and cost-effectively exploit that product with a range of derivatives which create a seamless overlay of a potential customer's needs. This "product family" is then *constantly* enhanced with features, functions, and improvements until a new original product begins to make the current product family obsolete. Skillful transitions and high-speed cutover to the new product family are then required to sustain market control and customer interest. As shown in *Figure 4-13*, patterns of introduction include the "down" strategy, where the new product family originator is first introduced as an upward growth path item (G) from the existing product line (A-F) from which replacement products are derived (H-L). Another strategy is the "up" strategy where the new product family originator is first introduced as a new, smaller area of the product line (A-F) from which replacement products (C,D,E,F,A) are then derived.

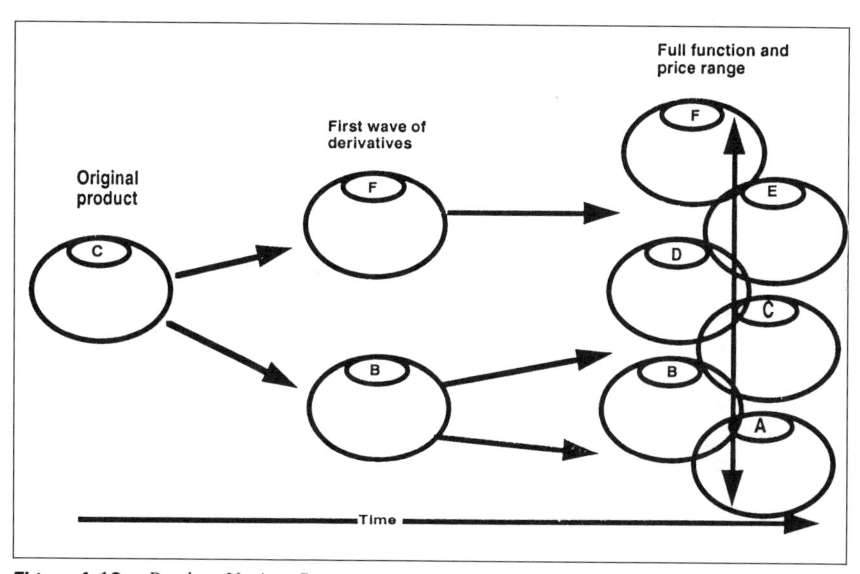

Figure 4-12. *Product Variety Strategy.*

In both of these circumstances, the best product approach is to have multiple product family members almost completely designed from the beginning. This

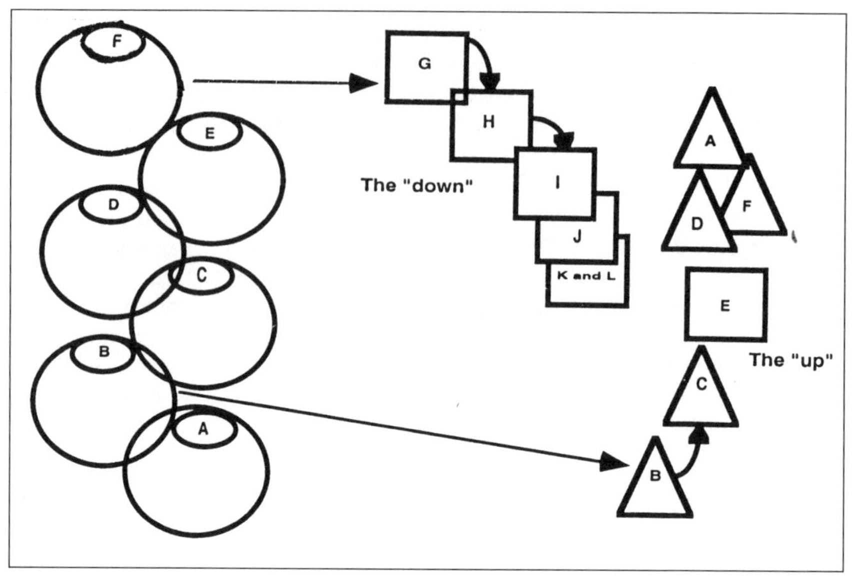

Figure 4-13. *Product Family Cycles.*

permits common design elements and functions which facilitate ease of understanding and use by the same customer with no new education or training required. This crossover of common design attracts customers in a complex product environment because relearning a new complex product is so difficult, expensive and time consuming. Product families sell each other and have a strong multiplier effect, in which it is sometimes unclear which individual product is the most profitable or important.

The broad interlocking range creates competitive advantage and strong barriers to entry because a competitor must equal or better the entire product family at once, a significant and daunting task in a complex product environment. Concurrent Engineering Design plays a key role in supporting the initial design, and in sustaining the evolution of product families. The simultaneous nature of CE Design and the electronic proximity created in the CE Design environment permit many alternatives and concepts to be developed and managed independently and concurrently. With proper coordinated design, many overlapping products and their common production processes, tooling, etc., are produced under configuration control using as many common elements of functionality as possible this commanality promotes reuse and lower costs.

Complexity in a Concurrent Engineering Design environment has three aspects: inherent product complexity, organizational complexity, and implementation planning and strategy. Complex products today are inherently more complex than they were 20 years ago. For example, if you count the number of gates in the circuits in a modern automobile as substitute for parts, the modern automobile may have a hundred times as many parts as the automobile of the 1950s. All one has to do is look at the engine, the dash, and the various additional electronic and hydraulic systems to appreciate this increased complexity. The radio (now an AM/FM stereo, compact disc player) is in itself more complex than the 1950s' automobile, with a host of technologies unavailable at that time. *Figure 4-14* depicts the rapid rise of computer hardware and software complexity now facing manufacturers today.

Managing product complexity has strained the organization, and it has, until recently, had to evolve its complexity, via specialization for example, to keep up with the product's complexity. The organization's structure has evolved in a pattern similar to that shown in *Figure 4-15*. In the 1950s, touch labor and materials dominated. By the late 1970s, product complexity had caused touch labor and materials to shrink, while middle management, which handled the

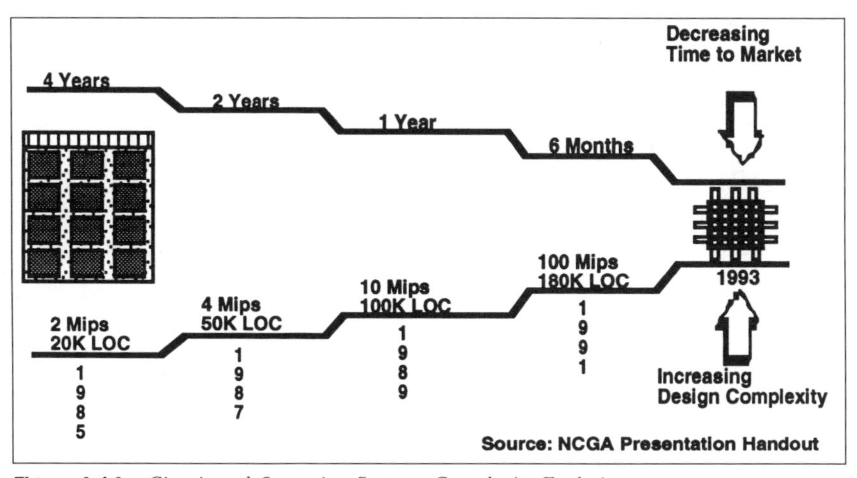

Figure 4-14. *Circuit and Operating Systems Complexity Evolution.*

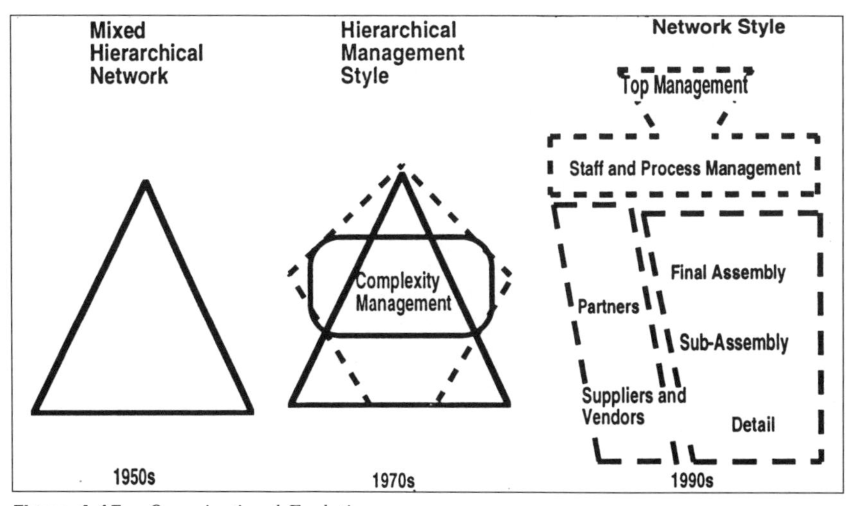

Figure 4-15. *Organizational Evolution.*

management of the organization's complexity, grew. The Concurrent Engineering Design organization (1990s) is evolving into one which is layered, and which marshals systems and knowledgeable workers, to assist in the management of the continuing growth of product complexity into product family and CE Design teams. These layers can be in one organization, but increasingly it is an interlocking confederation of suppliers and end-product organizations in related industry categories.

Figure 4-14 provides an example of how complexity is rapidly increasing throughout manufacturing. While the number of discrete parts may be stable or even declining in number, product complexity is actually rising rapidly if electronic circuits are being substituted.

Overall productivity improves when the number of activities used to manage this complexity stays the same or rises more slowly than the complexity. In the 1970s style organizaion, complexity management grew faster than revenue. The Concurrent Engineering Design environment is intended to improve the management of organizational complexity by at least several orders of magnitude. These improvements occur within the design process, configuration management, authority management, and resource management components of Concurrent Engineering Design.

Inability to keep up with or get ahead of complexity could mean the end of complex manufacturers who do not adopt Concurrent Engineering Design quickly.

Organizational complexity, as well as product complexity, is reduced using Concurrent Engineering Design. The Concurrent Engineering Design environment permits multifunction design teams to follow at least a portion of the end product from inception to obsolescence. Each team is empowered with responsibility for its portion's initial incorporation success, reusability, maintainability, etc. By closely associating the members of these teams and utilizing both physical and electronic proximity, the organizational complexity of getting something done is substantially reduced.

Organizational complexity results from functional specialization, which is a subject of debate. Some say that complexity is the result of too many organized fragments, too much staff and support functions trying to run the business. However, the large increase of overall technical knowledge probably requires functional specialization.

The process of communicating between disciplines and work groups is made difficult by the different terms and concepts used by each, and by the business systems and processes built to manage error detection and correction.

The integration of these complexities is at the heart of CE Design. CE Design's intent is not to eliminate the specialists but to eliminate unnecessary activities forced on specialists by current systems, policies, and practices. Concurrent Engineering Design communications systems and the technical enabling of easy-to-use electronic proximity should assist in greatly reducing the impact of complexity generated by functional specialization.

One of the most significant barriers to Concurrent Engineering Design success and to the achievement of world-class manufacturing is resistance to change. *Implementation* and execution *planning* and *strategy*, used to implement Concurrent Engineering Design, are critical to its initial and ongoing success. Initial success is impacted because appropriate use is not possible without proper education and training. Appropriate introduction and organizational change management techniques can allow the Concurrent Engineering Design process to sustain itself.

The change management process is aided and resistance to change is lowered by the CE Design process itself. There is an inherent rightness to Concurrent Engineering Design, which appeals to individuals participating in its implementation. In addition, the Concurrent Engineering Design process, by its inclusive nature, leads individuals into the expanded use of the process. It immediately begins to speed the pace of its own implementation. People are aware of competition and other business matters. This process enhances their perception of control over their organization's success, which is highly motivating.

World-class manufacturing can create permanent competitive advantage. Concurrent Engineering Design is an absolutely necessary enabler of world-class manufacturing in complex product manufacturing environments. The major characteristics of a world-class manufacturer depend on each other, and especially on Concurrent Engineering Design. As is true with any crossfunctional process in the highly specialized environment of the complex manufacturer, the introduction of Concurrent Engineering Design requires different thinking and plans, but aids in its own implementation because of its inherent desirability.

The enabling function of CE Design differs for each of the world-class manufacturing critical areas. For *Quality*, CE Design enables quality to be designed into the product. Its manufacturing process and quality control techniques, such as SPC, can ensure that the design process is performed in a quality-oriented fashion. For *Costing*, CE Design enables Manufacturing, Management, Marketing and Engineering to produce quality commensurate with cost. For *Time-based* competition, CE Design is absolutely necessary because its high-speed concurrent design process, as well as its manner of organizing and pre-planning the manufacturing process, are really the only realistic way to achieve necessary cycle reduction time.

Technology and Variety and Complexity Management and their involvement with CE Design come about in a different manner. These two critical areas of world-class manufacturing have reached the saturation point in many organizations. In general, technology and market-driven product variety requirements continue to advance at a rapid pace. CE Design permits technology and product variety, and the complexity they generate, to become manageable at substantially increased rates of introduction and in greater aggregate volumes. CE Design enables the organization to continue accepting and even advancing in these final two areas, while controlling costs to do so in the process.

SECTION II:

Concurrent Engineering Design Business

Process Framework

This section of the book is focused on the business, technical and managerial processes of CE Design. These major business process groupings are pictorially represented in *Figure II-1.*

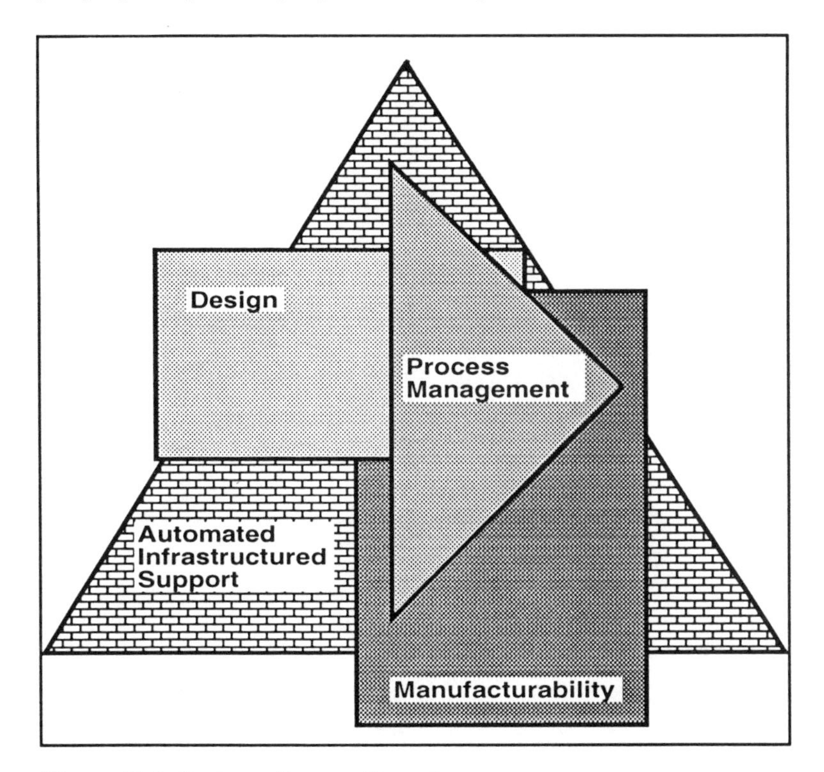

Figure II-1*. Business Process Groupings.*

Process Management is the highest level of the processes. It is internal to the organization and overlays the other processes. Design reaches outside the organization, and is interelated to

Manufacturabilitiy by Process Management. Manufacturability reaches out to suppliers. All these processes rely on automated infrastructure support, which is made up of many computing building blocks.

The "manufactured product of CE Design is information. This information includes many different categories, product images and specifications, the history of their production, and many instructional and informational sets of information in various presentations, including computerized data, text, image and various mixtures, as well as many different types of paper formats.

The major implication of the process "pull" and technology "push" combination to CE Design is the change from computerized systems as "record keepers" to these systems providing the framework within which CE Design is accomplished. It isn't yet all electronic; the CE Design process described herein could be done manually. But its power and elegance can only be fully recognized with automated support. *Section III's* chapters reflect this fact, and thus they, of necessity, have a "computer system's" orientation.

5

CONCURRENT ENGINEERING DESIGN'S PROCESS MANAGEMENT

CE Design is a multitiered set of business, technical, and managerial processes. These tiers of processes were first depicted in *Figure 2-10. Figure 5-1* shows these three tiers, identified as Process Management, Process Flow, and Process Execution. The focus of the Process Management tier is on managing the overall flow and integration of the various processes. Management's attention is typically not on an individual product, group, or process. Instead it

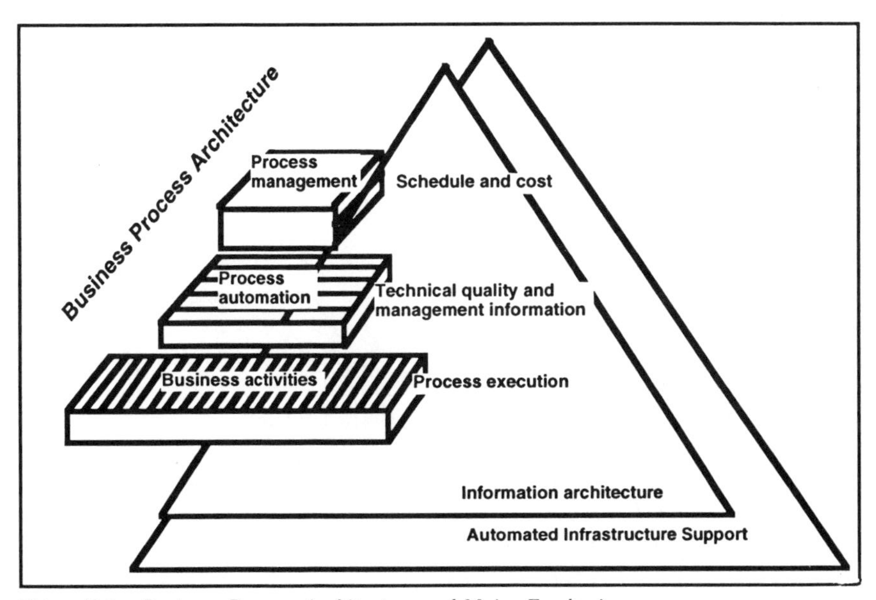

Figure 5-1. *Business Process Architecture and Major Emphasis.*

concentrates on their interrelationships, operational status and resource use (cost, time, people, machines, facilities, etc.). Information used in Process Management is not related to a specific product or process.

CE Design focuses on complex products. Complex products are products which continue to change after they are produced. Such products include airplanes, automobiles, computers (and many other products containing computer software), bio-genetics, etc. For these types of products, change is a constant. Accordingly, the process management approach for CE Design must be about change.

The Process Flow tier's focus is on the execution of an interconnected set of business activities and their associated business functions. These business functions are integrated into a flow of interconnected functions by defined function relationships. Management's interest in Process Flows is on process status, operational interrelationships, standard expectation exceptions (where normal activities are interrupted by error or other problems), schedule, and resource use.

The Business Activities tier has a functional execution focus. Management's attention is on status, use of resources, and schedule, but only on an as needed basis. Simultaneously executing three interrelated major processes of Concurrent Engineering Design (Design, Manufacturability, Process Management) is a key concept in CE Design.

Process Management depends heavily on CE Design's Automated Infrastructure Support capabilities. The Automated Infrastructure Support for Process Management also operates within the other two major CE Design processes, bridging all three, and permitting them to work together. The four major activities of process management are:

1. Routing and Queuing;
2. Configuration Management;
3. Resource Management, and
4. Release and Distribution.

ROUTING AND QUEUING

This activity of CE Design's Process Management enables CE Design to have an *electronic proximity* dimension. It is the most important feature of CE Design because routing and queuing enables and manages the execution of process models. The three most important concepts associated with CE Design's Routing and Queuing are Models, Authority Management, and Queue Management.

Routing is focused on sequencing the activities of a variety of groups and individuals. *Figure 5-2* shows that building the definition of product, process, and manufacturing support is accomplished by moving the growing ''package'' of information and work products from group to group and from individual to individual within the group, according to a model or route. This route often cannot be predetermined. The CE Design control group (if established) manages routing to each design manufacturability group. Each design manufacturability

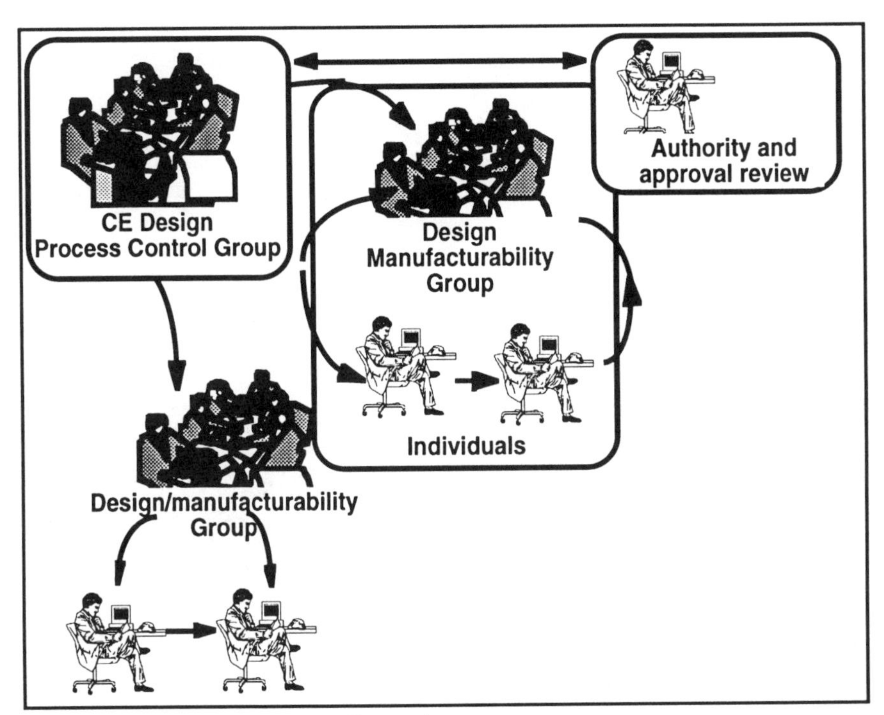

Figure 5-2. *Routing and Queuing Sequencing.*

group manages its own internal to the group routing. Routing authority and approval review is managed by such a CE Design process control group as necessary, and to the individual level if necessary.

Whenever decisions are needed, or work is waiting to be addressed, the "package" is queued until it can be addressed. Priorities can be assigned to work to manage queues as they develop.

The growing package of information must be organized. As the combination of work products resulting from design, analysis, manufacturability, etc., collects, it must be accessible through a variety of indexes and types of information management and retrieval. *Figure 5-3* shows how CE Design information should be organized into envelope sets, linked (or electronically "stapled") together into organized layers of detailed information. *Figure 5-3* also depicts an example of these layers of information. It shows a top level work statement stapled or indexed to a more detailed analysis envelope. This envelope also contains sets, one of which is stapled to a detail test and control envelope with various shape analysis results. In many firms, the design and work packages follow a Work Breakdown Structure (WBS). They usually use an extensive level of detail, based on a numbering scheme, to classify the elements of the information envelope. In CE Design, these electronic packages of envelopes are sent among groups and individuals with automated infrastructure support tools as a part of the routing process.

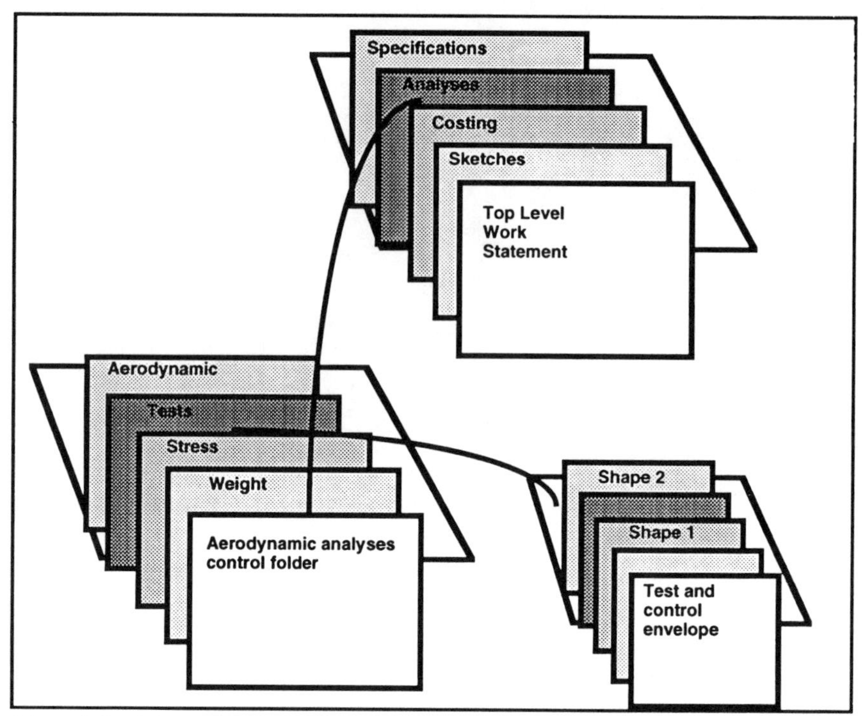

Figure 5-3. *Information Envelopes.*

MODELS

The routing between groups can be consistent for similar problems. Models of these routings can be built and reused. In the routing and queuing context, a routing model is a graphically presented sequence of planned or previously encountered activities. These activities are typically shown as boxes interconnected with lines of relationship. These lines of relationship show which activity boxes precede and follow, and show dependency and constraints. An activity is dependent on another when it must be at least begun before the next activity can be started.

How problems are addressed, the sequence of actions, and the collected *set* of information envelopes associated with each issue and its routing model, make up knowledge that can be reused. Continuing to accumulate these combinations of routing models and instruction envelopes in this fashion leads to capitalized intellectual property due to its process knowledge capture and reusability.

These "models" should be built in a fashion similar to PERT (Program Evaluation & Review Technique) using an information modeling technique called *sumentic nets* schedules. By using "a schedule-type" model, (which considers sequences, priorities, and constraints) several other elements of the CE Design processes are enabled. This schedule-type model should be composed of activities or "nodes" to develop submodels for both queuing and authority management purposes. As shown in the *Figure 5-4* sequence of diagrams, these models should "nest" or be built from submodels. In *Figure 5-4A* the major

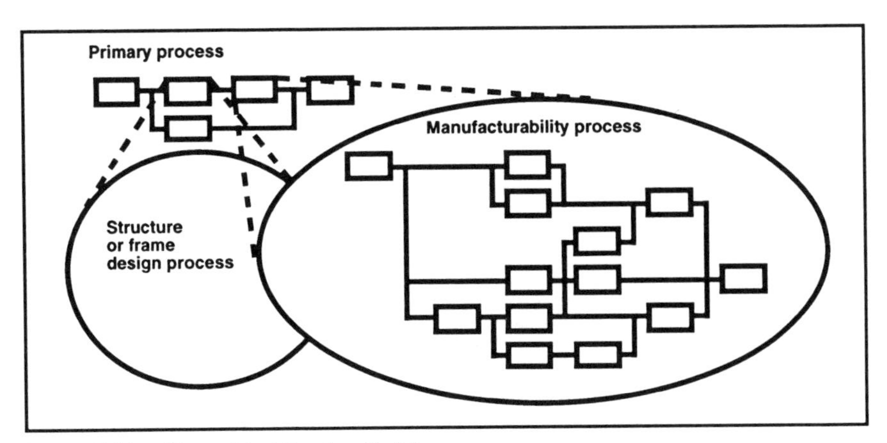

Figure 5-4A. *Hierarchical Routing Models.*

schedule and combination routing flow can be a design work flow model superimposed on groups involved in that product component design, with subordinate work flows for supporting groups, and further subordinate models for individuals. In the example shown in *Figure 5-4B*, several "concurrent" engineering teams are simultaneously "designing" elements of the product and interacting with various members of the "manufacturability" portion of the organization. The primary CE Design Process includes design process and manufacturability process models. In *Figure 5-4B*, the manufacturability model consists of other models, such as an electrical assembly process design model and a final release management process model.

The activities of these processes are represented by the model work flows, and documented by each model's work flow as the work product grows, and the groups and individuals meet. The model acts as a road map for what should be done next and what has already happened. It also helps relate all information developed to the model sequence of events within the context of that particular design issue and design intent. The concept of a "model" driven process thus becomes very powerful.

Typically, the sequence of events necessary to "get a design" through the entire design sequence is not well understood, especially in complex organizations dealing with complex products. Often, procedure manuals, memoranda, individual knowledge, and prior experience are required, in combination, to determine how to address design issues. By using a model to represent the activities required in each situation, the organization and the individual are *learning*.

Models can come about in two ways, generative and derivative. In the generative case, rules are used to compare the issue, circumstances, physical, and design properties of the item in question to generate a routing model. In manufacturing, attempts to develop systems that generate the actual manufacturing process (or routing model) for a particular part have been under investigation for some time, with positive results. In the area of design,

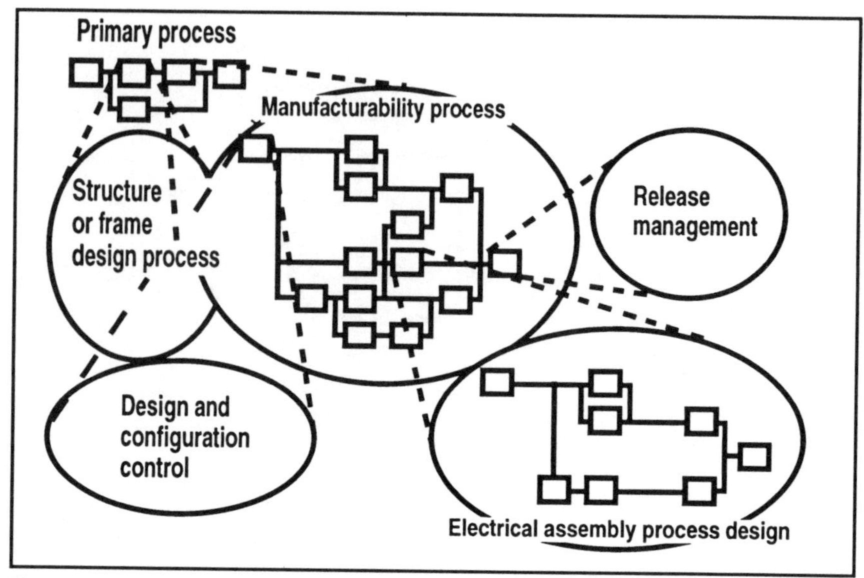

Figure 5-4B. *Hierarchical Routing Models.*

generative models have not had as much research. Instead, the derivative model, where a copy of a previous model is changed to fit the current situation, acts as a base for a new model used for the next design issue.

Models have become increasingly important to all aspects of design. They represent the CE Design process to be followed. In three-dimensional solid geometric representation using a Computer Aided Design (CAD) System, electronic models of the product can be used to simulate action and substitute for a mockup. According to analysis by the Boeing Company (reported in *Aviation Week and Space Technology*, June 3, 1991), this "digital preassembly" and mockup represent one of CE Design's main contributions. Models of the manufacturing process can be used to simulate the manufacturing process and its interaction with aspects of product manufacturability, serviceability, and operation. Models also can represent and store the exercise of authority.

A typical, off-the-shelf PERT scheduling and cost tracking software program is not adequate for providing automated support to the *routing* model construction, manipulation and usage process. Most of these software packages use their own proprietary database that is not accessible to other software. They also lack several key features. The right Routing and Queuing software tool has at least the following characteristics:

1. *Open database*—the ideal candidate system can use a database independent of the application software. This permits the use of schedule and resource information to drive electronic mail, movement of the information envelope, and additional information not used in determining scheduling dates and tracking status.
2. *Flexible graphical user interface*—the ideal candidate system allows

forward and backward scheduling, partial completion, and independent-of-completion later task initiation. It also provides several other flexible "model" building techniques through a graphical user interface that treats the model image as a layered graphical object in and of itself.

3. *Alternate Scheduling Logic*—the "back-to-front" PERT scheduling logic breaks down when the same resource is scheduled on competing tasks, or when the model changes during the process, or priorities interfere with standard scheduling. Logic for forward scheduling, and manufacturing shop floor scheduling, such as in OPT, which is priority and constraint oriented, should be available and insertable.
4. *Support of Additional Functionality*—the ideal candidate system permits a wide variety of "add-on" user or third party provided programming to interact with its supplied logic at many preselected logic points. The additional functionality needs to operate on the same information used by the standard scheduling and resource balancing logic typically found in such a system, as well as other additional information, as shown in *Figure 5-4A*. Many different requirements for CE Design routing and queuing support are met through this sharing of information.

AUTHORITY MANAGEMENT

A prime example of the need for such "additional functionality" in the Routing and Queuing model management environment is authority management. In many types of engineering, it is important that aspects of the design be reviewed and approved by competent authorities in the organization. This is required for good management, technical quality and legal liability reasons. In this technical and managerial type of review process, several variations from a standard scheduling model are apparent. They include:

1. *Electronic approvals*—In regulated industry situations, a signed drawing is required for review and approval. Many other organizations require this as well. When a digital model becomes the design control point, a secure method of assuring appropriate design "sign-offs" is required. One method uses a scanned image of signatures that can be applied to engineering drawings as they are printed on a plotter. The signature image can be protected by a "double blind" password, where the first password is to obtain the image and the second is to use it.
2. *Vertical processes*—the review and approval process is not just another activity or milestone event in the model. It has its own process rules and is not sequential. As shown in *Figure 5-5*, a separate process, requiring its own model, must be supported by additional functionality. This authority review process is dynamic. In a CE Design environment, the evolving product design or change and its impact on the product's performance and expected design specification is not always known at the beginning of a model-based set of engineering design activities. For management to monitor progress, prevent "design paralysis via endless

analysis'' and review for quality control, several standard released-for-review activities should be included in the CE Design process model.

3. *Applicability*—As the information envelope is released for review, the design's present impact needs to be determined. Referring to *Figure 5-5*, as the circumstances of the impact grow (cost, weight, performance, ease of use, maintainability, or manufacturability), then new people of higher authority may need to be included in the review. Depending on the type of impact (for example, additional manufacturing process steps and costs), these individuals may need to be scheduled in varying, but specific to that product's situation, sequences based on management rules (for this example, manufacturing, before finance, before product marketing, or management).

As these information envelopes containing the design are being moved about, various groups and individuals will have more envelopes ''passed'' to them than they have time and resources to address. As a result, there is a need for queuing management.

QUEUE MANAGEMENT

Queuing occurs when one of two events occur:

1. *WIP*—there is already work-in-process (WIP) open for the group or individual, and their previously committed work level indicates that they cannot accept any more work until their WIP queue is at least reduced. Sometimes, the WIP level may be acceptable normally, but a queue is

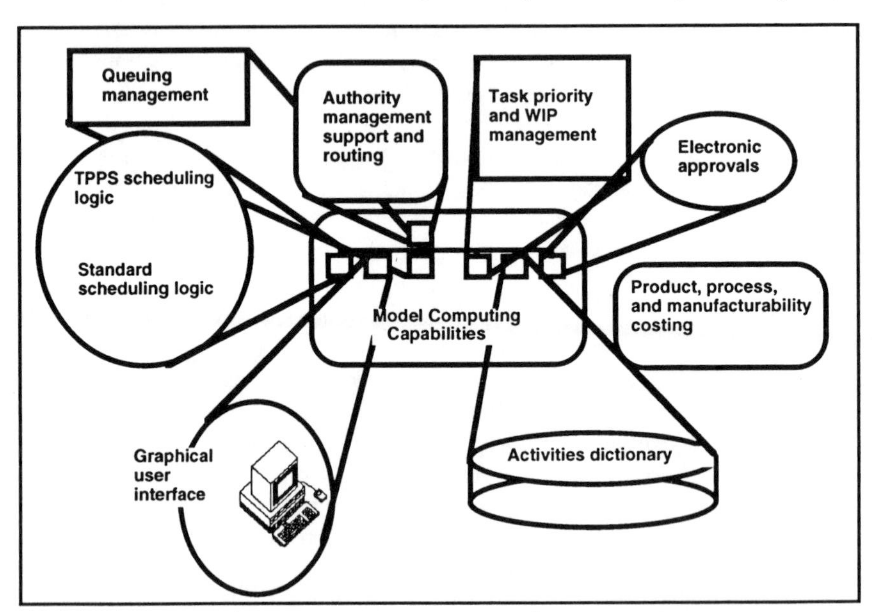

Figure 5-5. *Additional Routing and Queuing Model Information Requirements.*

needed now because the individual or group is unavailable (vacation, trip, illness, training, etc.), or the individual or group indicates they cannot accept more work (impending loss of personnel, anticipated high priority work, etc.).

2. *Priority Assertion*—in an idealized manufacturing situation, through the balancing of work across multiple groups, various engineering work activities could be handled in a "smooth flow" fashion. However, in manufacturing, a "balanced" design factory, where everyone is equally busy simultaneously, just doesn't happen. Especially when the "product" is information, and where invention and innovation are concerned, and variations in schedules, and new or unexpected problems and issues are occurring. Such "balance" seems elusive. A priority setting activity is necessary.

Priority Assertion has several aspects to consider in this context, including Schedule Management and Compliance Reporting, Security and User Profile Considerations, Aspects of Configuration Control, and Information Availability. *Figure 5-6* shows how important it is, from the Schedule Management and Compliance Reporting aspect, to have all CE Design activity authorized. There should be some "project," or cost accounting and collecting structure, against

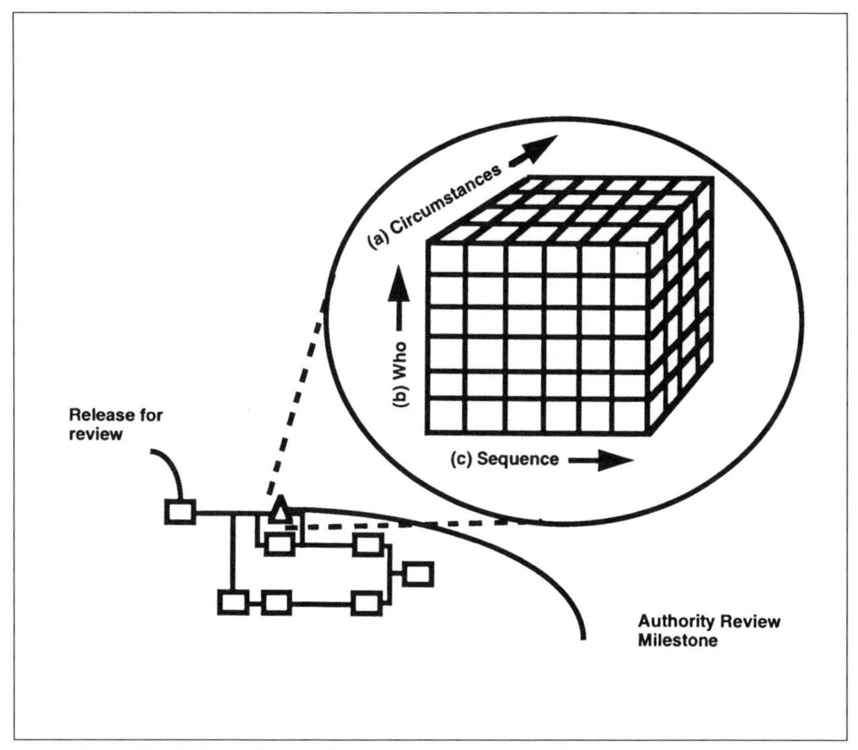

Figure 5-6. *Vertical Routing Model with Authority Review Matrix.*

which the schedules, costs, and resource uses are posted. The information envelope acts as an information storage repository of such information as the project's authorized routing model plan of activity. The information envelope is routed, or sent, from group to group and individual to individual. This information can be used by the routing and queuing model management system to monitor progress as the information envelope is sent to other groups, individuals, and systems.

Product introduction and product change *incorporation* schedules are first produced at a high level by the Program/Product Management Group. These schedules describe the overall plan of product development, introduction, refinement, and support. As the design is functionally decomposed and various teams work on elements of the design, an overall design is completed. The design, accompanied by the other CE Design Process outputs like process plans, support requirements, costs, etc., is released for production. Even in the best of CE Design environments, changes from the field, manufacturing, and other internal and external sources accumulate.

Figure 5-6A also shows how a Change Incorporation Board, together with the Program/Product Management function, agree on Committed Development Schedules (CDSs) which sequence committed, or accepted for incorporation into the product, changes into production. These CDSs reflect the various detailed CE Design activities necessary to get the change, new product, or new product family ready to be manufactured. It is in this incorporation process where most *priority assertions* occur. Safety, regulatory, immediate need, and significant cost improvement opportunities all compete for incorporation priorities and engineering design time. Schedule Management and Compliance Reporting via the CE Design process models to all interested parties becomes an important CE Design Control Group function.

The CDS schedule is a part of several different types of interlocking schedule types that must be supported in the CE Design environment. *Figure 5-6B* shows other types including the CPF, or Change Process Flow, which tracks design changes before they become committed for incorporation; MPS, the Master Production Schedule, which drives component production and orders off the TPPS, the Time Phased Procurement Schedule, or the rate-driven, final assembly sequence and customer related deliver schedule; and FCS, the Finite Capacity Schedule, which drives the local shop schedule based on priorities and constraints, similar to those of the design group.

Security and User Profile Considerations occur when the information envelope is routed from group to group and individual to individual. Each user and group is "profiled" to describe which portions of the overall envelope they can access, in order to copy, add, update, delete, or destroy information. If national security or company security is involved, additional "screens" of access and additional recordkeeping are needed.

In addition to the usual security access concerns, control over "who" is changing "what" and "when" is very important. The addition level of concern involves aspects of configuration control. Configuration control ensures that

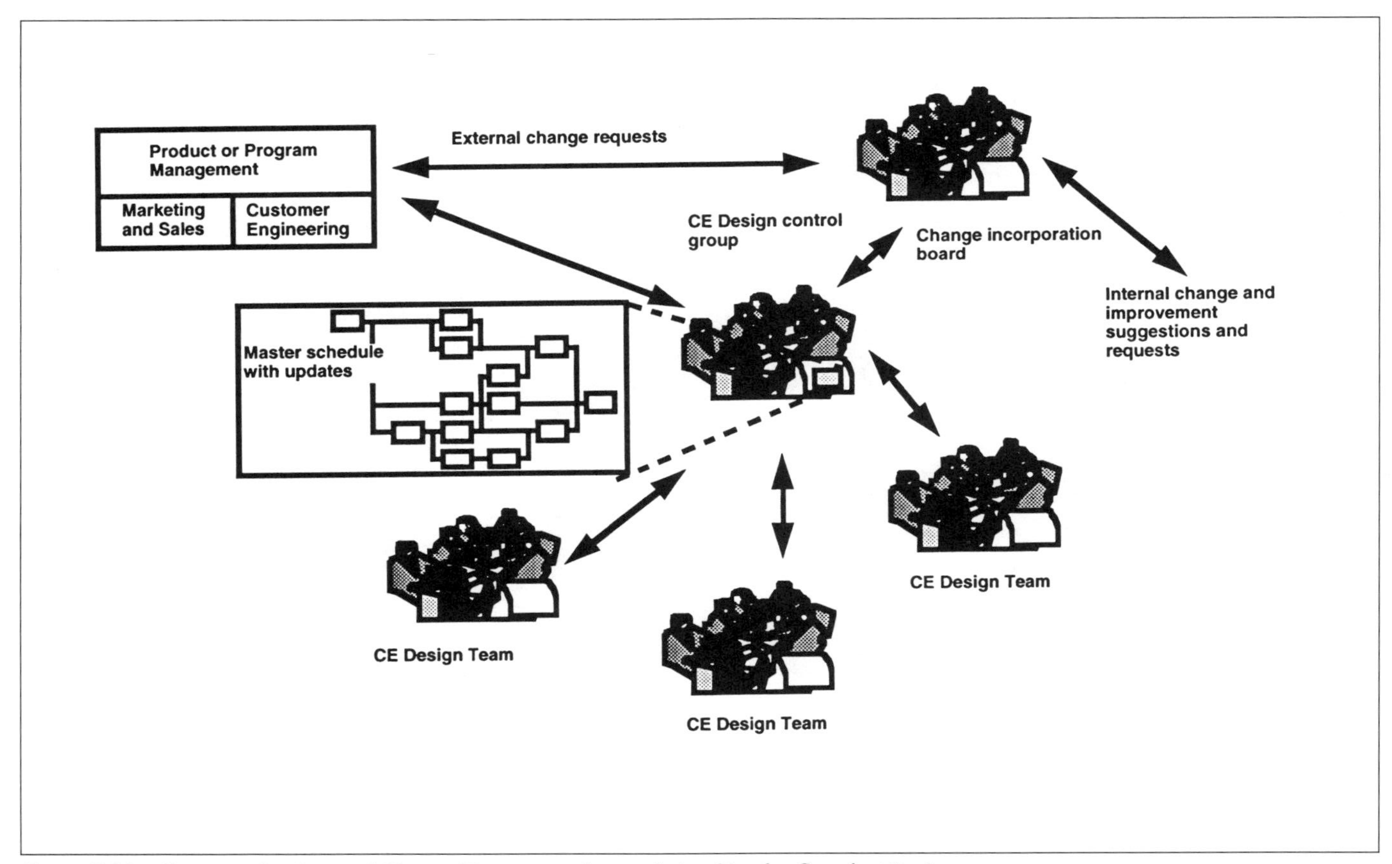

Figure 5-6A. *Program, Product and Change Management Inter-relationships for Complex Products.*

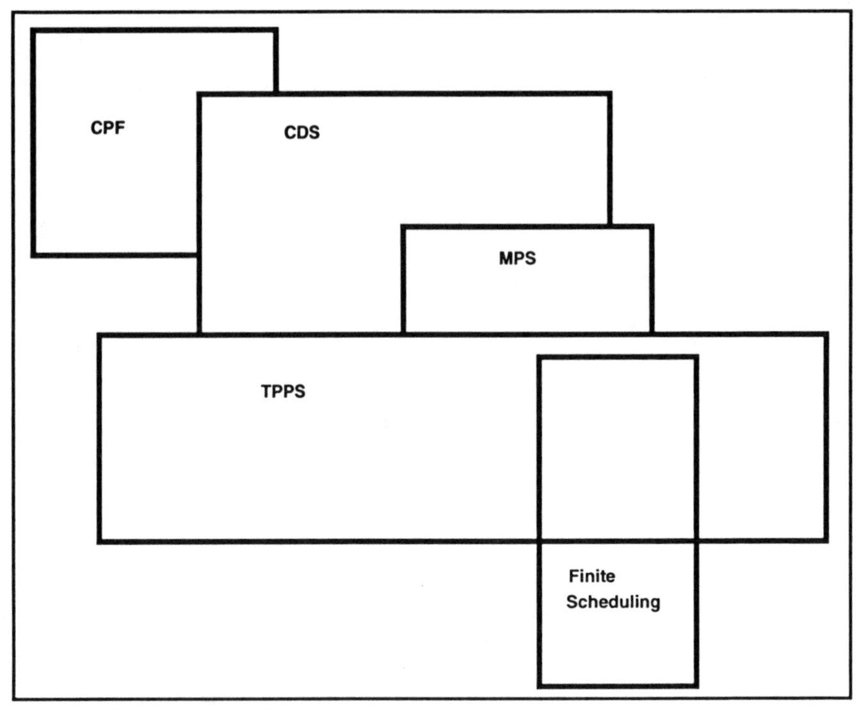

Figure 5-6B. *Scheduling Types.*

both the part of the process in which the group or individual is involved is both permitted and consistent with the current configuration "state." For example, the tooling group doesn't need to review, for manufacturability purposes, an early concept work statement about a potential product. That work statement still needs to be kept under the type of configuration control appropriate for this portion of the information envelope. Various sections of the envelope require different types of configuration control. This concept of different types of configuration control for different areas of the organization, types of information, and information status is a key concept of CE Design. Configuration control is illustrated in *Figure 5-7.*

Information Availability is a significant concern because of the current and evolving nature of the CE Design process. Most CE Design information is still on paper, in "off-line film" filing systems, or in a wide variety of different 2D and 3D CAD systems with incompatible data storage formats. As CE Design is implemented, CE Design Control Groups will have to deal with information requests that will be filled by: paper file retrieval; CE Design process information on a computer system; CE Design or other automated system information in other computerized systems, databases and files; magnetic tape storage; and other information storage mechanisms. Priorities, cost of accumulation, retrieval, presentation, and other technical, regulatory, and management considerations will be part of the decision-making on the speed and capability of

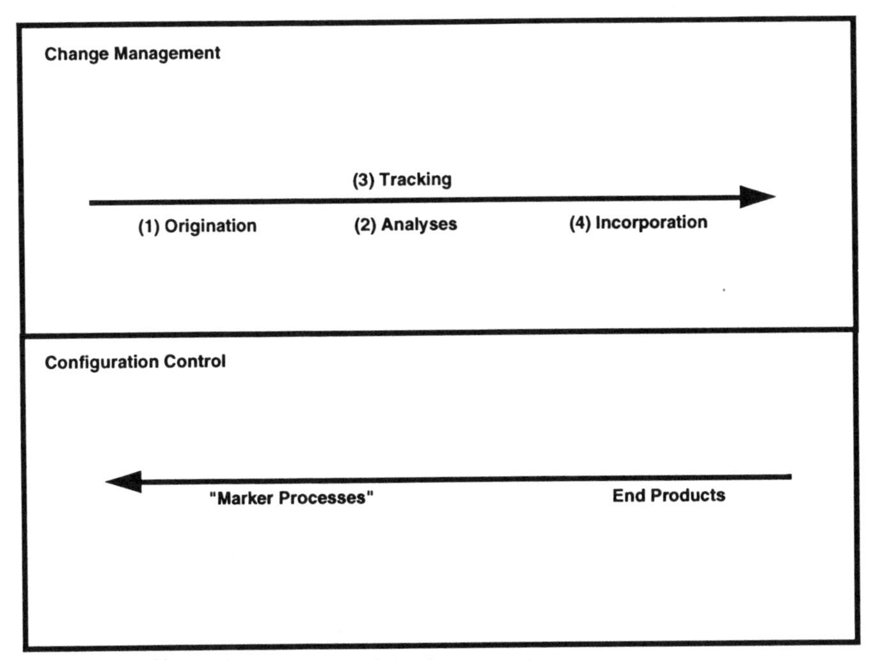

Figure 5-7. *Change Management and Configuration Control.*

information availability. For example, old drawings or filmed drawings can be ''scanned'' and stored as images for electronic review. These images have limited use because they cannot be changed. Standards for storing scanned drawings can be established (using standards such as SGML) after which technology can be developed to reduce (to geometric form or data) the product or part on the drawing. At that point, the drawing can be manipulated, just like any other 2D CAD product model, because it is no longer just an image.

CONFIGURATION MANAGEMENT

The configuration management element of Process Management is of importance throughout the organization in a complex product environment. The intent of the configuration management process is to get the correct component on the correct end-product delivered to the correct customer at the correct time and formation of these relationships.

Configuration Management is different from configuration control. The Configuration Management process is a cross-functional process coordinating several processes and various control techniques, providing a framework where appropriate configuration control activities can occur. Change Management is the management process that originates and manages the incorporation of anything from a new product family to a small revised product change. Configuration control is the application of various ''marker types'' to end-product related items, usually within a single functional area, to trace and manage changes approved during the Change Management Process for incorporation. Other aspects of the organization's activities may be kept under control using configuration control techniques.

CHANGE MANAGEMENT

The Change Management process starts what the overall configuration management process is designed to manage. Change Management has four main elements, as shown in *Figure 5-7*. They are: origination, analysis, tracking, and incorporation. The change origination process is focused on "channeling" the many types of ideas, regulatory mandates, customer requests, technological improvements, and internal improvements from both a necessity and desirable change perspectives. The origination process is shown in *Figure 5-8*. It begins with the Change Management Board's evaluation of various types of changes for consideration.

These considerations take on a variety of characteristics. Some changes requested come from the shop floor, requesting immediate change because of a detected error. Other changes include the introduction of a whole new product family. When a change is determined to be worthy of additional consideration, as seen in the lightest shaded area of *Figure 5-8*, a change consideration process involving preliminary engineering, management, and manufacturing analysis is begun. The process utilizes a routing and queuing "model" for reviewing that type of potential change. Various Concurrent Engineering Design teams are part of this preliminary consideration analysis process. Consideration factors might include cost, improvement impact, market impact, and other priorities.

If a change is considered acceptable, a Change Management Board decides on how and when the change will be incorporated into production. The medium shaded area of *Figure 5-8* shows that the Committed Development Schedule (CDS), its associated routing and queuing process model of the CE Design Team(s) activities, and the potential point in the Master Production Schedule (MPS) where the change is to be incorporated, are all developed. This Committed Development Schedule and the Master Production Schedule are also routing and queuing models. The consideration process, and the analysis and incorporation processes that follow, are all tracked through these models. These models are developed by people who do the work and are "in-line" or involved in the processes the models represent.

As the consideration, analyses, and incorporation processes proceed, the information envelope containing their work product grows. At the earliest point it might only contain a few descriptive documents and a sketch. By the time incorporation occurs, it may contain a wide range of information about all aspects of the change, including product literature updates. The incorporation finally occurs when the change is blended into the master schedule under the control of production planning, as illustrated in the darkest shaded area of *Figure 5-8*. Important elements of the envelope's contents, such as work statements and manufacturing process descriptions, are discussed in the design and manufacturability process chapters.

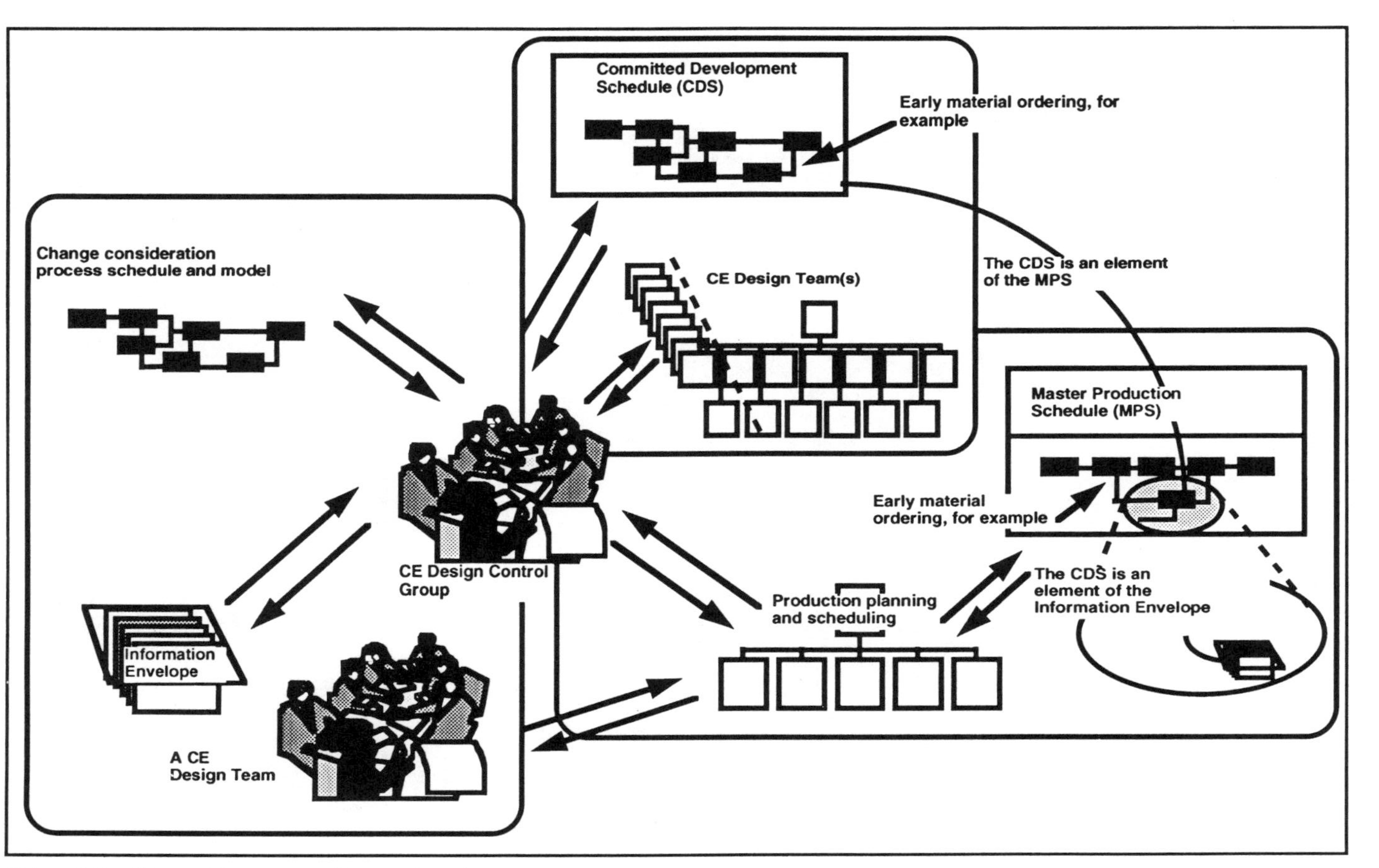

Figure 5-8. *Change Management Process Overview.*

Two aspects of change, (1) change variety and (2) adherence to change specification, reinforce the need for configuration management. The change management process is important because it establishes an orderly set of practices that promote improvement, quickly recognize and fix errors, and add discipline to the decision-making about what change to accept.

The biggest problem facing the complex product manufacturing organization is the pervasive rate and variety of change that must be constantly considered. Complex manufacturers have identified more than 150 different types of changes that might need to be considered. This change variety can create bureaucracies just for managing the many change types. Unfortunately many of the best and most experienced individuals in the organization are found in these bureaucracies. The routing and queuing "model" approach provides a common "system" of approach through which all of these varieties can be considered. These bureaucracies are limited or eliminated by this "model" approach and can be re-deployed to higher-value activities and needs. A new change type or subject means only another somewhat modified model of activities to be executed during the change consideration, analysis, and incorporation processes. People actually involved in the processes can initiate the models as well as the processes. As examples, three different change types will be considered. They are:

1. Shop Floor Detected Error–where a "touch" labor person attempts to complete the insertion of an Integrated Circuit (IC) into a board and the pins of the IC and the insertion holes don't line up;
2. Engineering Improvement–where an engineer finds that removal of an edge and the movement of an electrical power contact of a part permit easier manufacturing and reduces maintenance costs; and
3. New Product Introduction–where a new circuit board, fitting into a market niche not now occupied by the manufacturing organization, is selected to be produced.

The most important concerns of a shop floor detected error are speed and feedback to the originator. Unlike most other change requests, the manufacturing line may stop when this error occurs. Correcting this type of error, normally involving a temporary work-around, is the first level of concern. The second component of the correction process is to temporarily make the work-around a standard part of the manufacturing process while a permanent solution is developed. The third component of the correction process is to engineer a permanent fix of the problem, remove the work-arounds, and have the organization performing at as close to optimum as possible. This type of change is really three changes: initial work-around, temporary change, and permanent change. The configuration management process must record the initial work-around, identify and record the temporary change, and ease the planning for incorporation and recording of the permanent change. "Markers" are used to distinguish among these changes and the end-products that are affected by and incorporate these changes.

The Engineering Improvement change type has several potentially different characteristics. It could be a safety imposed change. For example, the movement of a pin may reduce the risk of fire on the board during heavy use. For this type of change, updates to units already shipped, through several generations of the product with that part, will be needed as well as the three category changes described in the shop floor error description above. However, if the change is a small improvement, only the incorporation process is required. This change type reinforces the need to record the details of each configuration for each end-product item, and for each individual unit if necessary.

Some of these changes can result in something other than an end-product change. Change requests of this or other types can result in modifications to the processes or documentation to which the product is manufactured. Configuration Control "marker" techniques must be applied in these circumstances.

The new product introduction type is similar to the improvement change type, but multiple changes (different sections of the end-product, and all other aspects of the product) must be incorporated simultaneously. This simultaneity of incorporation, and the consideration of several "changes" reinforce adherence to change specification.

Adherence to change specification requires that "proof" of change be maintained for products undergoing continuous change, and must allow for options and customer special requests. *Tracing* the change through the entire manufacturing organization and recording the configuration of each end-item ensures that the requested configuration is delivered. This is the overall purpose of configuration management. Many complex products are subject to at least some types of government review, standards and regulations. Others may require extensive, complete testing of the entire product after each change if traceability and change isolation are not maintained. Tracing the changes using configuration control markers permits the positive correlation of design intent and specification of product performance, i.e., what is designed is what is actually produced.

MARKERS

When a complex product family is conceived, it has a large set of features, functions, potential customer options, and customer special features. The configuration management process must support the process of taking a customer order and producing it. In the complex product environment, products are either built for distribution/inventory, or made to order.

A sample order and manufacturing cycle is shown in *Figure 5-9*. The complicating factor of building for inventory or distribution is customer demand. Considerable time and energy are spent on forecasting, predicting, and guessing "what and how many" customers are going to buy, and preferences can change rapidly. Typical planning techniques have included building a range of products with various sets of options to anticipate demand. Unfortunately, this mix is never quite right. Eventually some product must be sold at a discount to

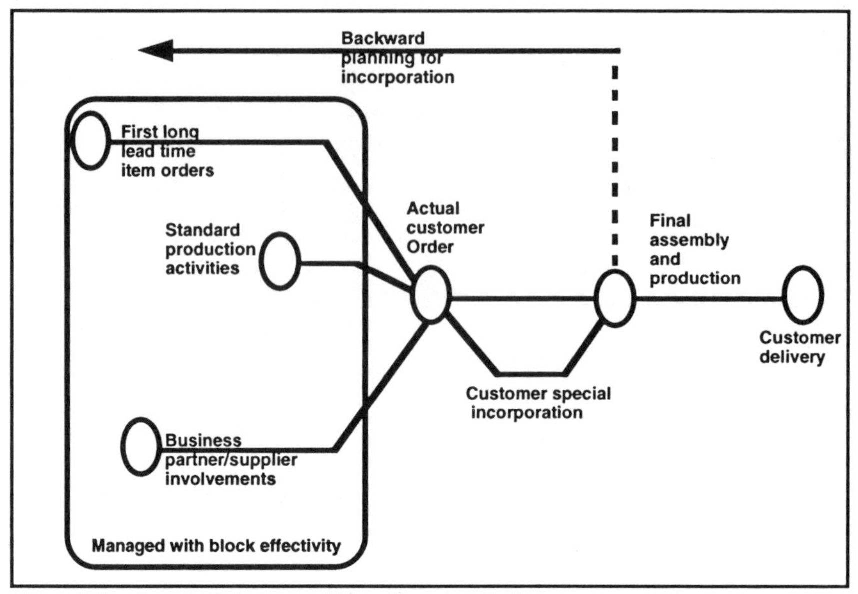

Figure 5-9. *Sample Ordering and Manufacturing Cycle.*

maintain adequate inventory levels and product mixes. Building for inventory has always been a compromise.

The manufacturing process requires time to take an order and produce the desired end-product. The customer's order-to-requirement time is short. Building for inventory satisfies the customer order cycle, but at real cost to the complex product manufacturing organization. Using proper CE Design and production manufacturability techniques, it is possible to build to inventory, for short delivery time, while avoiding overbuild. Effectivity is an important component of this capability.

Tracking the configuration for a complex product from the order point requires a marker technique called effectivity. For complex products built for inventory or distribution, there are two types of effectivity used to identify end-product configuration variations. These are date and serial effectivity. As design and manufacturing activities proceed, changes are identified which either produce a new product or product family (just a big change) or produce a newer, improved existing product. To differentiate products produced with these changes, or mark the point at which the change is to be incorporated, an effectivity (when the change is effective) is selected. If it is as of a certain time, then a date after which all affected products are produced with that change or a serial number after that particular serially numbered product is selected. The CE Design and manufacturing schedules are then backward planned to permit smooth incorporation of the planned change. Product orders for actual production, after considering the older configuration products, are matched with the effectivities of these products.

Certain types of process or complex manufacturing, such as drugs or chemicals, where quality, health and safety concerns always potentially exist, require recording the source or origin of their raw materials. This recording requirement also may affect composite materials, plastics, or even some metal parts or subassemblies. In these cases, a lot or batch number, or even the particular serial number of each important component, is attached to the end-product as additional piece(s) of configuration information. These items can also typically be built for inventory or distribution.

For the complex product made to order and only on demand, a third type of effectivity technique, the Block Effectivity approach, is used to identify, track, and record the product's configuration. The block effectivity approach uses an effectivity identifier for the basic framework of the product and for all the variable elements of the product (options, alternatives, and customer specials) to be added onto the basic framework to fill out the complete end-product as specifically ordered.

For the complex product, such as an automobile, a computer, or an airplane, produced on an assembly line and to a schedule with a build rate, which may involve different members of the product family being produced on the same assembly line, all types of effectivity are typically utilized simultaneously. As these complex products move down the assembly line, it is possible for individual end-products to be unique. *Figure 5-10* shows that as the product moves from final assembly station C to B to A, various additional items are potentially added to the end-product. Based on configuration control markers

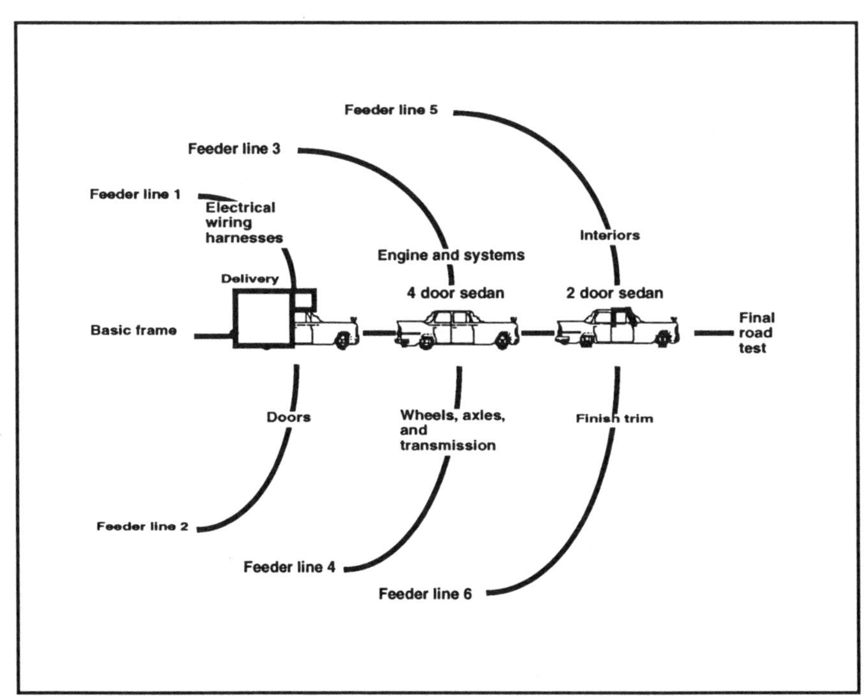

Figure 5-10. *Variable Products on a Single Production Line.*

assigned, the feeder lines could add basic and variable elements, options, or special/unique customer requests, all of which are changing based on their own configuration control markers. There may be a need for a "nesting" of these markers within the overall product structure.

However, the end-product produced is usually more inclusive than just the primary physical product itself. Other elements of the end-product might include spare parts, operating support equipment, maintenance and operations manuals, service equipment, operator training and education, and marketing literature. Part of the deliverable end-product items also may include providing drawings or schematics, used to design and manufacture the end-product item, for maintenance purposes. These items, in some instances, may be associated with a single end-product item in some unique way. These end-product related items, and the primary end-product, must be kept under configuration control and be tied together using effectivity.

Block effectivity is also important because it permits the collapse of the production schedule. To produce these effectivity controlled end-items, all the predecessor items that lead to the production of the end-item must be placed under effectivity control. Effectivity and configuration control via markers becomes a pervasive management technique, touching every part of the manufacturing organization very quickly. As shown in *Figure 5-11*, effectivity covers product descriptions and materials over the entire product design and production processes.

While effective date and serial number are obvious, the operation of block effectivity, crucial to overall complex product configuration management,

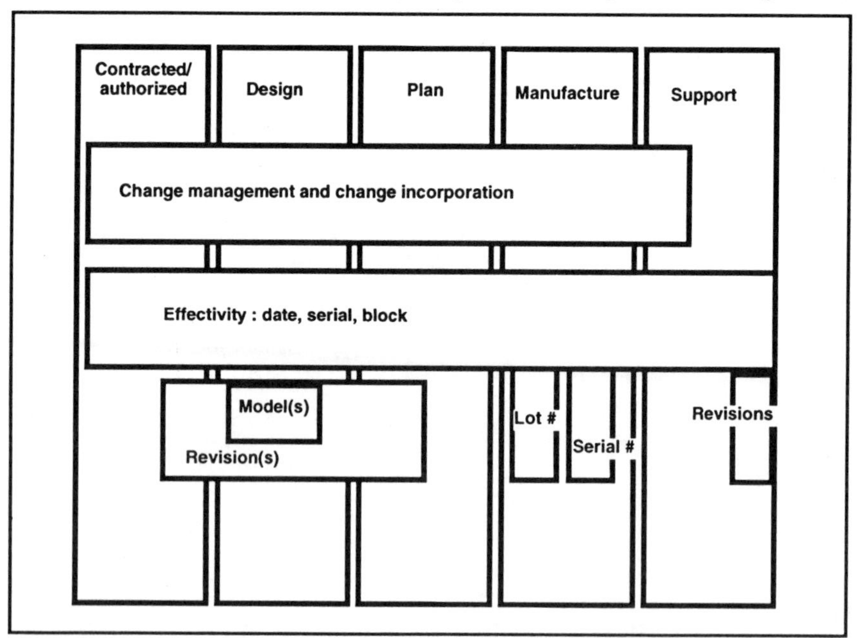

Figure 5-11. *Configuration Management Process and Marker Types for Complex Products.*

requires an example. A typical complex product has various physical characteristics, including a "frame" or other structure upon which most of the rest of the product's structure is placed. This "frame" and associated structure also might include most of the product which doesn't change frequently, even across the product family. This slow-to-change portion of the product, its definition, process, and support, are controlled by a Basic Effectivity block marker.

For example, a product family of elevating platforms might break into five products with increasing height reached and/or weight supported across the product family. *Figure 5-12* shows how the basic effectivity block elements of this elevating platform family might include the base "frame" and the platform. As the height reached by the platform increased, the strength of the lifting arms and the propulsion unit's power output would increase. Models of the product family would be designed to reach increasing height limits. Finally, at the top end of the product line, from a height-reached perspective, and for stability at increased heights, support extension arms to broaden the effective coverage of the base would become standard, but appear in the product structure as options. These options would be controlled by the Variable Effectivity block. For a special customer request such as "stabilizing guide wires" for attaching the

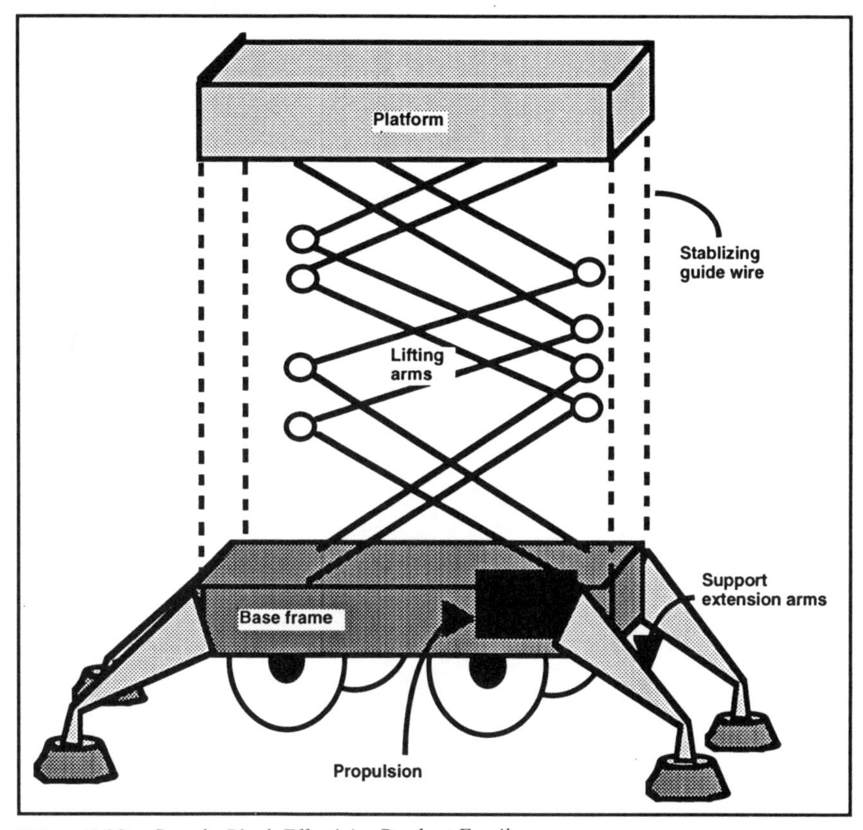

Figure 5-12. *Sample Block Effectivity Product Family.*

lifting basket to the base, a special subset of the variable effectivity block might be used.

The block effectivity markers for this example product might be "codes" constructed as follows:

1. Basic Effectivity Block–A nine-character symbol of AAAANNNNN where AAAA = model code and NNNNN = a range of end-product item markers which that model might cover. NNNNN might have to be much larger if the number of complex products to be produced is much larger. For this example, LIFT00001–LIFT00999 might cover the beginning of the line model; the next, middle of the line and largest selling model might be LIFT01000–LIFT39999. The other top-of-the-line, higher weight, height, and price models might use the rest of the range (LIFT40000–LIFT99999);
2. Variable Effectivity Block–a nine character symbol of AAAAPNNNN where AAAA is an abbreviated name or acronym for a customer, P (which provides for 36 products (A-Z, 0-9)) is the product model identifier, and NNNN is the range of orders now possible for that customer for that product. Other combinations and sizes are possible, of course.

For example, the latest order from the SAMPLE company for two models, one each, might be: SAMPI0003 and SAMPT0001, where SAMP = mnemonic for the customer, I and T = intermediate and top-of-the-line products, and 0003 and 0001 the serial number of the third "intermediate" and the first "top-of-the-line elevating" lifter ordered by this customer.

Regardless of the exact coding approach, the variable effectivity identifier and the basic effectivity identifier are both required to describe each end-product. An individual end-product's description is not complete without both. In combination, they act as a unique identifier for that end-product. As shown in *Figure 5-13A* (starting with the customer order), when the order is received, an element of a product or program management group responsible for effectivity control, consults the master schedule, considers the existing schedule, and assigns the basic effectivity block marker or markers to this order, as well as any needed variable effectivity block marker(s). Any changes to the prod-

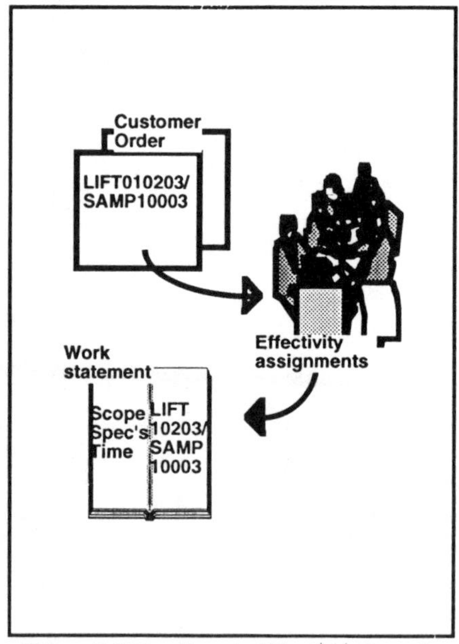

Figure 5-13A. *Sample Effectivity Overall Flow.*

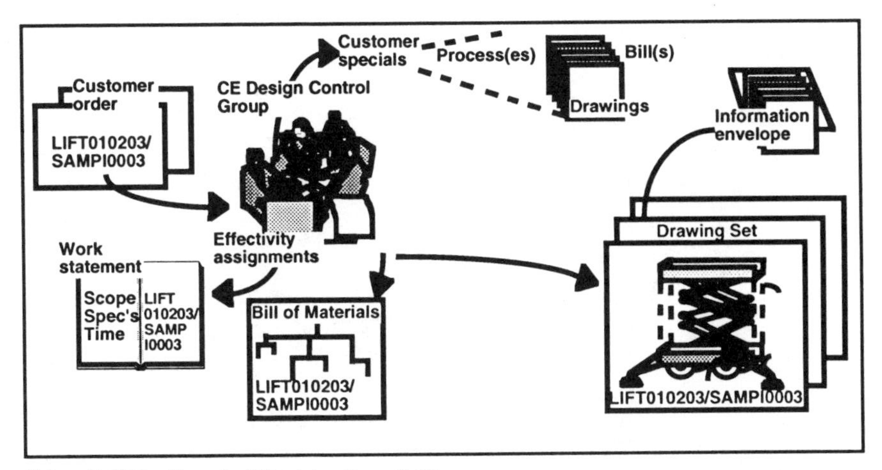

Figure 5-13B. *Sample Effectivity Overall Flow.*

uct design arising from this order are made using these effectivity markers, as depicted in *Figure 5-13B*. Once any design work is complete, the firm product design is released and the production manufacturing process occurs, as in *Figure 5-13C*. The effectivity markers appear in the MPS, as well as on supplier orders, shop orders, and in final assembly instructions.

Figure 5-14 shows that the master schedule, or final assembly sequence, is actually a sequence table with target dates. Every part, document, and other end-item that applies to LIFT01023/SAMPI0003, the intermediate lifting elevator, is marked specifically to be part of that specific end-product. For LIFT40201/SAMPT0001, the top-of-the-line unit, including the special request for braided steel wire, "stabilizing guide wire" ropes to be attached has also become a line in the sequence table.

Other categories of effectivity could have been included, such as an Engine effectivity block to cover various engine types, and perhaps a protection effectivity block to cover preliminary manufacturing for anticipated option packages for an end-product for a customer who either hasn't decided on a final configuration but is willing to pay extra to get started now, and/or in anticipation of the order that might be provided. These block identifiers are additive.

As the work flow proceeds, these effectivity blocks distinguish between work items intended for individual end-products. The block could have been a range, or even multiple ranges simultaneously for bigger orders. For example, LIFT01024/SAMPI0004 could be a continuation of a multiple intermediate order from a larger order of five intermediate models with the block of effectivities of SAMPI0003–SAMPI0008. The bills of material and the process plans for this end-product must reflect the effectivity. When supplier parts are ordered, their receipt must trigger a function to tie the parts to the correct subassembly. Effectivity is additionally used to relate included or not included components and subassemblies into the end-product. It is this technique of conditional inclusion that allows schedule collapse.

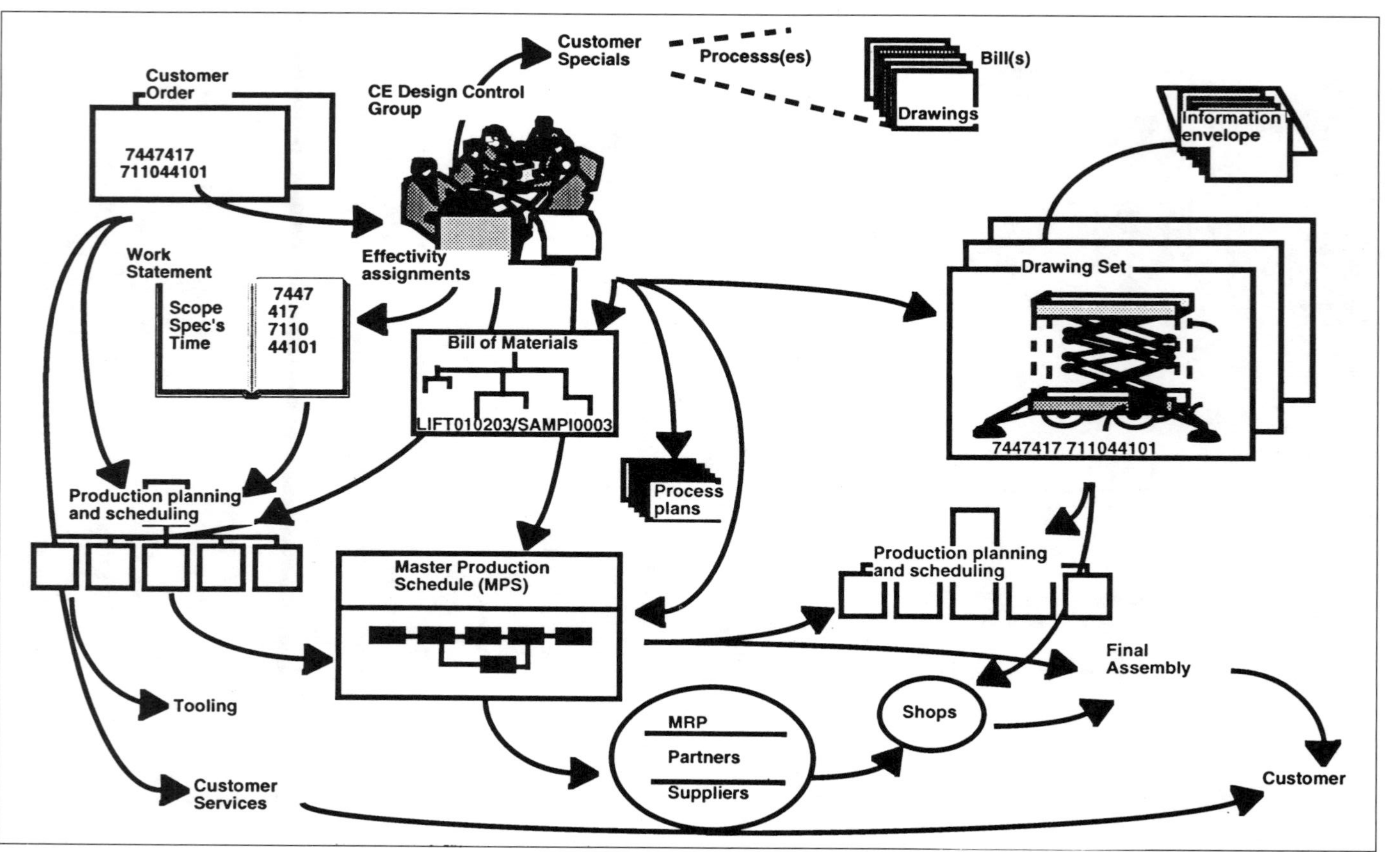

Figure 5-13C. *Sample Effectivity Overall Flow.*

Basic Effectivity	Model Designator	Variable Effectivity	Special Effectivity	Engine Effectivity	Protection Effectivity	Start Date	Load Date	Due Date
LIFT01021								
LIFT01022								
LIFT00099								
LIFT01023		SAMPI0003	—			5.5.95	5.7.95	7.4.95
LIFT00100								
LIFT10200								
LIFT01024								
LIFT10201		SAMPT0001	SAMPT0001			5.25.95	5.28.95	8.7.95
LIFT00101								

Figure 5-14. *Sample Master Schedule.*

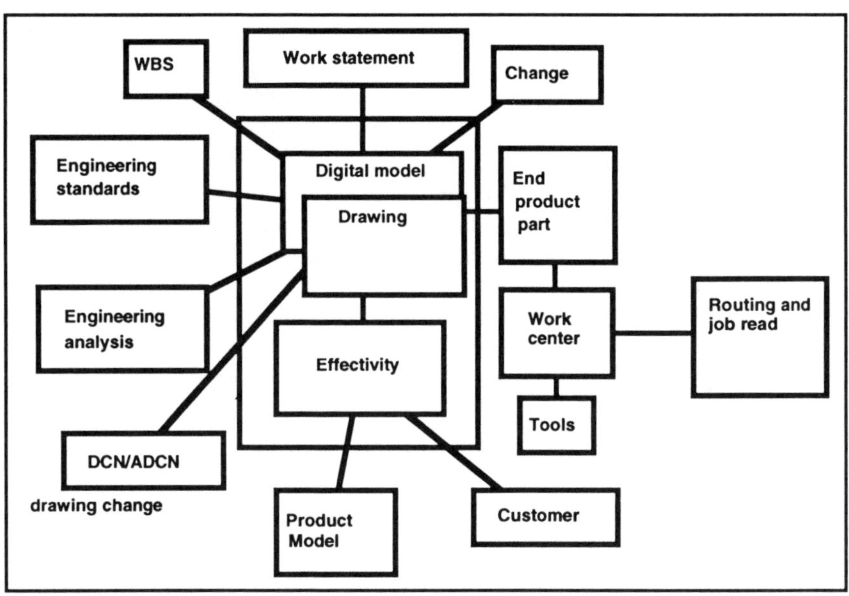

Figure 5-14A. *Composite Information Items for Configuration Management.*

Figure 5-15 is an example of a typical bill. Notice that the connections in the tree are not fixed. Their inclusion in any particular end-product's bill is conditioned by and governed by the effectivity block. Elements of the basic and variable portions of the end-product are "hung" off the bill tree. The net effect of this block effectivity approach is an individual bill for each end-product. This might be a "virtual" or an actual individual bill. The desire of most complex product manufacturing organizations is to track the individual end-product through the "as designed," "as planned," "as manufactured," and "as supported" bill structures. This tracking capability facilitates change management and configuration control activities, since each "state" of the product is somewhat different.

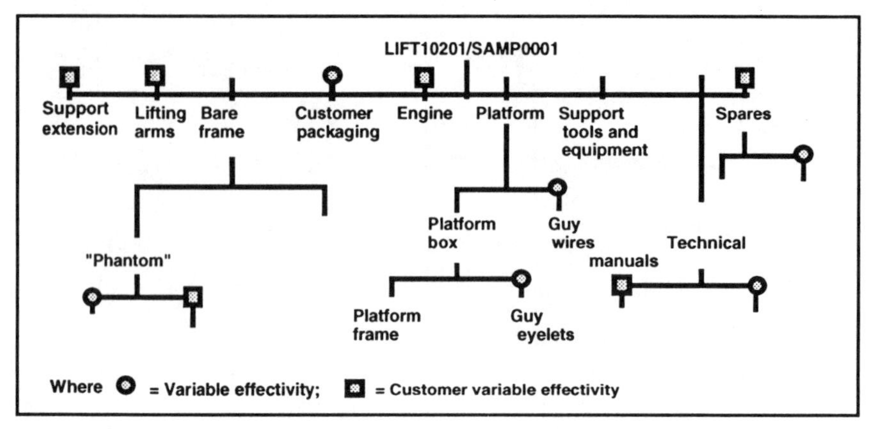

Figure 5-15. *Sample Bill of Material.*

From a systems perspective, the block effectivity approach is useful. It is costly to store and process bills of materials with 25,000 or more part numbers when explosions (a computer process to determine what parts to order), or implosions (a computer process to determine what parts are associated with each product or where each part is used), are required, and these bills are stored on a disk drive and the explosions/implosions are performed on a computer. The block effectivity approach permits a single bill structure with many conditional branches to replace these many bills if desired.

The technique, when combined with good discipline in how the end-product is planned for production, can be less confusing than the alternative, which is to constantly change part numbers on the design and the shop floor assemblies themselves. When appropriate major elements of the end-product are identified, effectivity as a component of the bill's parent-to-child part number relationships can stop at this high level. Using this technique, most of the organization is still working on a pure part number basis. Parts are tied to the individual end-product at the last possible moment.

SCHEDULE COMPRESSION

The power of the block effectivity technique is its ability to permit the complex product manufacturing organization to collapse the overall schedule substantially. The basic portion of the product, and a certain number of each model's component parts, can be started well before the customer order. The planner and scheduler can shrink the final assembly time, and use effectivity to tie together the basic and variable elements at final assembly time at the last moment. When the variable elements begin to get out of balance, scheduling and marketing can adjust these flow rates to reflect recent experience and anticipated demand. Safety levels of some items may be appropriate or not, as circumstances dictate. In any case, automobile manufacturing could be controlled in this manner and a "custom" order produced and delivered in 48 to 96 hours if all went well. Several operating examples of this substantial collapse can be found in industry including computers, aircraft, and electronic products.

This same effectivity control marker approach must be applied to the manufacturing processes, as well as the tools, the drawings or digital models of all components, including those for the special braided wire rope "stabilizing guide wire" ordered, the special processes used to add this customer feature, and the schedules themselves.

The relationships between effectivity and the other major information aspects of the overall manufacturing process are shown in *Figure 5-14A*. Effectivity and Digital Models are at the heart of the configuration control process. *Figure 5-14* depicts the various elements appropriate for configuration management concern. Notice that effectivity acts as an anchor point for these elements. The control point or authority for everything is the drawing, if there is no digital representation of the part or the digital model. The physical part itself is but a current physical manifestation of the control point, and is not the point of configuration control.

Considerable emphasis, in recent years, has been placed on part numbers as the key point of control by some in the industry. Drawings are being considered "just reports" of the digital database. It now appears that the *digital model* of the part, the subassembly, or the entire end-product, is still the basic configurable item. The design's Information Envelope and its various elements must be controlled by effectivity as well. This control must be extended to all elements affected by such effectivity.

Figure 5-11 shows other types of control techniques used on types of information evolved to meet the need of that individual type of information. *Figure 5-16* summarizes these. It is useful here to discuss the Digital Geometry Configuration control technique.

The Digital model of the Geometry Control technique has evolved to permit a team to work on the same related set of product and manufacturing artifact-related geometry simultaneously. This apparent simultaneity is a key feature of CE Design. *Figure 5-17* depicts control concepts for digital geometry. It describes the various "states" a geometric model can be in. The geometric

Information Type	Control Technique
Text	**Revisions**
Drawings	**Revisions**
Part Batches	**Lots**
High Expense Items	**Serial Number**
Traceable Items	**Serial Number**
Software	**Version / Release**
Digital Geometry	**Model / Variant**

Figure 5-16. *Information Types and Control Techniques.*

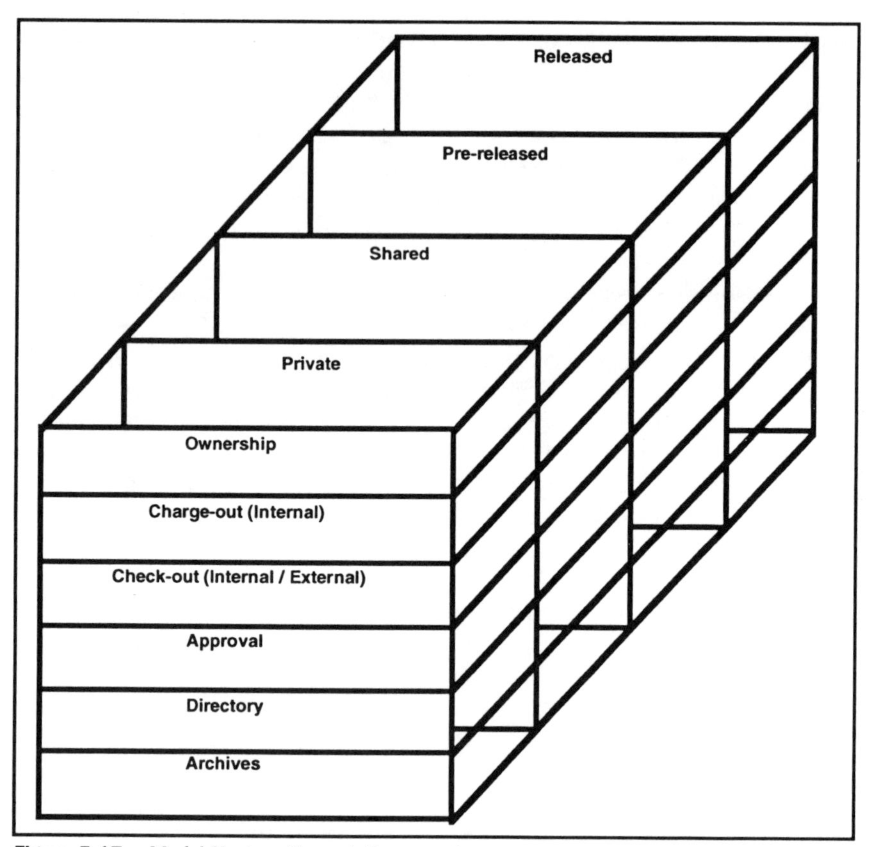

Figure 5-17. *Model Variant Control Framework.*

models are always "owned" by someone to insure design intent control. The models are "checked out" when actual changes are to be made. "Check out" occurs when the model is to be used, but not changed. When the model is being reviewed, it is in a state of "approval." Directories of models are maintained with these status indicators across these conditions, in one's private space, or when shared with others, or when they are under control for review, or when they are finally "released" to production manufacturing.

Figure 5-18 illustrates the control rules of this framework, providing the details of how to control the movement of the information envelope containing the digital geometry and product specifications between various "states" or status points (private to the individual, shared for review by co-workers, pre-releases during approvals and other CE Design Team reviews, and released for manufacturing and external organizations). Moving from top to bottom of the movement control arrows pictured in *Figure 5-18*, a model is "charged out" to a private environment model and "variants" are prepared. Once it is ready, various specific variants can be moved to "share" status. Each is reviewed, and compared to other proposed variants. Each may need to be moved back to "private"

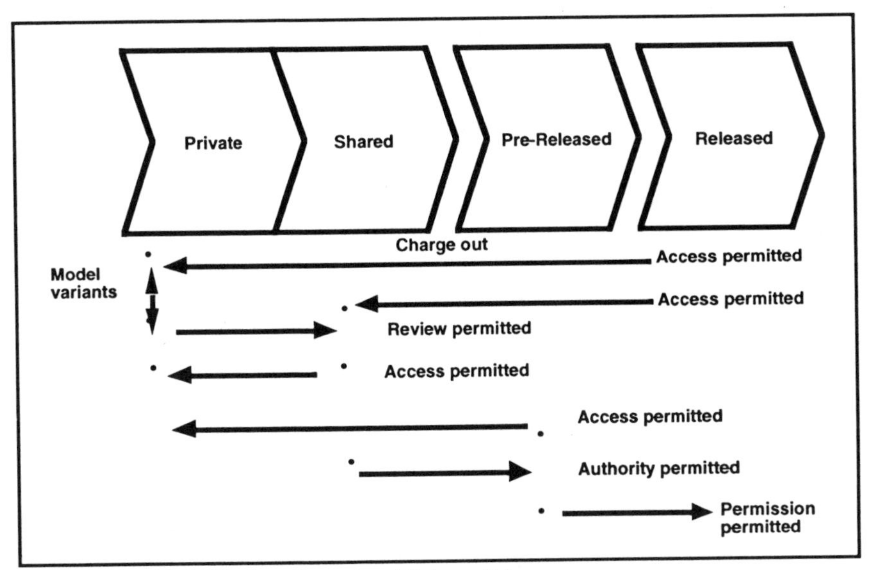

Figure 5-18. *Model/Variant Control and Release Scheme.*

for changes. After all changes are made, the reviewed model is proposed for final authorization for a change to be committed. When permission is obtained, the models are moved to release for production status. Effectivity permits the adoption of more than one variant at the same time.

Referring to *Figure 5-7*, the Effectivity process *is performed in the interest of manufacturing and product support*. They are the ones most interested in getting the right part on the correct end-product. Change Management, on the other hand, helps engineering manage and channel improvements, problems, and new product ideas. Concurrent Engineering Design is a natural reflection of these concerns. It puts together teams and processes to simultaneously address these complementary interests.

RESOURCE MANAGEMENT

This last element of CE Design's Process Management provides management of the resources applied during CE Design. These resources fall into several categories, including people, machines and tools, time, facilities, and outside firms. Techniques such as scheduling, cost accounting, balancing, OPT, and other load management approaches are also considered as aspects of resource management. The Resource Management process operates as another element of the process control logic using the Routing and Queuing process's model of activities seen in *Figure 5-19*, which must be executed to complete design and production responsibilities.

As the process model's activities are executed, the starting and stopping of these activities can "trip" or initiate the recording of the start or stop. In a computerized CE Design environment, the model acts as an electronic router,

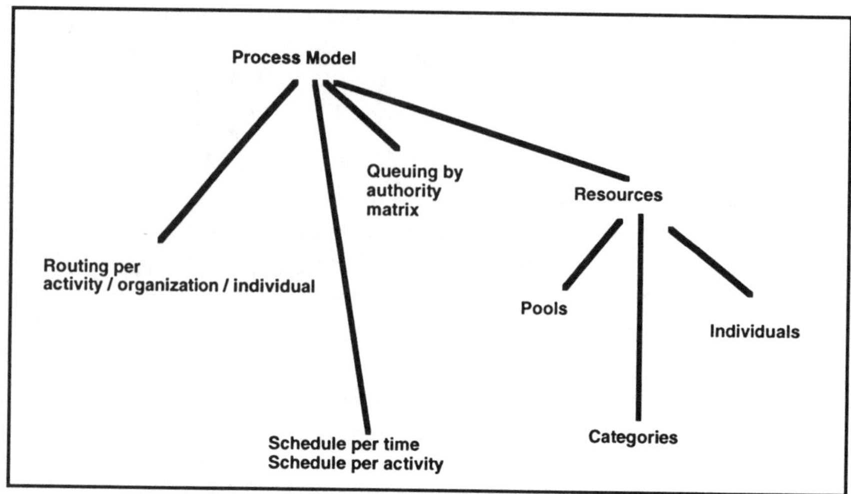

Figure 5-19. *Process Model as Base for Process Management.*

and the next work item appears on the appropriate individual's workstation. "Start" occurs upon "opening" the electronic mail, and "stop" occurs when the work item is sent with a "completed" tag. Elapsed time and the individual's and group's activities can be recorded.

The relationship between resources, the individual, and the group should be pre-established. An activity-based, resource management cross-reference process is used to establish these relationships. *Figure 5-20* is an extrapolation of the Activity-Based Costing (ABC) process originated by Cooper and Kaplan. Their concern was the traditional costing approach of manufacturing. The traditional costing approach was based on hours with overhead applied. This technique is discussed in general in Chapter 4.

Their technique can be broadened, and it works well with other activity analysis processes, such as those used to evaluate value-added activities and

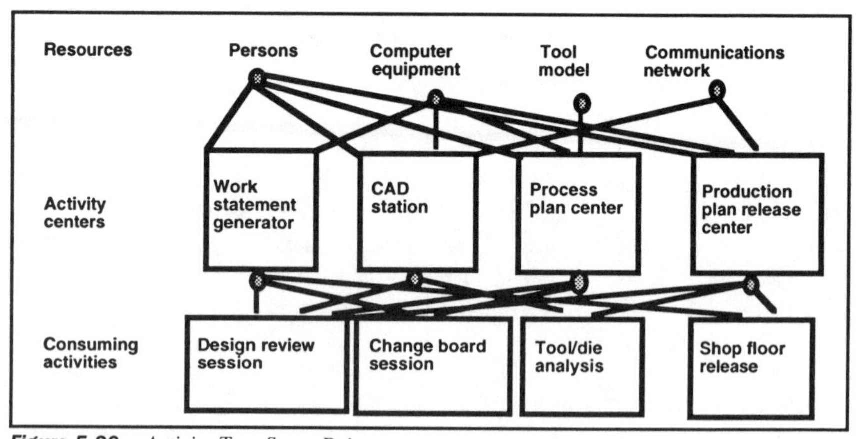

Figure 5-20. *Activity Two Stage Drivers.*

other process improvement techniques (see *Figure 4-3*). An important element of these activity-oriented analyses, including ABC, is that, historically, there has been no organized process to cost the activity-oriented portion of manufacturing. The dominant concern about nondirect labor is the unstructured nature of its activities. Many starts and stops, diversions, and a somewhat "reactive" style have created an environment in which measurement was very difficult. One important feature of the routing and queuing model, with automated infrastructure support, is its ability to measure a high percentage of these activities, including management's activities. See *Figure 4-7* for a conceptual model. It is not necessary to capture precisely or measure all, but to reduce the residual, nonallocated amount of all activities to less than 20%. The remaining time can be accumulated into an overhead pool and allocated across activity centers without substantial distortion of costs.

WORK STATEMENTS AS WORK AUTHORITY

The work statement process is discussed in the Design Process chapter to follow. One of its purposes is to create the cost accounting collecting structure or authority against which someone or something may charge expended resources. The authority is typically a project charge number against which all related activity costs are charged. This project number, in turn, is related to or cross-referenced to traditional accounting transaction-oriented numbers. This allows the integrity of the traditional accounting system to continue while project/activity accounting is being used to manage the CE Design portion of the business.

Most of CE Design involves activity-oriented management. *Figure 5-21* depicts some of the resources provided to an engineer to perform the design

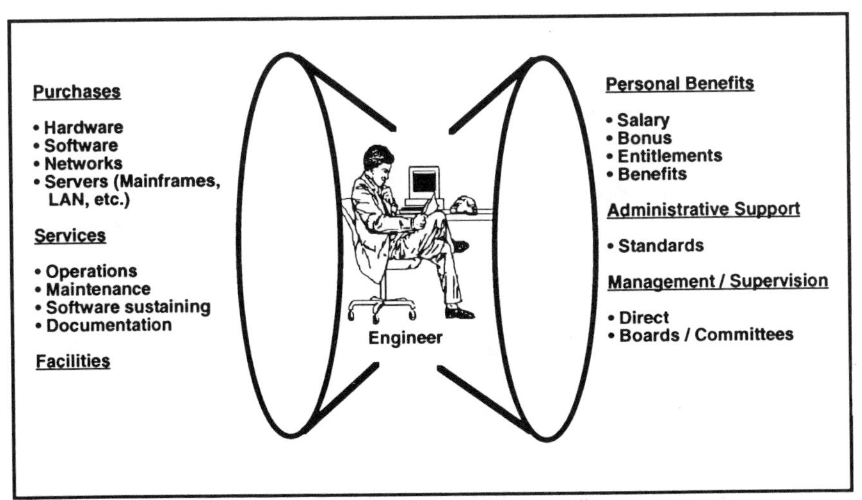

Figure 5-21. *Accounting (Project/Activity and Traditional) Sample Information Categories.*

process. These resources fall into two broad groups, "people" and "things." "Things" begin as purchased raw materials of some kind or equipment; "people" related resources include benefits, administrative support and management supervision, among others.

Figure 5-22 takes this broad activity and examines a next-level of detail activity. *Figure 5-23* takes the overhead accounting activity and examines a next-level activity. At this level, traditional accounting transactions are generated, as are project/activity accounting transactions. As shown in *Figure 5-24*, there can be a relationship established between even indirect action, such as purchasing paper for use in a printer used by an engineer, and directly measured

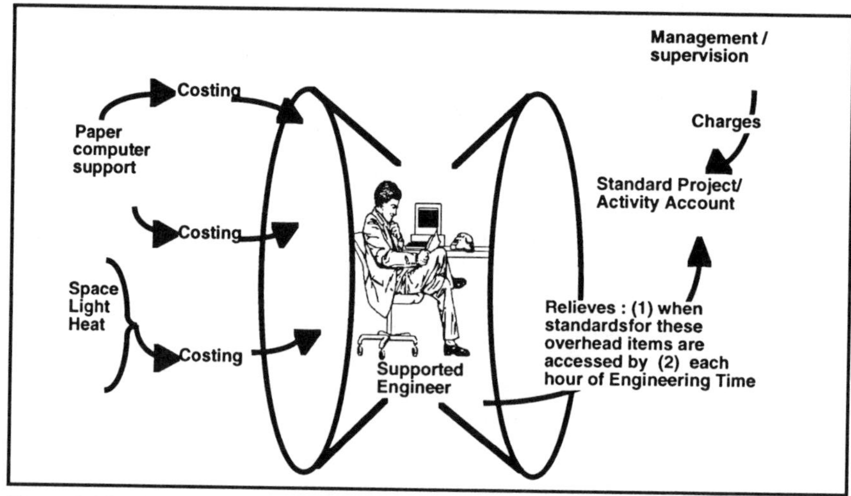

Figure 5-22. *Overhead Costing Example.*

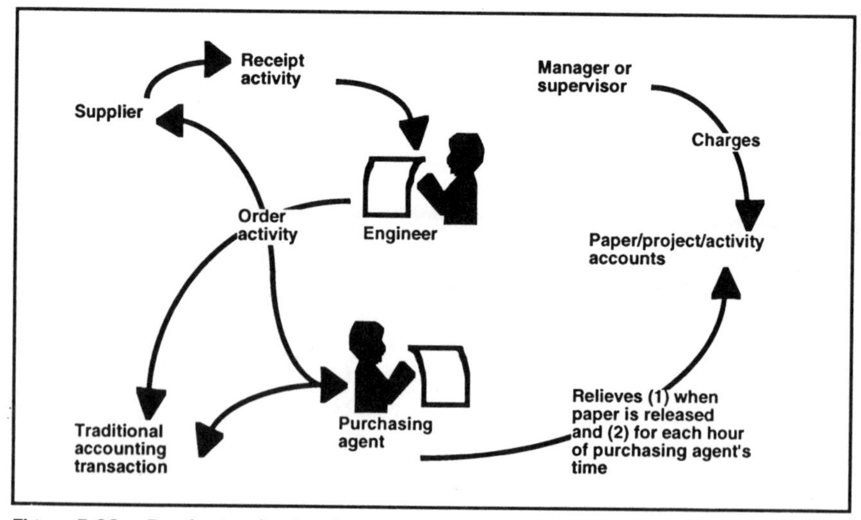

Figure 5-23. *Purchasing for Overhead Items.*

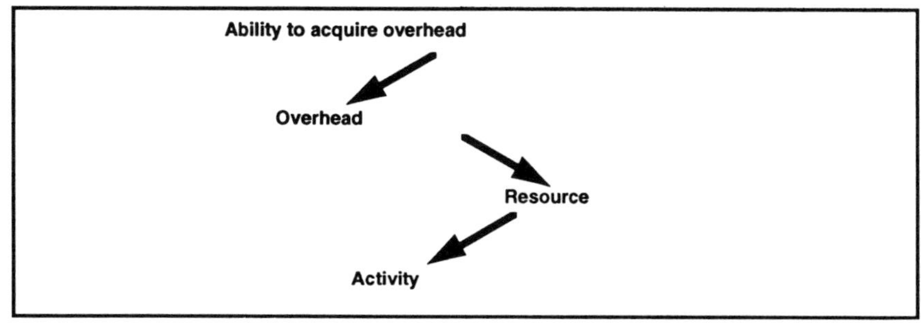

Figure 5-24. *Linking Overhead and Activities.*

activities. By establishing this rich infrastructure of costing, using activity *and* traditional accounting transactions, Resource Management accounting and usage/consumption information is produced. The starting and stopping of first level activities creates demand for second and third next-level activities, and it does so in the best ''pull'' fashion.

The historical problem with these nondirect activities, because of their unstructured nature, has been the difficulty of forecasting demand and determining the best mix of resources. This has created a constant interchange between senior and middle management to whom responsibility of process execution is assigned. Middle management wants a good mix of people and assignments, so varying tool availability (for example, computer systems, mockup facilities, space, communications) and priorities can be accommodated even in a very heavy work load period. Senior management wants a higher net productivity and lower cost, in a constantly improving environment.

Using the in-line model execution and activity costing approaches, middle management can more accurately show composite workload, forecast, and budget once in-line model-driven process experience information becomes available. These models must include resource usage patterns as an element of their stored characteristics. Because these models can include outside the firm activities and external resources, the models can be interactive with the activities and schedules of these firms. Their progress and cost parameters can be rapidly reflected in the complex product manufacturing organization's own schedules and budgets.

This model, when acting as a routing, can relate to a group or an individual's calendar. The widening use of electronic calendars, interacting with all routings to identify conflicts, prior allocations, and opportunities, also can aid in resource management. Most scheduling management software packages are now able to record activities against schedules, and can attempt to balance various resources within these schedules. The schedule, time, cost and resource analysis capabilities are valuable within this context. Unfortunately, this automated process has also been very people, technique and resource-intensive, because the tool is a ''stand alone'' management tool, unless it is an extension of the ''inline process'' model described earlier.

The activity database, shown in *Figure 5-25*, should be absorbed into the

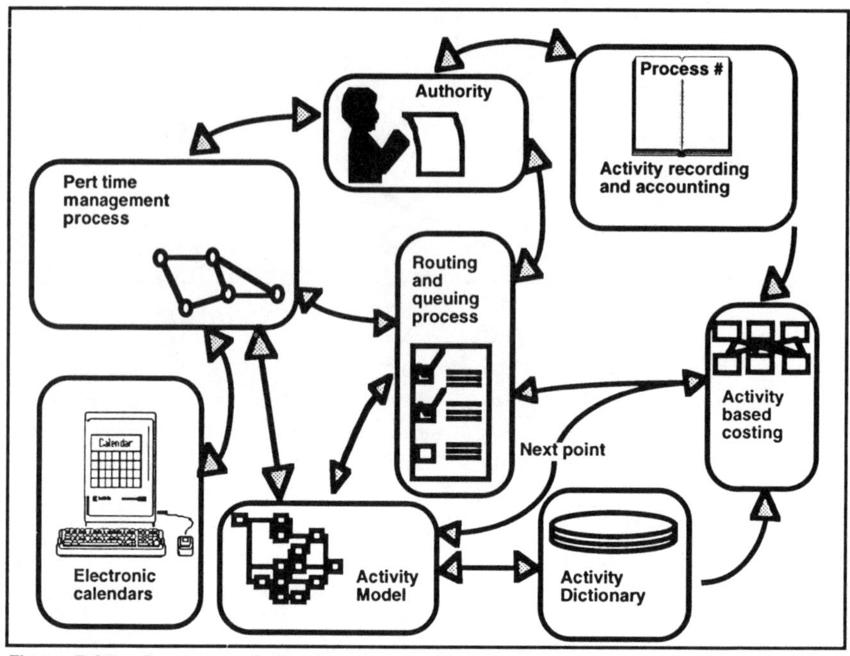

Figure 5-25. *Routing and Queuing, Activity Model Relationships.*

routing and queuing model data structure. This permits the scheduling process to be integrated into the other processes by using a common base of information. For each point on the routing and queuing process, an activity model is developed. This model is based on the activities of the activity dictionary, which included the ability to support costing. It is also the focal point for time and calendar management.

This common base of information does not have to be a single database. Using a database system, an object-oriented architecture, a message-oriented architecture, and other sets of computerized tools to provide functional integration is all that is necessary. Chapter 8 discusses computing support more extensively.

Once the model has been established and activated by an appropriate authority, the model-driven management process can provide visibility into progress, resource use (planned, actual, and variant), and electronic calendar interface and interference analysis. As processes are started and stopped, costs and consumption can be tracked and requirements projected. Once this overall process has begun, and more models covering more CE Design activities are built and used, the wide variety of models can be *combined* within a high level, milestone-oriented model-driven process. This rollup of projects provides for overall resource use and projected budget analysis at the executive level.

The temptation in the resource management process is to attempt to *balance* work load and resources. As has been shown on the shop floor, and in any unstructured work situation such as CE Design, a balance of this type is impossible to sustain. The scheduling process across all the groups involved in CE Design should be focused instead on managing "elasticity." By having a good mix of people and tools, and having more resources than needed, middle management has built excess elasticity into their organizational structure. As work across these various organizational elements ebbs and flows, this excess elasticity also goes up and down. In the competitive market, this excess elasticity adds product costs that cannot be sustained. In the model-driven environment, better information and experience should accumulate about which elements are being over- and under-utilized, and adjustments can be made as appropriate. The key difference is that these adjustments are not the "5% across the board" increase or decrease-type approach, but are specifically directed against known requirements, and against activities, not people. The results include increased throughput and reduced costs, and is more sensitive to people.

One of the challenges of the CE Design approach is its teaming concept. All elements of the product must have teams, and if all teams had to be physically co-located, this would require additional staff because a minimum of a person for each function for each team is required. Some duplicate assignment of manufacturability and design team members would be possible, but inefficiencies would occur and ineffectiveness (too many, not enough) would still be a factor.

Electronic proximity or "electronic teaming" provides a better mechanism, especially for the more static elements of the product in a change, rather than new product development environment. Resources can be partially assigned. Routing and queuing models of past changes and new product development have been kept. Intellectual capitalization of what was required, of whom, with what cost and time parameters have been accumulated and are easier to use and understand. In this environment, it is easier for someone, perhaps even unfamiliar with the product, to take on the next change. The improvements in these models and the tools and techniques of design can be cumulative. Improvements in resource use and effectiveness can continue.

6

CONCURRENT ENGINEERING DESIGN'S DESIGN PROCESS

The CE Design process has several objectives. It is most important to facilitate the *completion* of product design. In spite of all the attention on engineering, there has been little focus on the design process itself. Yet, design has always been perceived as important. Design Process importance, historically, appears limited to activities associated with initial product development.

This inconsistency can be observed in a recent survey. It was conducted with the cooperation of a number of manufacturing organizations, worldwide. As seen in *Figure 6-1*, design quality was ranked as their most important manufacturing concern in over 88% of those responding. Yet, *Figure 6-2* shows only a little more than 10% of those surveyed have manufacturing involved in a design effort on a joint basis. A fully deployed CE Design process, with its disciplined, integrated design methodology, would have made the percentage of involvement much, much higher!

This chapter addresses the Design Process of the CE Design. It examines the "mental model" in which a concept becomes a product. This "mental model" is mapped into a general design process. The approaches to design which address the initial product development process are described. The complex product, with its long product life cycle and duration as an operating product in the marketplace, is also discussed. Various engineering design processes, comprised of activities necessary to complete each approach, are presented, analyzed, and the implications discussed. The role of the Design Process in the overall CE Design process is also described.

An interesting aspect of CE Design is the need to develop a model of the design process to complete a product design. This chapter describes design process elements within the context of CE Design. Such a design process has several different perspectives. These perspectives include the intellectual process, the engineering method of analysis, physical and system product designs, and the complex product design and matrix teaming invoked in the complex product manufacturing organization.

THE INTELLECTUAL PROCESS

The intellectual process, which each group and team goes through as they consider a problem or develop the idea of a product, can be described as having three general stages (described in *Figure 2-7* series). These stages are conceptualization, visualization, and realization. The conceptualization determines *what*. Examples of *what* include the product's characteristics and operating parameters, identifying tools required to make it, developing the content of the process for making it, and identifying the documents needed to describe its operation. Visualization of the *what* occurs when the characteristics, ideas, and parameters identified are reduced to usable information. Examples of visualization include:

- A 2D and 3D set of images of the product which provide details of characteristics and operating parameters first conceptualized,
- A model of the manufacturing process with instructions, or
- An operating manual developed on a computer.

Realization of the visualized product comes when the physical manifestation of the *what* occurs; e.g., when the hard part is produced, the manual printed, or,

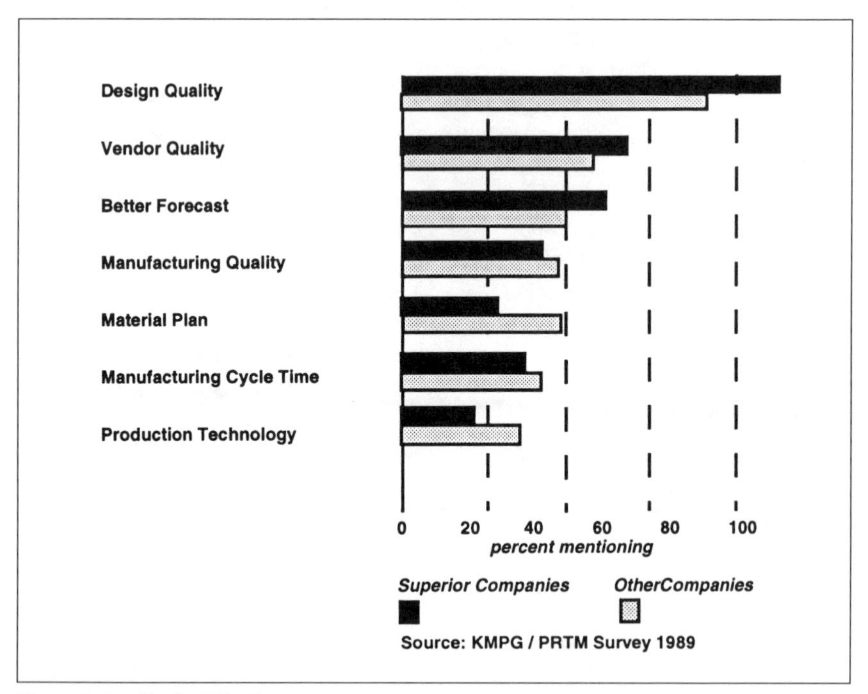

Figure 6-1. *Profitability Impact.*

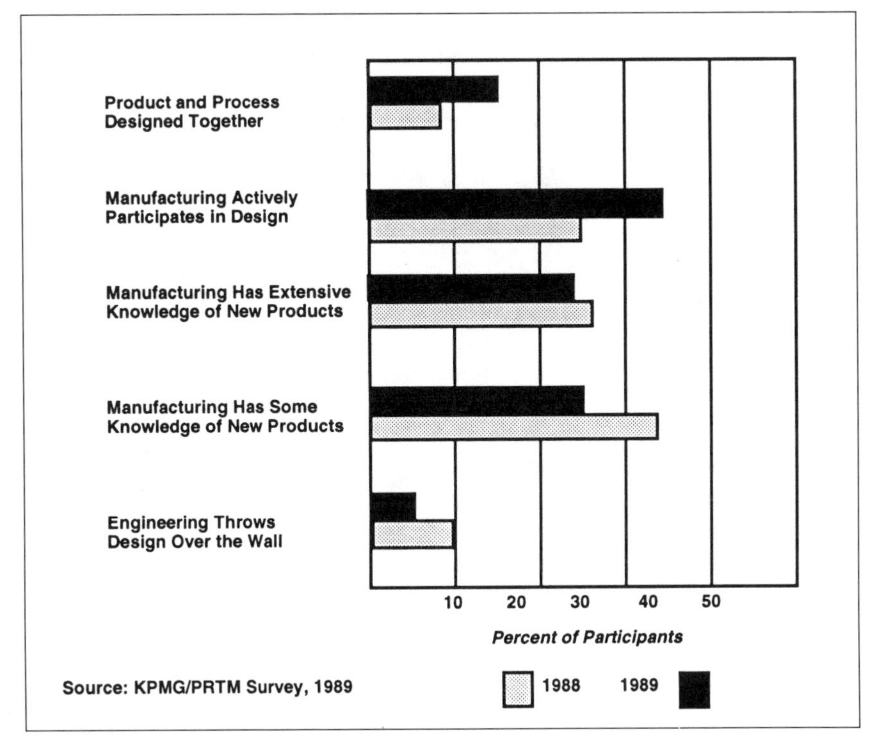

Figure 6-2. *Manufacturing Involvement.*

in the case of an information system, when the input/output screens and reports can be seen, used, and results based on internal computations and data provided are achieved.

A discussion of these intellectual process steps is important. If the automated infrastructure support for CE Design is to be effective, the information associated with this process must be captured quickly and integrated as necessary as the stages progress. A challenge of CE Design's Design Process, and its supporting processes, is to support conceptualization so it becomes as easy or easier to work within the CE Design Process while capturing appropriate process step information, compared to a traditional design environment.

The chances of a successful CE Design Design Process increase dramatically if the tools ''induce'' use. Using those tools as after-the-fact recorders of already fully developed ideas is a redundant activity. Also, the thoughtful, analytic process is not captured well after-the-fact; there is a natural tendency to ''short-cut'' to the ending description of the product. The best environment captures design intent and process to facilitate organizational learning and the accumulation of information useful to manufacturing.

Utilizing visualization tools (such as CAD) to support the conceptualization process has limitations. Many times, the speed, responsiveness, and complexity of these tools is an inhibitor to creativity and conceptualization. Other potential barriers, such as management styles, cultures, and team psychology, are discussed later in this chapter. For these reasons, the best approach to ensuring that the intellectual process is quickly and successfully executed is to stress small groups of two or three people conceptualization activities. *Large groups or teams will tend to stress organizational harmony over creativity.*

There are competing strategies about how, when, and what size teams should be used at each point in the design's evolution. A variety of approach-oriented trials have yielded two recommended teaming approaches. If the intent of the project is to develop a new product, the new innovation design teaming approach in *Figure 6-3* should be used. The early conceptual process is begun with the small team of two to three lead designers. As the design matures, more individuals and teams are added. The teams decline in size as the product reaches manufacturing and realization.

For a derivative product, such as a new automobile of the same type but with different characteristics, start with an almost completely staffed team as in the existing Derivative Design team approach in *Figure 6-3*. Conceptualization at the entire product level is not as important. The "full-up" team can move immediately to the details of the design in smaller subteams of specialists where most of the innovation and conceptualization is to occur. The team can complete the process quicker and still derive the benefits of smaller teams. The small team can be innovative if necessary, while the large team can quickly get the derivative accomplished. It is important to use each of these teaming approaches as described.

ENGINEERING DESIGN APPROACHES

Since the Design Process intends to manufacture information (its product) used to manufacture a product which is used or consumed by others, there is a

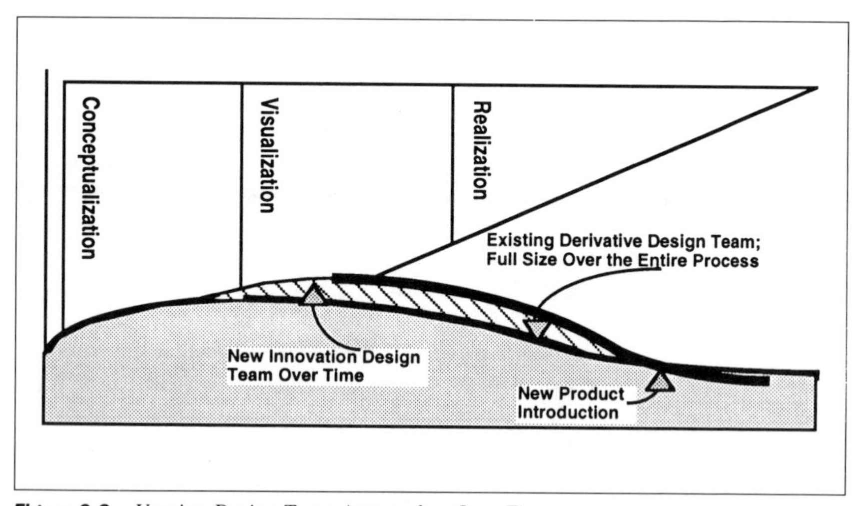

Figure 6-3. *Varying Design Team Approaches Over Time.*

constant concern about how the product will work in practice; i.e., in the field by the user, consumer, or buyer. While the intellectual process is being executed, the individual or group is simultaneously concerned about the characteristics of the product or process. The Design Process is applied by management as a guidance and control mechanism. The teams adopt the design approach used during the Design Process, based on perceived technical product characteristics, team interactions and requirements, and the perceived best match. Various engineering approaches all apply a variety of techniques to the execution of the intellectual process. These different approaches are represented in *Figure 6-4*. The design approach is also related to Design Processes and analysis techniques. The design approaches include:

Step-wise Refinement. The most common design approach. During step-wise refinement, decomposition of the potential product is developed, various "layouts" or images of the potential design are prepared, and performance and manufacturability options are analyzed and tested. Layouts communicate how the product appears by using a paper drawing as the primary method. Other aspects of product performance, assembly instructions, and remaining characteristics of the product which need to be communicated to production manufacturing are contained in "notes," or supplemental drawings. Many organizations, when referring to a drawing, are actually referring to the entire set of papers comprising a complete communication about a product or component. This concept of layouts or images, supported by supplemental information, has been carried forward into most current CAD systems. This has resulted in systems operating as drafting or drawing support tools, instead of design support or digital product definition support tools.

As design proceeds, a "trial and error" method is used to analyze each design variant. Each variant is considered, and as the intellectual stages proceed, a repetitive set of activities (design, analysis, test...design, analysis, test) are executed. This approach designs from the "inside-out," where components are designed, and then "rolled up" into larger and larger components via a *design integration* process.

In stepwise refinement, a general design is still prepared first. This design is done for organizational and design responsibility assignment reasons. The general design is usually quickly out of date once actual design starts, is not to scale, and not fully maintained.

The step-wise refinement approach provides good discipline, is good for inexperienced designers, and produces gradually improving designs. Its primary disadvantages are: it tends to have a result which comes from compromises occurring as the process proceeds and it is difficult to execute many of its small steps concurrently.

Concurrency is achieved only by considering variants through the design and manufacturability stages simultaneously. This multiple variants evaluation speeds up the process, but creates waste when variants are eliminated. In addition, some components can't be designed until others are complete.

Sometimes the compromises finally become too onerous. Unfortunately, the

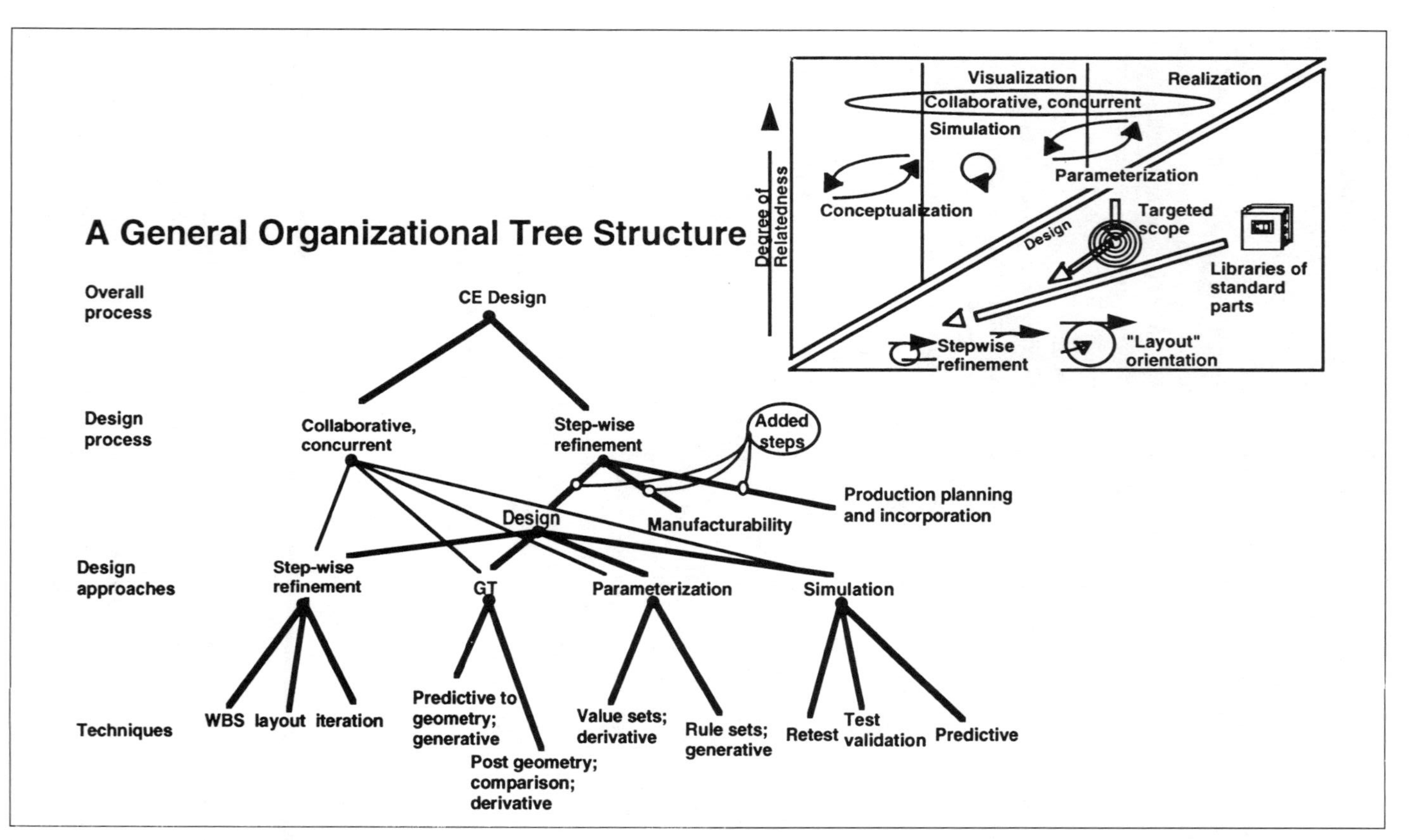

Figure 6-4. *Intellectual Process Stages and Various Design Approaches.*

decision to go back to an earlier design step is usually encountered late in the process, and an expensive, time consuming restart is required. Since restart is avoided if possible, an average or worse quality of design is usually the result of this approach.

Adding Work Breakdown Structures (WBS) to step-wise refinement adds additional structure and order to the complex product design process. Work Breakdown Structures organize a product's structure by using a differentiation technique. The product is broken down into detail. During this differentiation, different teams are assigned to different product components. The WBS drives the design process to an interface-driven integration technique. The WBS also serves as the drawing index structure, and helps keep the paper, or the digital equivalent, organized and easier to cross-reference with other indexes. It improves the step-wise refinement approach by providing a structure similar to the eventual production manufacturing structure required to produce the complex product, as represented by the structured bill of materials.

Group Technology. GT is another approach used in larger, more complex organizations. A natural way for an engineer/designer to address product opportunity or change is to ask if it has already been addressed. In a small organization, personal experience and knowledge can provide the answer. In a large organization with thousands of component elements, individual or group memory is not enough. Generally, GT is a technique used to classify previous designs using product or component features, performance attributes, and characteristics so that previously designed components can be easily identified and derived. There is a small but growing effort to generate designs using rules which shows promise in this area. This rule-based, generative approach and predictive simulation, discussed next, are similar in concept.

Once a generally applicable previous design is found, a derivative design can be developed by reusing existing information and modifying it to achieve a quicker, less costly to engineer information end-product. GT is still evolving. New compatibility with new requirement analysis techniques are being developed to ease the "fit" to these new requirements' screening processes. Because of its use of existing parts or components, perhaps with no change since they are already represented in physical form, Group Technology is an example of "reverse engineering."

However, if a new idea is required, GT may not be useful. In a new product situation, it is well into the design when enough detail can be understood to determine a GT fit. Design intent and other decisions may have been made precluding those of a seemingly good fit component. This approach retains promise, but the capabilities of CAD and Design Process Support information systems need substantial enhancement for this approach to be broadly successful.

Parameterization. Another approach, parameterization, is a combination of GT and the step-wise refinement design approaches, with an additional element. Parameterization works best where most of the design problem can be stated as a set of relational values, and the design results of the application of these values has been proven many times. As an example of the use of parameterization, consider a cowling for an airplane engine, suspended by a strut from a wing. When the design for the next cowling is to be developed, and it will be similar to others, then a set of parameters of the cowling, previously proven to result in the best design, can be reused. Previously, the cowling design had been reduced to a value set capable of being expressed mathematically. The parameter values and mathematical formulas can then be used to produce a visualization of the cowling. This visualization can be further manipulated, with these rules and formulae, until the optimum design for this particular situation is determined.

Parameterization shows greatest promise in change to, or improvement modifications to component design. Embedding the parameterization model into the CE Design Design Process Model to produce design work automatically without intervention is the key feature of parameterization and its most significant benefit.

However, if a new idea for a cowling or some other product is required, parameterization may not be useful. For example, a cowling which is internal to the main airplane body and is an element of the structural framework of the body would be so different from the suspended type that its parameters might be of little or no help. Parameterization has been limited to specific domains of applicability; however, this is an active field of development and this approach also has promise for the future.

Simulation. Complex product design teams are increasingly using simulation as a design approach within their design process. Simulation has traditionally been used to design, analyze, and test product functionality. Simulation is going through a revolution relative to actual physical testing and the step-wise refinement, design-analyze-test, basic approach.

Historically, series of tests were performed to validate a product design's performance. These tests were then simulated via derivative simulation for later, small changes to avoid the cost of a physical test in similar circumstances. Simulation techniques, computers, and simulation tools have become quite powerful in many areas. Now, predictive simulation is being performed. Testing is only used to validate the simulation. In certain cases, a test anomaly results in the test itself, rather than the simulation, being questioned as to accuracy.

When the design domain reaches the predictive simulation stage, design by simulation, as a design approach, begins to be possible. In this case, simulation tools have advanced to the point where the product's performance characteristics are *first* determined by simulation, and *then* reduced to a statement of design. This design by simulation first is unlike the first two types of simulation, where the design is done first and then tested within a simulation.

An example might be a logic chip or electronic component specified as a result of the simulation of the overall product. Its detailed circuitry is "built" using a simulator and an integrated test suite. Its structure might be stated using a special language, such as VHDL, whose logic can be simulated on a computer. All these simulation tools are based on computering. Design integrity and product quality can be tested during the simulation process itself.

In this case, physical product testing is not related to the product's design or simulation of actual use, but is intended to test the quality of the manufacturing process. Product performance is already theoretically assured by the simulation of operations, verified before committing to actual product production.

A substantial portion of step-wise refinement design can be performed as an integral part of the design by simulation. Design by simulation also supports CE Design's preferred Design Process, the collaborative-concurrent design-oriented process, to be discussed later in this chapter, since an "outside-in" approach to the product's design can be employed. The management of the individual areas of the product and their interfaces, usually specifically managed and controlled, are ameliorated by simulation since they are considered as integral to the Design Process, even though these issues must still be formally addressed.

DESIGN PROCESSES

Step-wise refinement was introduced as a design approach and process. Step-wise refinement is also a design process because its iterative, trial and error approach is a set of inter-related activities. Another design process, and the preferred process for CE Design, is collaborative-concurrent design.

Collaborative-concurrent design considers design on a holistic basis. Referring back to the *Figure 6-4* tree structure, notice that Manufacturability and Production Planning and Incorporation are not added steps, but are an integral part of the process. The dashed line from collaborative, concurrent to design indicates that these approaches and techniques are all still valid. *In collaborative, concurrent design, the fully designed end-product (the entire product or significant assembly, at least) is produced simultaneously.* This is very appealing to those interested in concurrency because it permits many parts of the product to be designed simultaneously. Design variants can be considered quickly, and the cycle-time, cost, and quality disadvantages of step-wise refinement can be largely avoided.

There are two significant issues within this approach. The first is *risk.* Because the fully designed end-product occurs simultaneously, holistic design can result in spectacularly successful, innovative designs, but sometimes can also result in designs which must be completely discarded. The other issue is *control.* It is difficult for management to understand status, schedule, and budget when the final result is a "big bang" or an all-at-once deliverable. Having "nothing to show for the effort yet" while "being 70% complete" are status reports that make managers who cannot affect the outcome of the design process very nervous. To reduce *risk* and maintain *control* when using this approach, a number of preliminary design releases, at the 20%, 40%, 60%, and 80% of completion points for example, are scheduled. Testing for interference, product simulation for performance, and GT examination are all possible within this holistic approach during these interim releases. These "artificial" releases add work and somewhat "step-wise refine" the holistic approach process. But these additions preserve its advantages while compensating for its current disadvantages. As technology improves, these disadvantages will be overcome by rules of analysis and review which will evaluate design quality as the design is being prepared.

In spite of the *risk* and *control* issues, the collaborative, concurrent design process appears to be the best basis for a CE Design process because it is philosophically consistent, can absorb many design approaches, and can be transitioned into from step-wise refinement.

For the truly complex product with many thousands of components, suppliers, partners, etc., a combination of these approaches can be used to transition from step-wise refinement to collaborative-concurrent, the preferred Design Process for CE Design. For example, work breakdown structures plus the collaborative-concurrent approach at the high end of the design and in various component areas can be employed, perhaps initially within an overall step-wise refinement process in order to reduce risk and insure control. During design, all these different risk abatement management and implementation techniques might be used as transition stages to full CE Design. These transition strategies are discussed in Chapter 9.

COMPARING DESIGN PROCESSES

The prevailing design process today is step-wise refinement. This process has been in use in various forms for several thousand years. As products have become more complex, the step-wise refinement process has evolved to accommodate new requirements. Its power is derived from its long use, stability, and adaptability.

However, a CE Design design process model based on the collaborative-concurrent approach, driven by process management's routing and queueing facility, is also very powerful. In addition, the use of concurrency is not very effective in the step-wise refinement process.

In a CE Design process employing the collaborative, concurrent design process, all the engineering tools can be connected into one or several process flows which, in turn, can be linked to each other. With provision for the automatic interchange of data via parameters for the ''electronic envelopes'' in which they are stored, these model flows can also be interconnected with other model flows into an overall model-driven CE Design process. By combining this computing architectural capability with the collaborative, concurrent design process, significant actual concurrency can be achieved.

The CE Design collaborative, concurrent design process model is very adaptable. It can be varied to the design approach used for the problem to be addressed. However, these two design processes (stepwise refinement and collaborative, concurrent) also need to consider the intellectual process and how it might vary for each approach and the design process in which it is embedded.

Figure 6-5A through *Figure 6-5C* depict a simplified view of the step-wise refinement approach. In *Figure 6-5A*, the approach begins with preliminary analysis. The initial product design concepts, its WBS, and its overall project schedule are established. Based on the WBS, the product is then divided up into areas of the product. For example, the motorized lifter discussed previously might have been divided into the engine associated systems, base frame, lift basket, and lift system. These areas, in turn, might have been divided into specialty areas for assigning design team responsibilities. The WBS is used to assign responsibilities across various groups, either functionally or discipline organized, along the speciality areas previously identified for more detailed design.

Detailed Design is depicted in *Figure 6-5B*. Each area assigned group goes through a design-build-test cycle for their individual area of the product according to their scope of work. This may go through several iterations until each design team is satisfied. Substantial engineering analysis may occur during the detailed design stage. The traditional emphasis during detailed design is on meeting schedules for design release to the final stage, integration, and analysis.

In *Figure 6-5C*, the individual product areas are brought together for integration. The integration process, done using mock-ups, testbeds, or prerelease product or prototype production manufacturing, leads to feedback to the preliminary stages of the approach. Trade-offs are made about what to fix, what to replace, and what to leave alone in the evolved design.

In step-wise refinement design process, there is little differentiation between a product design approach and the process of producing a design. This creates confusion in engineering and manufacturing. The *design approach* is related to the product and its requirements. The process refers to how the act of creating the design is accomplished and managed. The tree structure of *Figure 6-4* is important to communicating the "architecture" of design. For example, designing a new product is not a good idea with a GT design approach. GT is best used on evolved product, based on existing technology or previously designed standard parts. A new Lego® product based on existing parts is a good example. Lego® assemblies can be reasonably complex, involving hundreds of different part numbers. A new type of assembly system, however, is probably best performed using a "clean sheet of paper" or within a "brainstorming" series of meetings involving a small, innovation-oriented team.

The design process includes the business, technical, and managerial activities which must occur as the design is prepared for release. The chart in *Figure 4-7* describes the general categories of these activities, including those associated with design. These other design activities need to receive additional attention. They have never been separately identified and managed before.

Figure 6-5D shows the overall step-wise-refinement process includes manufacturability analysis, incorporation planning, and design related elements of actual production. This definition of traditional design is not the accepted view. Most engineers and designers think the job is done when the product design is released. Actually, their job is just beginning. As problems are identified, they are looped back to the design teams. For most complex products, the product design is in a state of change for a considerable period of time after the initial product design release. Change may be constant for months or years. The changes can add up to many times the costs and time associated with the initial

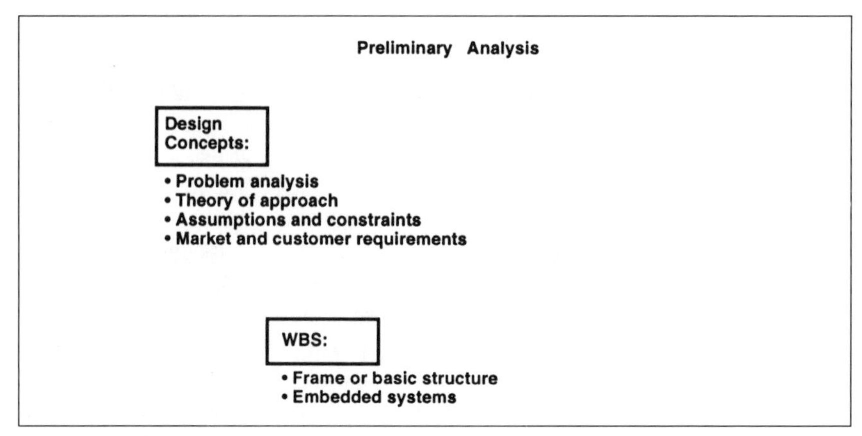

Figure 6-5A. *Stepwise Refinement Approach.*

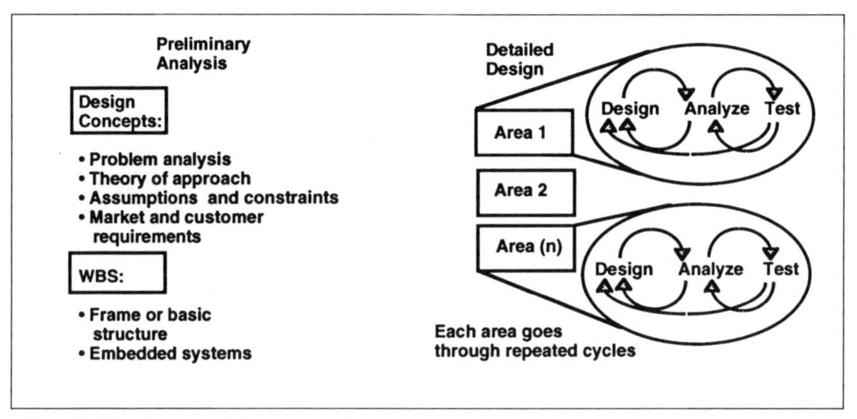

Figure 6-5B. *Stepwise Refinement Approach.*

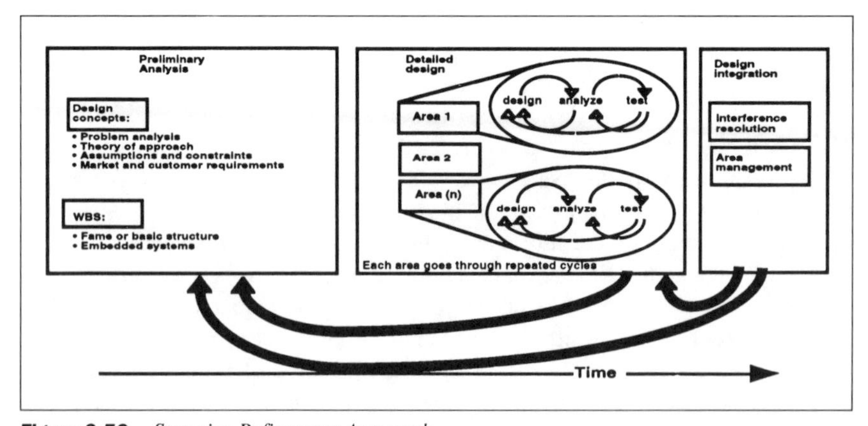

Figure 6-5C. *Stepwise Refinement Approach.*

product design release. Thus, the need to focus on the entire process, and the product design's complete life cycle.

In the overall step-wise refinement design process, some design concurrency can be accomplished. In the major tasks of manufacturability, analysis, and incorporation planning, some concurrency is also present. However, these tasks generally follow each other in serial fashion for control and assurance of completion reasons.

Achieving the benefits of the overall CE Design process can be partially gained using this serial, step-wise refinement design process. These partial benefits are a key element in the overall CE Design implementation strategy discussed in Chapter 9.

When using the Group Technology (GT) and parameterization approaches within the step-wise refinement process, the overall flow and concurrent-serial combination of activities are essentially the same. The difference is in how some of the details of design occur. GT and parameterization are design approaches, not full processes.

Using prior design process flow models, and prior associated activities, makes the next execution of a similar step-wise refinement design process quicker. These techniques can be applied to the step-wise refinement design process or CE Design's collaborative, concurrent design process. Their use improves the quality of either process and their work products to the extent these techniques apply in each individual circumstance.

Figure 6-5E depicts the application of these process management techniques to a complex product, step-wise refinement design process situation. The diagram proceeds into greater detail and calls for five levels of detail from a management perspective. The scheduling is based on a deliverables-oriented perspective, where one must demonstrate the end of an activity with evidence of completion. Milestones are used to record and measure progress against major project goals. A WBS is used to develop teams and establish schedules. Three design releases, at 25%, 75%, and 80% are established. Various design issues (2) and manufacturability issues (3) are considered. Finally, various issues regarding the support of the design process and deliverables (4,5) are also considered and put on the activities list.

The collaborative-concurrent design process is quite different. As shown in *Figure 6-6A*, WBS activities are at the center of this CE Design Process. It uses the holistic approach to the product. In addition to the design inside-out and outside-in steps, design control ensures interference analyses and management and interface definitions are occurring concurrently. Interference management predefines the content of various areas to preclude interference. Interface definition predefines the interfaces between system elements that cross area boundaries. *Figure 6-6B* depicts how multiple teams using the holistic approach, and the design control-interface-interference set of activities, monitor and correct potential issues before the design becomes more "fixed." This simultaneous activity must have an end, and it must be kept under control. The end, or schedule adherence, is principally managed by having multiple design

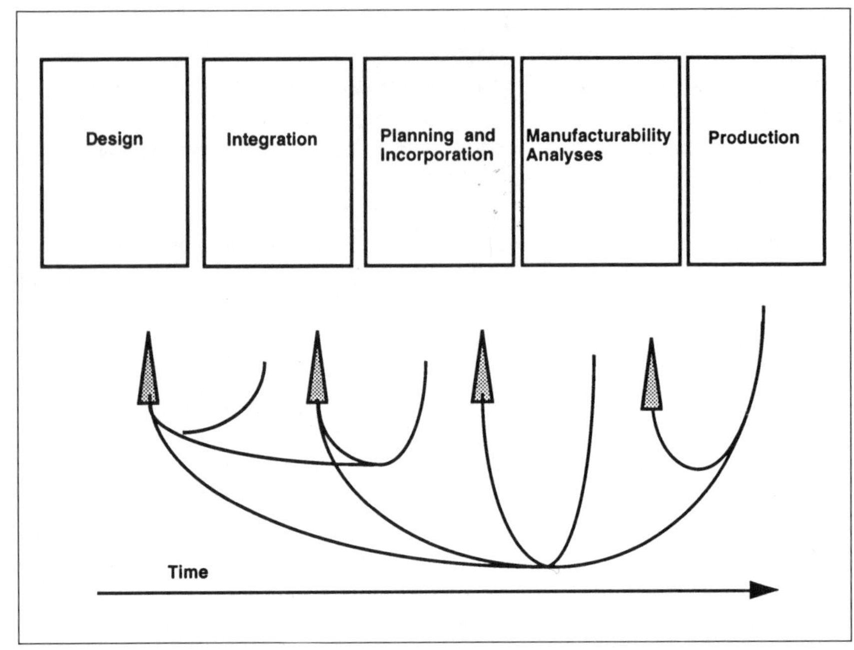

Figure 6-5D. *Overall Stepwise Refinement Design Process.*

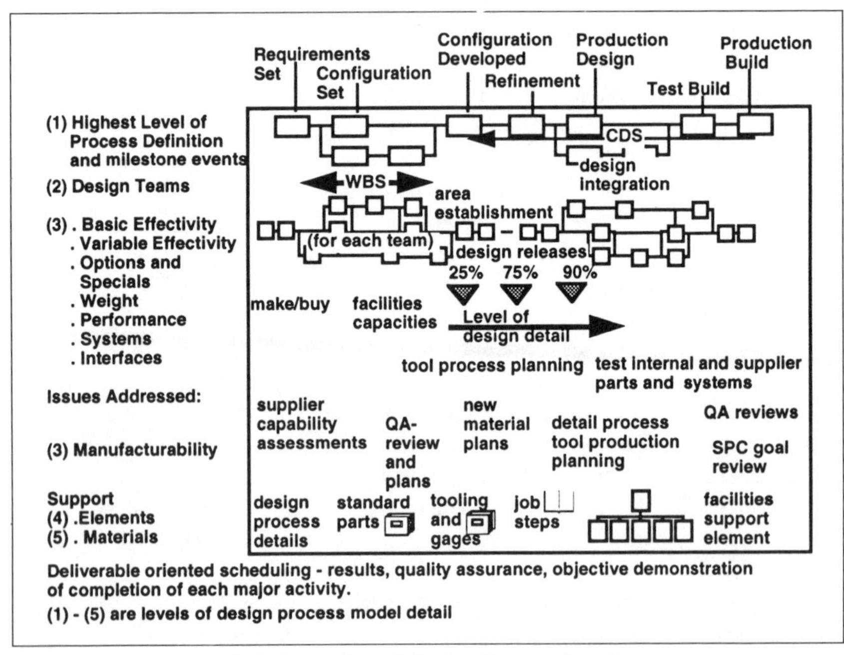

Figure 6-5E. *Stepwise Refinement Overall Design Process Sample.*

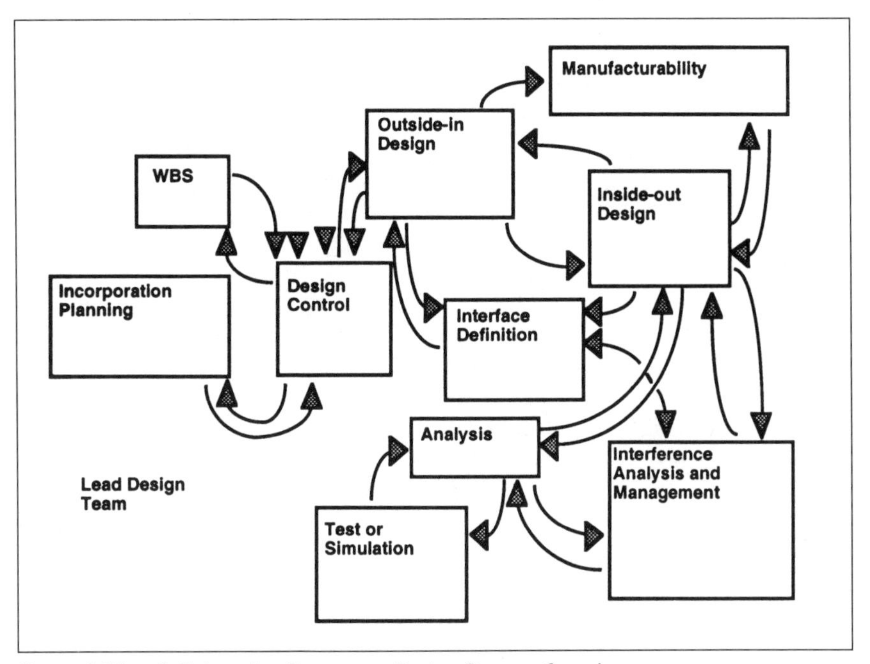

Figure 6-6A. *Collaborative-Concurrent Design Process Overview.*

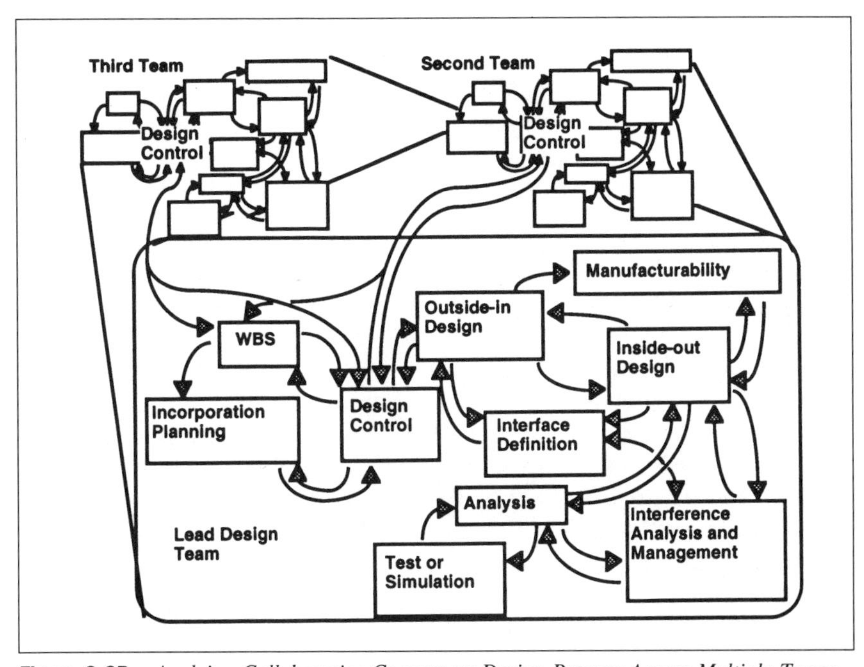

Figure 6-6B. *Applying Collaborative-Concurrent Design Process Across Multiple Teams.*

releases. Control is the responsibility of the design leaders, as supported by the process management, "in-line" process model concept described in Chapter 5.

OTHER DESIGN PROCESS RELATED ISSUES

There are other issues which must be considered by the manager, the engineer, and the designer when determining the design process for their environment. The first is the physical product itself.

Physical Products. The physical products produced by complex product manufacturers have "hard part" results; examples include automobiles, engines, buildings, and airplanes. A number of important factors go into the design of a physical product. The first factor in physical product design is the materials from which they are constructed. Materials science is not in the scope of this book. However, support within either CE Design design process for assistance in selecting raw material, supplied parts, subassemblies, or internally manufactured items should be provided. Promising approaches to material selection include GT and product standards information systems, using a combination of rules of use and selection to assist the engineer/designer in materials choices. Selection is included in the design process model of choice.

A second factor is how physical product characteristics prevent decomposition of its structure. This nonsimple decomposition is reflected in the activity necessary to complete a design. As shown in *Figure 6-7*, the activities of the design process model of choice differentiate into submodels and executable activities. "Differentiation" is the term used to describe this breakdown because the product does not break down into smaller areas or sections of the product on an even or gradual decomposition basis. For example, the differentiation may be into systems and the physical parts through which they pass, a very common differentiating approach.

At the finer levels of model granularity attained through differentiation, access to the types of systems are included as support materials to that portion of the overall design process. At this level of granularity, these process models should also contain directories of who to call in the organization and what outside-the-system reference materials to consult for material and other issues and questions.

A third factor in physical product design is its performance characteristics and parameters. Basic performance information about materials in use can be stored with other material information. The information about combinatorial performance, or the anticipated performance of the parts when combined into a subassembly, is also important and not available directly from any source. Combinatorial performance is what the product is being designed to accomplish; it may also include design intent. This is what testing, simulation, and mock-ups attempt to predict.

A fourth factor in physical product design is mock-ups. The issue of mock-ups actually varies, in approach and expected results, between design approaches. In the step-wise refinement approach, design *integration* brings the detail parts together to make subassemblies, which are in turn assembled, and brought together into a final assembly process which produces the end-product. Mock-ups (for this step-wise refinement approach to integration simulation using actual products) are brought together in the same pattern and at the same time during mock-up and integration portions of the step-wise refinement design process.

This similarity is a disadvantage because it goes through the same iteration process as design. Judgments are continually being made about whether to go to a complete, full-size mock-up (stage III), a lesser mock-up with simple models (stage II), or simple illustrative mock-ups (stage I).

The disadvantage of this similarity, in the digital environment, is that the digital mock-up process can be very computational intensive, people intensive, and time consuming. As the number of parts to be combined rises, the capability of the computer to store, combine, and display these now huge CAD data sets becomes a limiting factor.

This similarity is an advantage when, in a digital design support environment, the product and process information is available electronically in the design area of interest. It doesn't have to be an all-digital product definition. In this digital environment, the process of product definition *integration* combines these digital models (typically stored in mono-detail or a single part per digital model

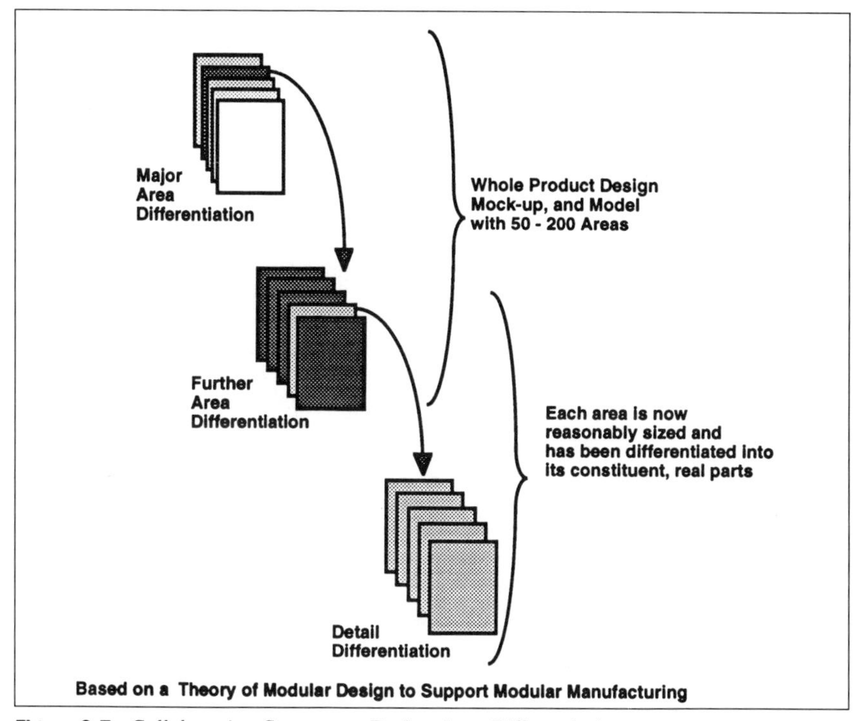

Figure 6-7. *Collaborative-Concurrent Design Area Differentiation.*

form), into subassemblies and then larger assemblies. An electronic mock-up provides an excellent benefit from a cost-of-errors-prevented perspective. It also saves time compared to building the physical mock-up.

The collaborative-concurrent design approach provides an evolutionary response to the issue of mock-ups. Collaborative-concurrent design uses a set of functionally differentiating, collaborative-concurrent models of the complex product. These differentiations increase in detail, accuracy, and specifications as the decomposition proceeds.

The focus of the design teams during the initial design process is on the holistic elements of the design. The intent is to design the mock-up first, as the product is being designed. In automobiles and airplanes, for example, it might be aerodynamics, weight, strength, and aesthetics. As the differentiation proceeds, the emphasis might change to interfaces between cross-area systems elements and areas themselves. Additional details, increased accuracy, and the other aspects of overall CE Design are then considered.

This collaborative-concurrent design approach precludes almost all need for post-design physical mock-ups. Natural elements of differentiation activities would include interference analysis and area management concerns as a part of the holistic approach. The holistic approach may also use mock-ups, either digital or model, and perhaps even go to stage III for portions of the design, but for the purposes of design, and during the initial product design itself, and not as a pre-manufacturing stop.

The beneficial difference between the collaborative-concurrent and step-wise refinement design processes appears to be in earlier, more complete holistic designs with carefully controlled differentiations. This has reduced errors, improved quality, and reduced the cost of the design process and manufacturing costs. Step-wise refinement and collaborative-concurrent can coexist; step-wise refinement can be an evolutionary stage leading to collaborative-concurrent.

This sort of collaborative-concurrent design by holistic differentiation can be very effective:

- in a complex product manufacturing environment where business partners share design responsibility,
- in which the product is a system, or
- in which modular manufacturing and production is to be utilized.

Shared product design can be facilitated by an early focusing on interface analysis and area management. That amount of design, at the whole product level, supporting the more detailed design is all that is required. This saves time, reduces design cost, gets the design partner involved early, and does not prematurely commit the product to a high cost or less-effective business partner design. In the step-wise refinement process, the precommitment problem comes through indirect design intent or design artifacts imposed on this area of the design. The design artifacts are imposed by other areas of the design having made choices before the partner can become involved, or after the partner begins. In contrast, the design is not as firm when using the holistic approach in the collaborative-concurrent process; the partner completes the design ''within'' the product model. This completion can be done outside and integration tested on the computer, if desired.

When the product is a system, the in-line design model and the processes it drives can be used to assure system-level performance *before* components impose performance characteristics. In modular manufacturing, this holistic differentiation permits entire areas to be designed and produced by a separate organization. In this approach, the final assembly process is one of "mating" various completely assembled product sections. For example, if the product was a bus, the bus might be divided into three sections: (1) the interiors, (2) the body, and (3) the drive train.

For the bus example, in a flow type of final assembly process, components are brought to the body during a series of final assembly stops. These components may be brought individually or in subassemblies. In modular manufacturing, the fully complete (1) interior is assembled to the fully complete (2) body and to the (3) engine and drive train at the same time. In certain circumstances modular manufacturing offers benefits compared to a typical flow-type final assembly process. Examples include joint ventures, large-size products, and when the product is a system.

System Product Design. The second design process-related issue is the design of systems as a product. The manufacturing progress can be seen from many different perspectives. These include the quality of the products, manufacturing processes, and others. Another measure of the progress of manufacturing—defined as the orderly and repeatable construction of something which takes on operating characteristics not apparent from an examination of the parts—is the manufacturing organization's increasing ability to deal with complexity. This is most obvious when products are systems.

The most important characteristic of these "system products" is that *the whole is greater than the sum of its parts*. In fact, it may take on secondary and tertiary sets of characteristics (operational characteristics) which could not be predicted from an examination of the parts. These emergent operational characteristics are normally why the system is manufactured. Usually, these characteristics can only be observed *after* the system is in use. This "only-after-manufactured" aspect is what makes systems manufacturing so difficult. Different systems products are at varying levels of design and manufacturing sophistication and understanding. Some systems are still almost custom products.

In *Figure 6-8*, systems products have proceeded through several levels of abstraction, or layers of functionality, from their constituent parts. Each of these levels involves greater complexity. The first level of abstraction is physical control and energy movement systems. These systems are hydraulics (fluid), electrical (power, energy, and control) and light (a new field—lasers, masers and the like). A second level might be computer hardware logic chips and genetic engineering. At this second level of abstraction, the system can be "pre-programmed" to control or direct a variety of activities and systems. These activities and systems are at first levels of abstraction. Genetic coding may drive altered protein or other natural organic manufacturing processes (natural system) which results in an altered biological entity, for example. A third level of abstraction would include computer software and biological entities (biologies).

At this third level, biological entities, including humans, exhibit the unique systems characteristics of "life," including adaptability, change, growth, and death. This is a little understood phenomenon. This level includes computer software because of its potential to eventually exhibit some characteristics of "life." Computer software might be characterized as going through its own "stages of evolution," including differentiation into components which have their own emergent properties.

Figure 6-9 provides a possible view of such evolution. The stages described include:

1. the business and technical tools in use today;
2. person-critical systems such as aircraft control, nuclear reactor safety, medical, and spacecraft;
3. intelligent systems such as expert systems, learning systems, and neural networks;

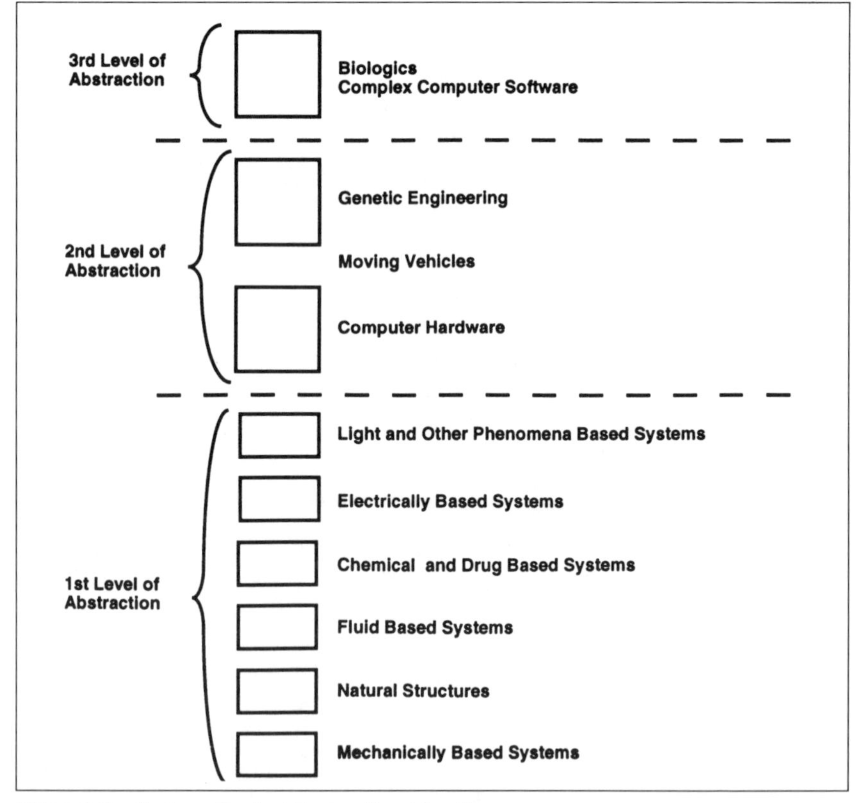

Figure 6-8. *Systems Product Design Considerations.*

4. complex systems such as weather forecasting, the proposed Strategic Defense Initiative (SDI) and android-type robotics, and
5. human and beyond functionality.

Manufacturing software at certain of these levels is now possible. For higher levels, many aspects of software development is "craft," custom, or dependent on individual skill knowledge and innovativeness. However, a theory of software manufacturing is emerging and it depends on CE Design.

The Ability of CE Design to Extend Complexity. The third of these design process-related issues is the growing impact of continuing increases in product complexity. The concept of the product as a system reflects the need for any manufacturing organization's world class operation to be fully integrated. *Integrated doesn't mean centralized.* "Integrated" means the integration of cross-functional processes across distributed organizational units, including CE Design, the cross-functional process which is the subject of this book.

This discussion of system products and their production is for the manufacturer who is the intended reader of this book. It should be important that the design process, and CE Design, be *extendable*, that is, continue to handle increasing levels of complexity and product integration. The commitment required to change the culture, individual behavior, and thousands of processes in a complex manufacturer is extensive. Conversion to CE Design must be worth it!

Step-wise refinement design can be an all manual process. It mirrors the first

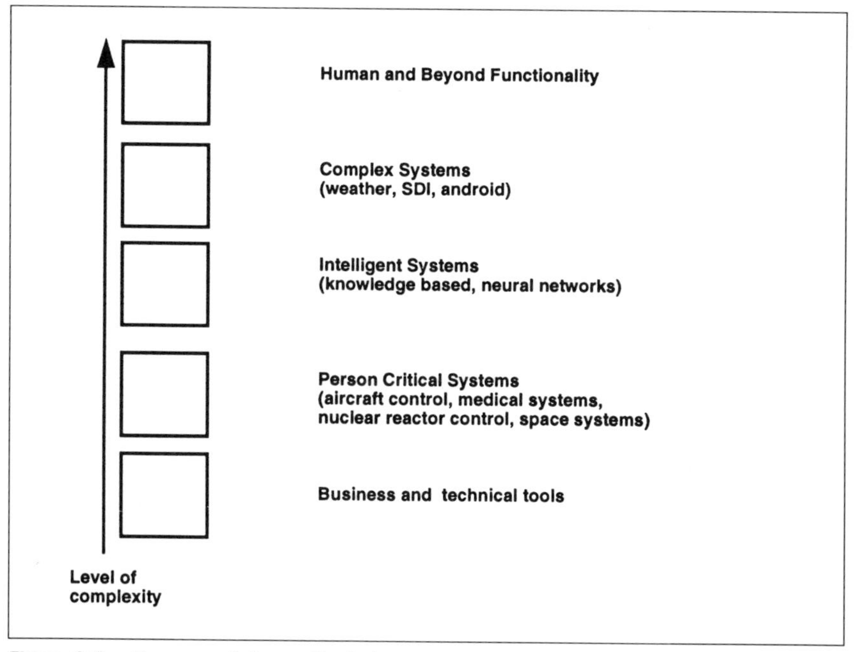

Figure 6-9. *Computer Software Evolution.*

level of abstraction physical systems "action-reaction-control" characteristics. By using "design-analyze-test: design-analyze-test," this basic engineering design technique reflects that, until after 1950, probably over 90% of design in manufacturing was focused on mainly physical property-oriented products. Control systems were used. But these control systems were based on physical control techniques such as hydraulics or mechanical leverage. These first level products could be designed by envisioning a physical product and then designing by doing. As these physical products became more complex, more pre-planning was required. As more complexity in the product, the process, and the organization occurred, steps to consider these complexity implications were added. This pattern of adding steps to the step-wise refinement design process continued. These add-ons continued to lengthen the time required to complete the design cycle by:

- adding to the loss of design intent communication by putting more functions between the production portion of the organization and the designer;
- increasing resultant errors and cost, and
- adding "rigidity" to the design cycle.

The *holistic* design approach requires automated support to be successful. It can be performed manually. However, without automation it would be very cumbersome at any significant scale of product and operation. That is why it has been used on a small scale and not widely deployed in large manufacturing organizations building complex products.

Collaborative-concurrent has not been a widely accepted design process in the complex product environment for the entire product for the same reasons. However, the use of the "in-line" model ameliorates the impact of the reasons not to use this design process and changes this process's level of abstraction. The model acts:

- as a primary process management tool in the collaborative-concurrent process;
- as the basis for the requirements for automated systems support for CE Design's process management, and
- as a change agent which "seduces" participants to make changes in the manner in which process management is performed and reported.

This book contends that the model-driven CE Design process is a second level of abstraction system since it produces manufacturing information, i.e., it manufactures information. It learns, adapts, and capitalizes on its learning. Additionally, the tools to manufacture information, at each level of abstraction, must be at the same level as the process these tools support.

Pre-defined models of behavior govern the collaborative-concurrent design process through "programming," or at least through models which guide or suggest prior experiences in like work areas. Since these models require other Level 1 automation systems to support the generation and management of these models, then the automated support systems supporting process management can be characterized as second level of abstraction systems.

Since CE Design is made up of second level of abstraction processes, it might be characterized as at least a high-end second level of abstraction process. Perhaps one day it can evolve to the point of being a self-perpetuating system and become a third layer of abstraction process. *Figure 2-15* first introduced these levels of abstraction. These levels of abstraction also are consistent with the opportunity evaluation introduced in *Figure 6-4*. Thus this level of abstraction concept, the roles of more complex systems, and the processes which use them and produce them are interrelated.

The point is that the complex and leading edge organization which intends to design complex systems needs a new design paradigm. Step-wise refinement as a design process, and not just a design approach, is an inhibitor, or this book would not be of interest. CE Design, with step-wise refinement as its transition design process, is a large improvement. However, using the collaborative-concurrent design process within CE Design appears capable of creating a ''paradigm shift'' within the organization and in its competitive position. The experiences that led to this book created the belief that CE Design, using the collaborative-concurrent design process, is a new paradigm. But what are some of the characteristics of systems where CE Design is required to effectively produce the design?

Systems Characteristics and Collaborative-concurrent Design. The fourth of the other design process-related major issues is the relationship between system products and the choice of design processes. One of the advantages of the collaborative-concurrent design approach is its focus on dealing with the holistic aspects of the product throughout the design process, instead of only at the start and during integration at the ''end'' of the process. This corresponds to a constant focus on the working characteristics of systems when they are products. Such a focus can be added to step-wise refinement by adding review steps for this purpose. These are likely to be only somewhat successful since they lengthen and add additional complexities, as compared to collaborative-concurrent's inherent focus. They can be added as part of the transition strategy.

A second characteristic of systems is the need to interact often with physical products. Today, systems are prevalent in complex physical products. Their use includes such purposes as guidance, monitoring, power generation, and environmental control.

One of the drivers for this need for a new design paradigm is the need to increasingly intermix higher level of abstraction systems with physical products and their first level (physical control systems). For example, the classic approach to physical product design and manufacture is to break down the complex product into smaller areas. Then they are understandable and manageable by humans. However, these same complex products have become increasingly complex. An example might be an airplane.

In the 1940s, an airplane about the size of a B-17 bomber (now about the size of a Boeing 737) had 10,000 to 15,000 part numbers, including the engines. Systems were very basic and were focused on communications (internal handsets), fuel, control (mechanism with some hydraulics), and some miscellaneous capabilities (bomb sight control, instrumentation, and the like). These systems were *added* to the airplane as they evolved and were found necessary. The airplane was built to fly a certain distance, at a speed and altitude, and carry a payload to its destination. To a certain extent, all other aspects of the design, including systems, were added afterward.

As shown in *Figure 6-10*, for today's aircraft environment, the same size aircraft has *over 90,000 part numbers*. The largest aircraft (three or four times larger) has upwards of 275,000 part numbers, and flies three times as fast with more weight at higher attitudes. This is at least a four to 10 times increase in part numbers. The next generation of aircraft may have even more systems and thus even more part numbers. All while the size of the aircraft has not changed by more than 200% or so. In Chapter 4, a similar story was described for computer logic, now microprocessor chips. The complexity level for very complex products like airplanes, automobiles, factory equipment, buildings, and the like is even higher. Now these complex products have hundreds to thousands of these computer chips imbedded in them. The mechanical, first level of abstraction, equivalent of these added systems might result in 1000 times (or more) equivalent part numbers, if such mechanical equivalents were even possible.

Obviously, one of the trends affecting this shift has been the microprocessor-size computer. As shown in *Figure 6-11*, the cost of enough computer power to control complex product operations has been dropping rapidly. It has reached the point where *embedded systems* (as they are referred to in the computing industry) have become a very large part of the computer industry's business. While

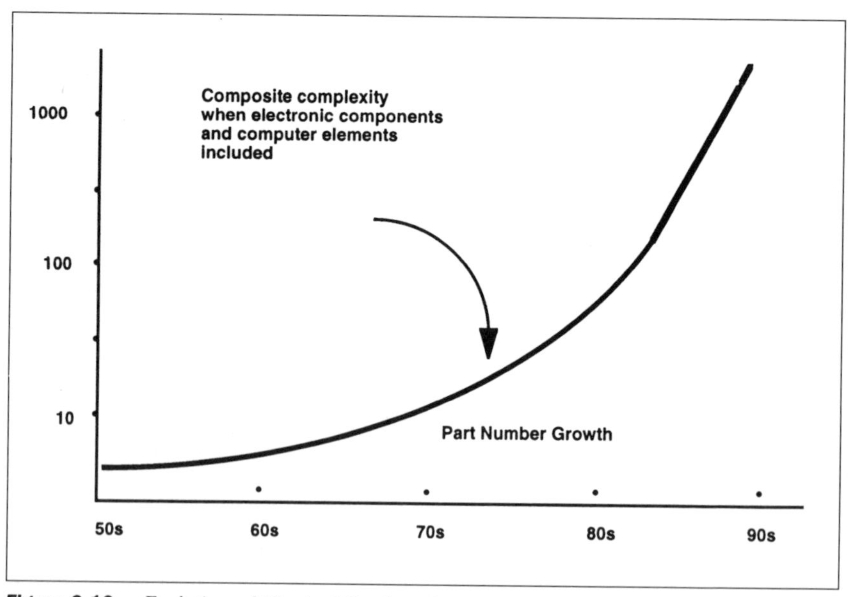

Figure 6-10. *Evolution of Physical Product Complexity.*

software's productivity and cost performance ratios have not been reducing in cost at this rate, improvements are being made. For total product complexity to be manageable, important breakthroughs in software are required. If one combines the electronic logic complexity with the other system complexity (perhaps represented by part number equivalents, as previously discussed) of the aircraft, the combinatorial complexity may have reached 500 to 1,000 times as great as the B-17. For the next generation of commercial airplanes, they may become computer systems which also fly. This same complex growth process might be true for a bus, an automobile, a large computer software system, a building, a ship, a personal lifter, a communications network, a biological monoclonal antibody, an oil and gas refinery, or any of the other thousands of complex products of the near future.

Step-wise refinement and collaborative-concurrent design processes manage the design approach for the "system" within the product differently. The issue is whether the design process is "outside-in," "detail-up," or "inside-out" (step-wise refinement) versus "holistic" and "interface oriented" (collaborative-concurrent).

When systems are part of physical products, there exists a natural potential conflict. Systems extend throughout the various elements of the physical product. The "step-wise refinement/inside-out" approach focuses strictly on the physical product's Work Breakdown Structure and not on the product as a whole for a very long period in the design cycle. Everyone is anxious to get to "parts" so the real design process and releases for details can begin. This emphasis can come from production. They must actually produce the detail *first*, though detail is described *last* in the design process. This "reverse urgency" results in premature releases and early material orders. Production Manufacturing expects the release of the detail production design quickly in order to allow detail production activities to begin without delay.

In the step-wise refinement design process, the product is broken into more detailed areas of interest. While this is occurring, the design management team can take one of two decomposition approaches:

1. Separate systems into segments or portions of the system by area of the product as determined by the WBS.
2. Apportion the product into its standard WBS areas, but keep the systems separate. The systems are assigned their own branch of the WBS.

As summarized in *Figure 6-12*:

1. the systems component design groups must coordinate and communicate across their areas in addition to coordinating within their assigned physical product design area. These inter-area teams coordinate via interface and performance specifications. A separate system design team also may be utilized, since many systems components may be partially or completely manufactured by a supplier/partner. Coordination among the supplier and the various area design teams may also be necessary; or
2. the coordination occurs between the systems design team and the supplier. The coordination between the system design teams and the

physical product area teams focuses on *interference analysis*. Coordination also is required for *design integration* and to ensure *design intent*.

Collaborative-concurrent Design also utilizes two approaches, as summarized in *Figure 6-12*:

1. The total product design approach considers the issues as an intrinsic part of the design process. There are no subsequent steps; or
2. In the collaborative-concurrent/WBS approach; the design differentiation process described earlier facilitates the overall design of systems as well. The cross area systems are designed as elements of the high-level design (cross), and their details are designed as the more detailed areas (intra) of the design are considered.

While both approaches will work, the collaborative-concurrent process is less artificial. It is less artificial in that fewer additional, separate steps for coordination and communication are required. Collaborative-concurrent should occur fast enough so that while it takes longer to release detail, the overall process is shorter and production's error rate will go down because considerable overall flow time is saved. Early material and production planning occurs at both the higher and lower levels of the design process. Production Manufacturing is not left with reacting to a mostly completed design which still contains some potentially significant anomalies.

Product Packaging. The fifth other design process-related issue facing managers, designers, and engineers is the packaging of the product in today's increasingly complex systems environment. In some types of complex manufacturing, the system is the product. Even in this type of manufacturing,

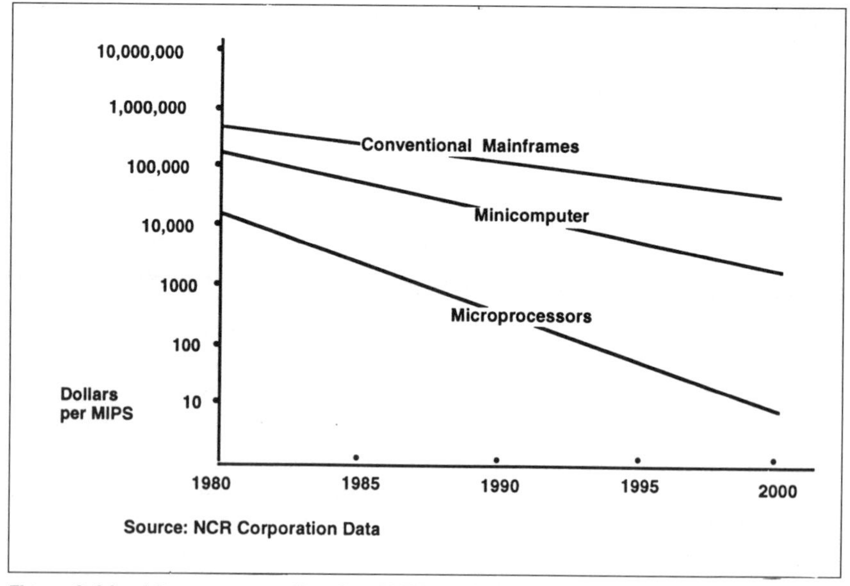

Figure 6-11. *Microprocessor Cost Per MIPS (million instructions per second).*

physical packaging is still required. As shown in *Figure 6-13*, there are potentially successive layers to the systems packing process. These include:

1. interfaces to other elements (such as other processor chips, system software elements, buses or wiring);
2. interfaces to external systems and support products, including test equipment, other system elements, and communications systems;
3. manufacturing packaging elements which are utilized to produce the product, such as designed artifacts for use in manufacturing but which are removed before delivery, as well as manufacturing technology interrelationships such as robotic machinery for handling, AGVs for movement and ASRSs for storage, and jigs and temporary framing or attachment points to assist in the assembly process, and
4. delivery packaging in which the systems components are placed for delivery and installation by the customer.

Many systems products are placed in product packaging for protection against the elements, electrical interference, cooling, and for other purposes. These packaging issues must be considered as the product is designed, and must be included in its WBS regardless of the approach. The important difference between system products with physical packaging, and physical products with embedded systems is their intended use. As complex products and systems continue to evolve, this distinction will continue to blur.

The collaborative-concurrent design approach, due to its *extendibility* through information system support and its high degree of concurrency, should rapidly increase as CE Design itself is accepted by the complex product manufacturing organization. Its acceptance also will be aided by the serial design approach which was largely used previously. As product *complexity* constantly increases the design cycle's time, resource and cost (both directly, and in the product itself), plus the already achieved partial concurrency will be enough to induce

	Interface	Interference Analysis	Design Integration	Design Intent
(1) Stepwise - refinement				
• WBS	Distinct step	Area management	After 1st design	Communicated
• WBS / systems	Area	Distinct	After 1st	Communicated
(2) Collaborative, Concurrent				
• Total product	Intra-intrinsic	Intra-intrinsic	intra-intrinsic	Intra-intrinsic
• WBS	Cross-intrinsic	Cross-intrinsic	Cross-intrinsic	Cross-intrinsic

Figure 6-12. *Design Adaptations.*

the organization to adopt collaborative-concurrent. For collaborative-concurrent to be successful, some different organizational business processes, and information systems support capabilities and characteristics are required.

OTHER CE DESIGN PROCESS IMPLEMENTATION GUIDELINES

If the decision has been made to begin the change to the CE Design collaborative-concurrent design process from a current step-wise refinement design process, the management team should consider some additional guidelines in their implementation plans. These include:

- complex product matrix design teams;
- teaming psychology;
- team management and measuring and rewarding success;
- technical quality;
- schedule compliance, resources management, and cost management, and
- documenting the design process model.

Complex Product Matrix Design Teams. Adopting the collaborative-concurrent design process means evolving a current step-wise refinement design process. Using this book as a guide, planning the evolution to collaborative-concurrent is an important first step. While this evolution is beginning, however, one must simultaneously be concerned about the people involved in the process. How they should interact, to be as effective as possible in this constantly

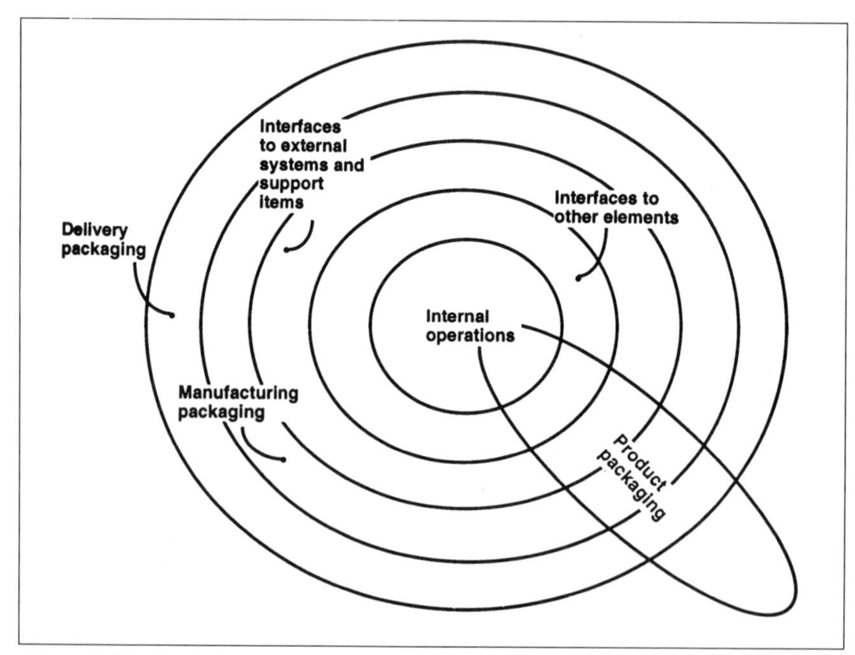

Figure 6-13. *Packaging.*

changing environment, is a critical success factor.

The organizational aspect of the CE Design process is critical.

It is *not* just putting people together.

CE Design must be operating with new guidelines, procedures, and information manufacturing tools. The CE Design process is a set of inter-related, integrated, cross-functional processes. These processes are integrated across organizational units. The successful CE Design process operates across various organizational units, both inside and outside the organization, with very little administrative bureaucracy. It operates as a relatively ''flat'' network, as shown in the *Figure 6-14* series. In *Figure 6-14A*, the initial organization of the teams is built around three different types of responsibilities. These responsibilities are:

- lead design teams,
- engineering administrative tasks, and
- supporting design teams.

The lead design team has overall product design responsibility. In the case of systems products, they have responsibilities including insuring that the product requirements and associated emergent systems performance characteristics, which only appear at the whole product level, are delivered by the design.

Engineering administration is another important element of the teaming process. They have responsibility for the design process itself and supplying the tools and techniques to be used by the design teams. This eliminates redundancy and confusion. They also have processes, internal to their teams to execute. These internal processes include researching and setting standards, establishing design work-product definitions, and coordinating automated infrastructure support.

The supporting design teams provide more detailed work products, as first discussed in *Figure 6-7*, in support of the lead design team's requirements.

Figure 6-14B introduces more teams and design partners into the network. The additional design teams must relate to each other, from an interface perspective where necessary, as identified in area management. The design partner, if in a partner role, acts in concert with the lead design team, or the proper supporting design team or teams to develop its portion of the product's overall design. Issues of control are discussed further in Chapter 8. *Figure 6-14C* depicts the more complete situation in the complex product design environment. In this figure, multiple design teams are both inside and outside the organization.

Networked Design Teams operate on a collaborative basis. Collaboration is an important concept to CE Design. When discussing collaboration, there are several personnel and team-oriented perspectives to consider. These include:

1. teaming psychology;
2. team management, and
3. measuring and rewarding success.

Teaming Psychology. Putting people together should be done with a purpose and a plan. There are several factors which should be considered when establishing design teams. These include:

1. the role and activity surrounding the designer;

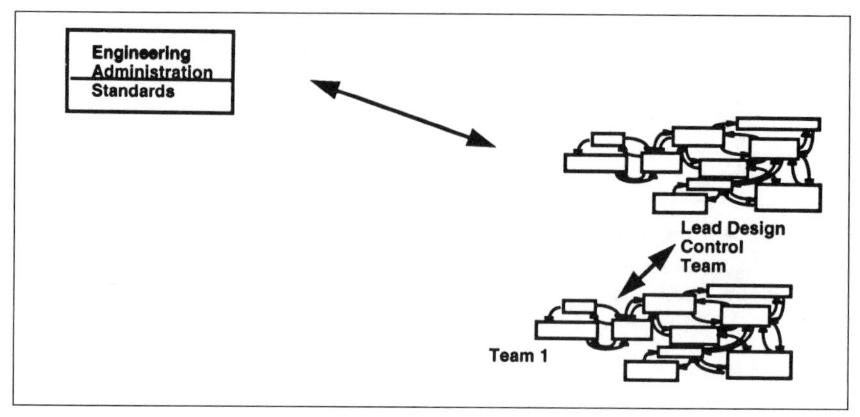

Figure 6-14A. *CE Design's Networked Matrix Design Team.*

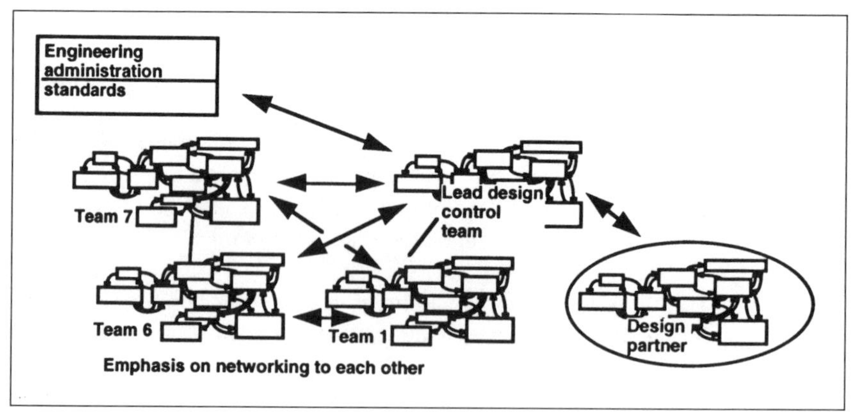

Figure 6-14B. *CE Design's Networked Matrix Design Team.*

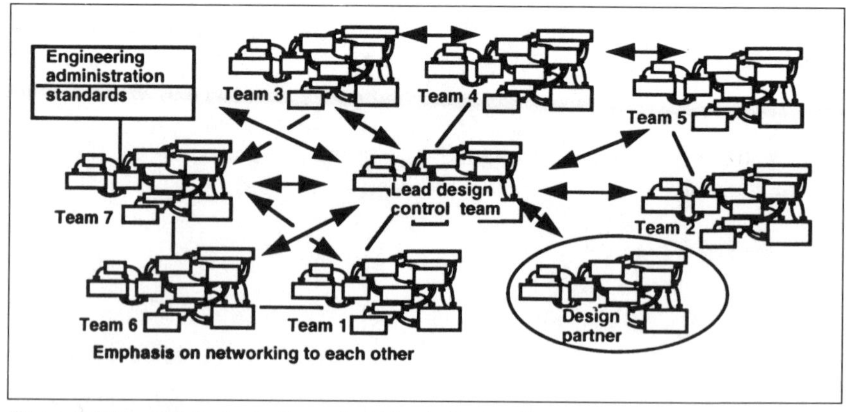

Figure 6-14C. *CE Design's Networked Matrix Design Team.*

2. the team selection process, and
3. the matrix interaction process.

One of the principal factors in how the management team goes about establishing teams is determining the role of the lead design engineer or team. To a certain extent, this may depend on the personality of the individual, the group, or its leader. However, it also depends on how the team should be operating and on the type of design process the team is to employ.

When selecting individuals to make up a team, some attention should be given to the personalities of the individuals involved. Some individuals are optimistic, others tend to have and generate concerns quickly. Some are natural leaders. Others are more at ease when another takes the lead but provide significant quality information and ideas once the interaction between people ensues.

Consideration also may have to be given to the relative "stature" of design engineers, manufacturing engineers, sales and marketing personnel, finance personnel, manufacturing production experts, and others who comprise these matrixed teams. A good team has at least one of each type of individual characteristic, a proper mix of engineering technical specialists, and adequate representation from each of the disciplines. The teams also need some training in how to work as a team. They must develop an understanding of the mutual respect which must be forthcoming for each team member's contributions and needs. In the complex product teaming environment, different styles may be necessary and appropriate.

In *Figure 6-15A*, the Lead Designer for this element of the product is intended to act as a *leader*. Most significant decisions are made by this individual. This individual typically has a natural leader-type personality. This individual, to operate effectively, should be a proven innovator who has the confidence and respect of the rest of the team. He or she should be comfortable with being "the first to take a position" *without* significant "pride of authorship" problems. In an organization with fewer individuals, or with fewer individuals of proven high caliber, this approach may be necessary to accomplish design and to schedule objectives while at the same time educating and teaching more junior individuals the design process, the products, and their respective roles.

The strength of this team style is usually that it can get started quickly. There is no confusion about roles. The various early parts of the process can be executed efficiently. A strong individual can "pull" the team through the entire process. On a team such as this, if possible, it is important to have another well-respected individual who is a "concern generator." This person will add balance to the team's perspective.

There are, of course, issues to watch for in this type of teaming arrangement. In the multiple design team environment of CE Design, using a Team style employing the Leader approach puts additional pressure on the Leader. All significant decisions and actions flow through, and are made by this individual. If there are sicknesses, personal distractions, or just a "mental block," the overall process may be quickly affected. The *direction* may be set too quickly.

Because of the strength of the Leader, it may take longer to recognize and deal with the need for redirection.

Additionally, as the emphasis shifts from the conceptualization of the design to its manufacturability, the need for other team members who provide this perspective to take a more assertive role about their contribution may become a problem. This is particularly true if the Leader is more of a designer than a manufacturing engineer. It is this *matrix*, the interactive process of design, manufacturability and product requirements/architecture which is the strength of CE Design. It must not be circumvented. *Figure 6-16* represents the relativities of these tradeoffs inherent in the matrix approach. The figure includes the relative strengths of each approach if these are purely new product teams, or are product change teams, where design is still important but where design restrictions are more prevalent.

The *Team Process With Collaborators*, as depicted in *Figure 6-15B*, has the opposite/strengths and concerns. A somewhat "softer" designer or lead design engineer will try to gain more of a consensus.

The team will get started slower. More patience and management support is needed early.

If the other participants are somewhat less experienced, there will be some "drift" before a direction is determined. The main focus of the collaborator in this teaming environment is to gain and to maintain consensus. If this can be maintained, then this team's overall performance can exceed all but the strongest Leader teams. This type of team executes collaborative-concurrent design better. The risks associated with this approach include the team mix of individuals, distractions, and the level of commitment necessary to seek and maintain consensus and common *design intent*. Management must be patient and allow the design to emerge. During the change to this approach, if a transition is

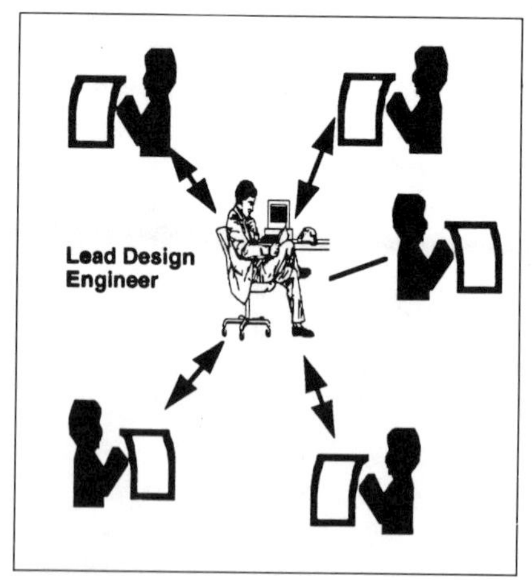

Figure 6-15A. *Team Process with Leader.*

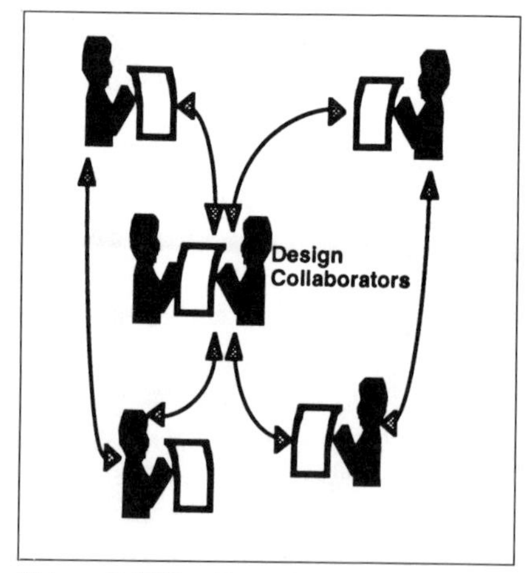

Figure 6-15B. *Team Process with Collaborators.*

required, particular attention should be placed on achieving early successes. Also, early releases can keep attention focused and encourage a consensus without undue influence.

Team Management and Measuring and Rewarding Success. As the teaming process proceeds, the management of these design teams also proceeds. Because these teams are operating in more of a network manner, an attempt must be made to make the teams more "self-managed." While empowerment is an important CQI concept, and can be particularly powerful, one must not "abdicate" control and management to the team. Good vision, direction, support, and conflict resolution are still overall management responsibilities.

One of the principal strengths of the in-line model-driven Process Management approach described in Chapter 5 is the embedded management process which is built into the model and which is executed by the automated support systems as the teams do their jobs. Because most of the *administrative* issues are already handled by these automated support systems and supporting engineering administrative processes, the management team can focus on:

- technical quality,
- schedule compliance,
- resource management, and
- cost management (product related and organizationally related).

Technical Quality. This is not a book on how to achieve technical product design quality. However, there must be some discussion of how management can recognize technical quality in product and process. Recognizing technical quality should focus on at least two approaches, product requirements validation and design robustness. There are many techniques used to gather, prioritize, and

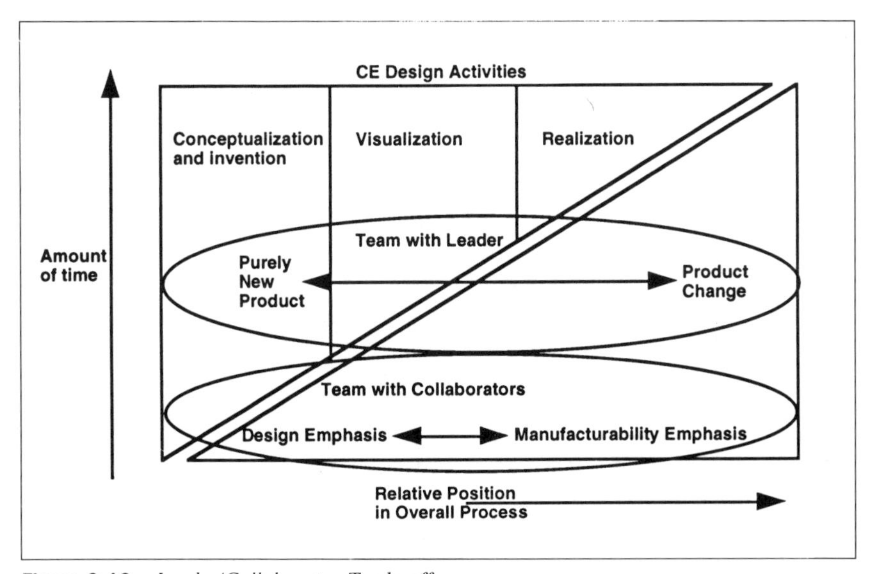

Figure 6-16. *Leader/Collaborator Trade-offs.*

track the achievement of requirements.

Quality Function Deployment (QFD) is a good example of a technique which helps in this regard. Whatever one's choice of technique or tool, however, it is important to have a good understanding of how and why this product is being designed or changed.

As depicted in *Figure 6-17*, the product may be a new idea. If it is a new product, then the *market need* may be difficult to determine because there is little awareness of this product. The new product's desirability and performance requirements may be hard to measure. This new product may need to be test marketed, for example. Another type of new product is one which arises out of a new technology (technology push). Like the new idea, it may be difficult to establish desirability and requirements.

Both types of products require sales and marketing support during the design process. They may require product education to potential buyers. Such design support personnel can, at least, provide insights and can act as selected potential clients. A product architecture with planned enhancements, and a family of products should be considered during this type of design process execution. Product requirements also may originate in engineering. However, these requirements should be validated to the extent possible by those closest to customers and clients.

Market need-originated products should be based on requirements. In this case, requirements are set by the marketplace, documented by sales and marketing, and executed against by engineering. The improved derivative can reflect any of these other three situations, and should be handled accordingly.

Another issue to consider as potential products are considered during requirements validation is the core competitiveness or competencies of the organization. This reflection takes two perspectives. The first perspective is what will the market believe; that is, if your organization brings this product to

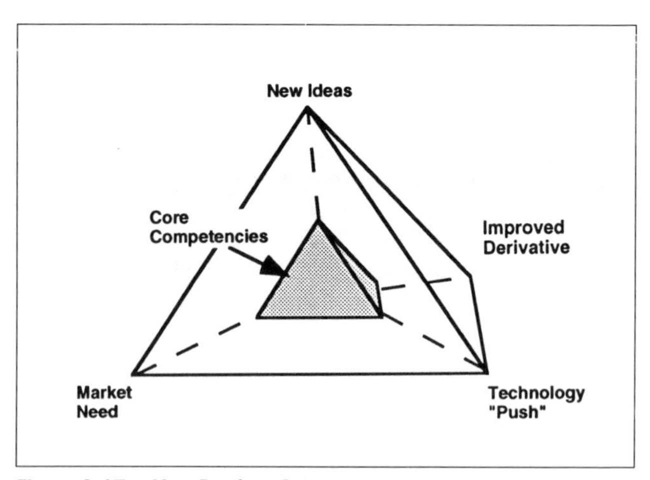

Figure 6-17. *New Product Origins.*

market, can it relate the new product and the existing organization and its distribution channels to each other. A great product which is difficult for the existing sales and marketing organization and your customers to relate to requires special planning and perhaps a longer, more extensive, and more expensive product introduction process. The engineering and manufacturing to support this new type of product should probably emphasize the ability to make changes frequently and rapidly.

The second perspective is what is the organization good at? Does it do a great job of engine design, or plastics and composites, or network communications? Usually, these core competencies are relatively small in number. While it is not necessary to focus only on these areas, consideration should be given to two controlling concepts.

1. Look for products which use these competencies and in which *substantial* differentiation from competitors is possible.
2. Consider a different mix of design teams to include new and existing teams and new and existing members. Perhaps different management should be included when an attempt to develop new competencies is planned. This emphasizes to the teams that this product is different. It may require everything about the product, the teams, and the organization to be different. The differences should be clearly communicated.

Requirements validation is especially important when a new product area outside current core competitiveness is to be developed. Extra time, budget, and resources probably are necessary. Acquiring a new lead designer/Leader or two (perhaps even from outside the organization) also may be appropriate if the teams can absorb such an additional change.

One final factor to consider about core competencies is the CE Design process itself.

After several iterations of products using CE Design, the fact that your organization is operating CE Design successfully is, itself, a core competence. The psychology of Collaborative Teams in a collaborative-concurrent design process promotes flexibility. This permits a wider range of problems to be addressed successfully. While not in itself determinative because other factors such as cost of manufacturing must be considered, the CE Design process is a highly significant and leveragable competency.

The other technical quality management approach should be to ensure that the teams follow the self-correcting design aspects of CE Design's collaborative-concurrent design process which focus on design robustness. This is one of the key objectives of the manufacturability process to be discussed in Chapter 7.

Schedule Compliance, Resources Management, and Cost Management Process Management, Chapter 5, discussed schedule compliance and cost accounting data gathering as an integral element of CE Design's Process Management. These issues should be considered, along with the growth and maturity of various teams and their individual members, as new products or derivatives are considered and new changes to existing products are contemplated. Management should consider the flexibility provided by the Process

Management process and develop occasional teams. These teams should use individuals from existing teams for special projects. *Figure 6-18* depicts the concept of temporary teams.

The concept of temporary teams provides variety, an opportunity to demonstrate growth, and perhaps new perspectives. Obviously, this third-dimensional teaming should be used only as appropriate. The flexibility of CE Design's automated support will make temporary teaming seductive.

Documenting the Design Process Model. The Design Process Model to be adopted by each organization should first be documented as a process model. This is something of a paradoxical issue. The in-line model-driven CE Design process should be self-documenting. Once the process is executed, a model to be followed now exists. But how is the first model developed? How does the organization transition from its current processes into this new design process?

The author's experiences recommend that the development of the first model should use a collaborative-concurrent design process. A first and second level of detail design process model is developed by a temporary team. This group is trained in teaming and in collaborative team execution techniques. Subsequently, product design teams are formed, and this second "wave" of teams use the actual tools and simulate via a "conference room pilot" the proposed process. This is done several times, and successive levels of detail are developed. *Figure 6-19A* depicts this overall approach for the development of the first three levels of process details.

It is probable that four successive levels of detail will be needed to fully develop and document a usable design process. Because of the additional level of detail (level four), and the differences between a new product introduction

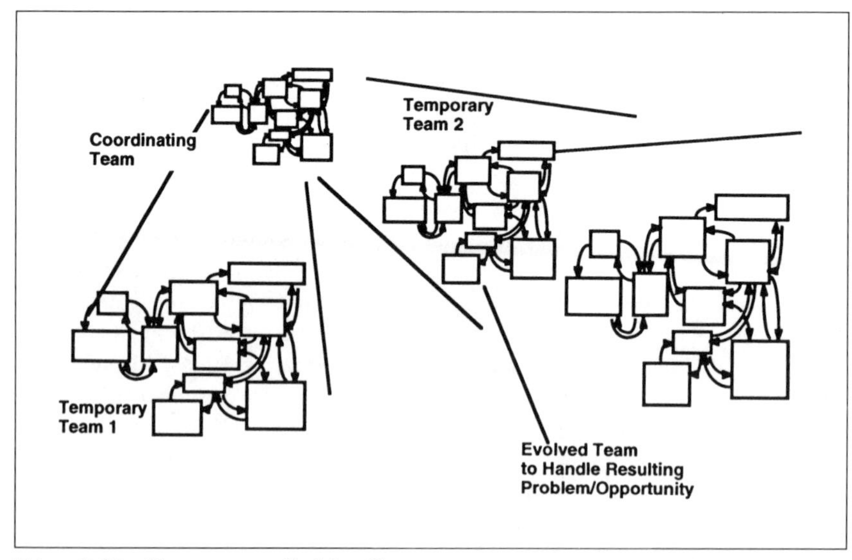

Figure 6-18. *Temporary or Evolving Teams.*

and product change processes, *Figure 6-19B* suggests that these additional details and differences be developed in a second iteration of the conference room pilot's testing activities. It would be best if these first process models can be captured directly as "in-line" process models using an appropriate computerized tool. Once the first design process model developments are completed, the formed product design teams can proceed into their first actual product design assignment.

Constant risk assessment is appropriate during the design process's evolution. This first process model design activity may need to be performed several times if the organization transitions from step-wise refinement to collaborative-concurrent, with varying levels of interaction, as described in Chapter 9.

Design Process Artifacts

During the CE design process, there are two sets of tools which capture the information which is being manufactured. The product of design is the

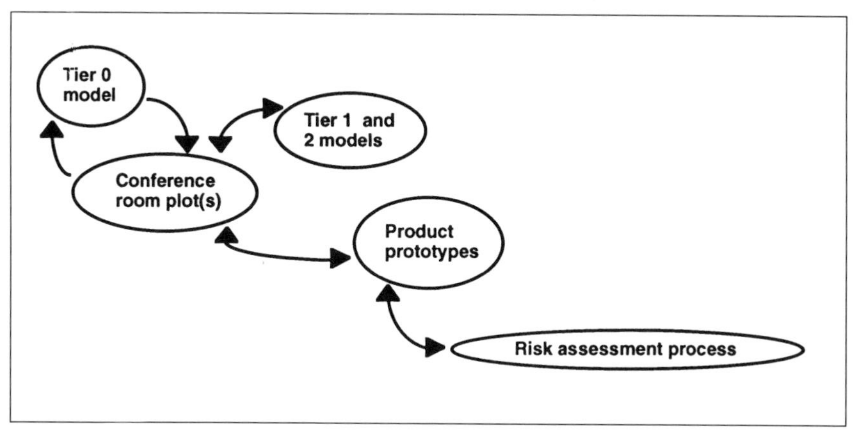

Figure 6-19A. *Design Process Model Evolution.*

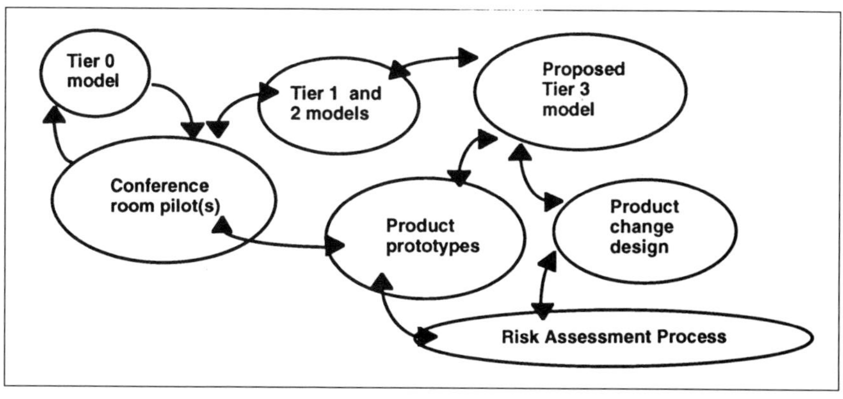

Figure 6-19B. *Design Process Model Evolution.*

information necessary to design, manufacture, and support the actual end product. For Engineering, these artifacts are normally divided into three types of information:

1. work statements;
2. geometry and models, and
3. engineering analysis information.

The intent of this portion of the chapter is not to discuss each of these items in great detail. There are many publications and much actual work in these areas. However, their use within the context of the CE Design process is discussed in the following paragraphs.

Work Statements. Work statements are useful in the engineering and design environments because they can act as a single point of control for the information end products to be produced. In a complex product manufacturing organization, there are many types of work statements. They represent the authority to do something, and are the mechanism to account for control of the application of resources (people, money, equipment) to accomplish a work objective. In addition, they describe the scope, approach, budget, and work plan of the effort. Thus, the work statement can become the information accumulation point for all product and process design and engineering activities.

In *Figure 6-20*, there are three major types of work statements depicted. The Product Family Plan and individual product plans, influenced by product requirements, product features and functions, and the preliminary product architecture, are separated into design work statements. Typically, there is one per matrixed design team. These may be further broken down into additional work statements for analysis, test, simulation, documentation, and the like.

In addition to this first type of work statement, other types of work statements may be used to document other efforts throughout the organization. Many types may not be related to the end products of design and engineering or manufacturing at all. For examples, work statements for support of environmental cleanup may be developed. Thus, any process for work statement development and maintenance must recognize that not all work statements need to be under configuration control, or need to contain all the complexities of a product design work statement. In addition, work statements can lose their identity and traceability. It is important to track the achievibility of requirements, and how they are being addressed; work statements provide the linkage between requirements and the design artifacts.

As the design process proceeds, the product's features and functions "move around." Work statements for each team may drift. Additionally, suggested improvements from customers, the production floor, and from other teams may cause drift as well. It is common for a design team responsible for a high change rate area of a design to have several suggestions, requests and/or production design revisions affecting the same area, and sometimes the same component of the product.

If work statements are prepared and not kept "up to date," and are not an intrinsic part of the design process, individual change requests and idea identity

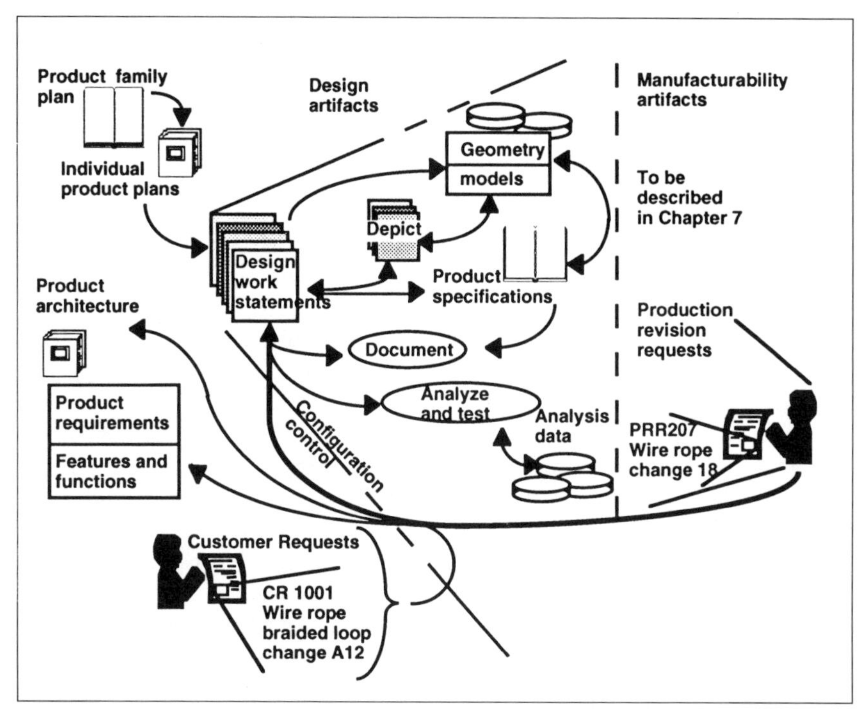

Figure 6-20. *Work Statement Hierarchy.*

and traceability can be lost through combining changes, breaking them apart, only using part of suggested improvements, and other well-meaning but lacking-in-precision actions.

The automated support for CE Design must use the work statement as the information accumulator. It must have those features and functions necessary to permit the identity and traceability of each proposed requirement to survive the process. The computing architecture necessary to deliver these types of computing features and functions is summarized in Chapter 8.

In addition to the design work statement depicted in *Figure 6-20*, there are two other types of work statements. These are Customer Request (CR) and Product Design Inquiry (PDI). The CR is used by sales and marketing to identify customer product design interests. These CRs may result in products particularized to a single customer or to product options or additional features, if these work statements are accepted for actual design and implementation. The PDI starts on the shop floor when a problem in production manufacturing develops which appears to come from the product or component design. The severity of these problems will vary, and might include ones which stop the production line to those which just require a clarified assembly instruction.

Geometry and Models. The geometry area of design and engineering is the subject of an entire industry. Many books and articles have focused on the features and functions of 2D, 3D, and 3D solid geometry systems. The discussion of geometry in this book, within the context of CE Design, is on:

1. management of the geometry once produced;
2. configuration control of the geometry;
3. manufacturability and other aspects of the CE Design Process in which geometry plays a part, and
4. relating geometry to itself and to other design and manufacturability information artifacts.

There are many important features and functions for a geometry computer system; these include accuracy, the ability to design a variety of physical, electrical, and other system parts; and the ability to combine parts to check for fit, interference, and utility. Combining parts for purposes such as packaging design, tool fit, and mock-ups is also important. There appear to be several potential useful systems in the market.

The focus of this book is not on the geometry itself, but on the design process within which the geometry package is utilized to first conceptualize the product, if possible. It focuses on how the geometry relates its components, and how CE Design's collaborative-concurrent design process manages the relationship of the geometry to those processes which produce and support the product.

The modern geometry package is so complex and laden with features that it may not be usable for early conceptualization activities. Early work statements may contain ''scanned-in'' hand sketches, product requirements as text, sample briefs of material from similar products, performance simulation data, and a variety of other materials in addition to the work statement. This rapid accumulation of information is why the use of an ''information envelope,'' which operates much like a ''manila paper envelope'' is so important. The work statement may also contain a high level geometry model of the design team's area of responsibilities prepared with a simpler, less precise, but easier to use design-only CAD program by a lead design team.

As the team goes through its design process, a variety of potential versions of individual components and their combinatorial models are developed. The model/variant configuration control process of *Figure 5-17* is used to manage this versioning. However, in a complex product environment, with multiple products with many thousands of components, it is the storage, indexing, and retrieval of these versions of geometry images which must also receive strong attention. A number of Product Definition Management (PDM) systems which store and retrieve geometric models have emerged in recent years. The approach most likely to prevail in this environment is one in which the geometry is stored across many different heterogeneous computer systems and computer hardware types and PDM is but a component of the larger set of systems which comprise CE Design automated in infrastructure support.

As shown in the *Figure 6-21* series, geometry information can be found on various media (tape, magnetic or optical disk, et al.) which are components of various hardware and software platforms. The computing architecture appropriate to CE Design is discussed in Chapter 8. From a design process management perspective, it is important that the computing architecture recognize these elements and concerns:

- as shown in *Figure 6-21A*, the most likely configuration from design automation is the use of powerful engineering workstations connected to a supporting computer acting as a temporary storage point for geometry, also called a server;
- as shown in *Figure 6-21B*, additional storage elements and more powerful computing for analysis, test, and to act as a central storage point for geometry are also connected to the workstation networks; and
- as shown in *Figure 6-21C*, additional elements can be added to provide the engineer, designer, manager, and others a single point of entry to the overall architecture.

Once this basic architecture is assembled, its operation, from a management perspective is composed of several important concepts which are as follows:

- in this architecture a directory locates the desired geometry. Typically, this directory contains not one, but several alternative indexes into the geometry. These indexes include:
  1. the work breakdown structure (WBS) used to plan the engineering work and to index the differentiated product components as they are developed,
  2. a bill of material index, based on assembly part numbers,
  3. a maintenance manual index, perhaps based on the indexing scheme of the manual itself, or how the product is assembled and disassembled in the field, and
  4. a group technology or standard parts indexing approach.
- the network linkage establishes a data transfer setup with the hardware/software platform where the actual geometry is located,
- the user profile and security capability determines if and how the requested geometry can be provided to the regulator, and
- the translator to/from converts, if necessary, the stored information into a form which the current geometry system can utilize.

If the geometry is for an older, more stable product it may still be stored in paper form, or perhaps on microfiche. The computerized, automated storage of all product and process related information will require a large amount of digital storage capacity, so alternative storage approaches may be appropriate.

Data management and the Data Directory together identify other pieces of information and geometry to which this geometry relates. All of this capability may be reached through an information envelope which appears to actually contain all this information.

Digital models, within the context of this discussion, are intended to be combinations of detail parts in a single digital representation. These combinations could also be called assemblies, higher levels of the bill of material, mock-ups, or in aerospace, for example, installations. These combinations of detail parts, or process elements, also have part numbers.

In the collaborative-concurrent design process, there is a need for constant downward-differentiation and upward-recombination activity, as the product is designed. The release and distribution from the collaborative-concurrent pro-

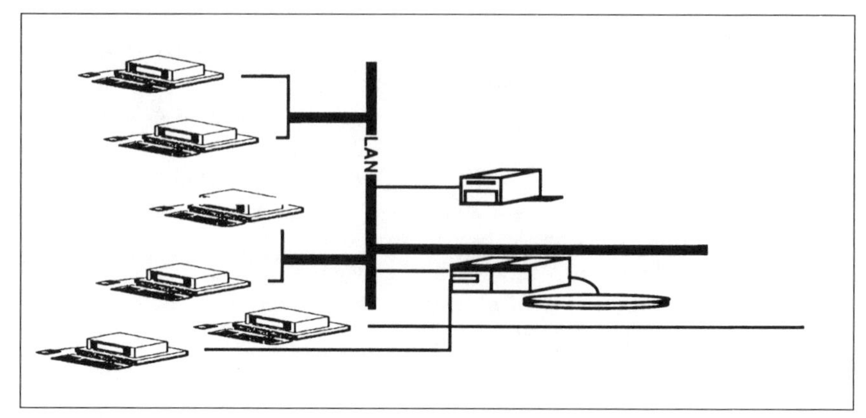

Figure 6-21A. *A Generalized Approach to CE Design Information Storage and Retrieval.*

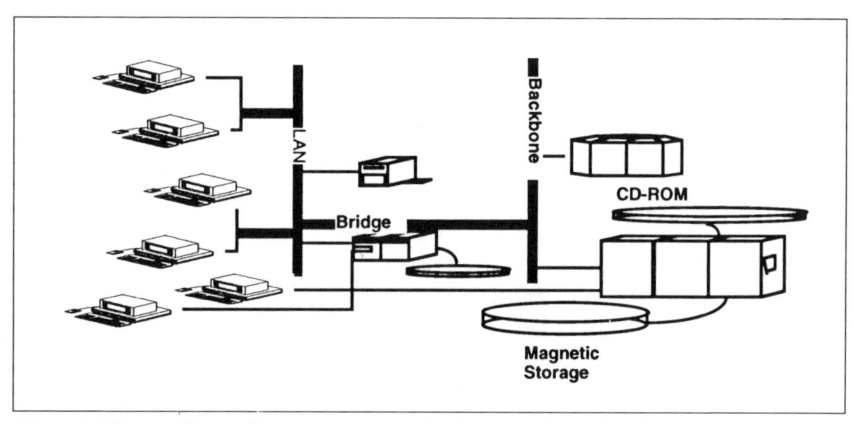

Figure 6-21B. *A Generalized Approach to CE Design Information Storage and Retrieval.*

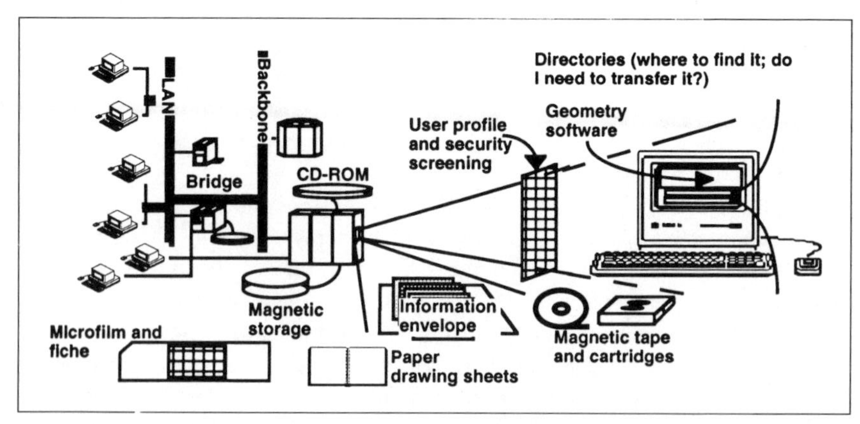

Figure 6-21C. *A Generalized Approach to CE Design Information Storage and Retrieval.*

cesses should be at least one level above the detail from a bill of material perspective. No work statement should be limited to just one detail part. The lowest level model, or higher levels if possible, are the central point for configuration management. Release at the effectivity level of design is best.

At the end of the CE Design process, the releasable geometry and other manufacturability information artifacts are stored in this same architecture. The function of Release and Distribution takes control of this design process information end product.

Engineering Analysis. As in the case of geometry, there are a variety of tools and techniques used to perform engineering analysis, testing, and product performance simulation. The issues for engineering analysis as they relate to this book and to CE Design are the same as those of geometry and models:

1. management of the analysis information produced;
2. configuration control of the analysis software in concert with that information for reproducibility reasons;
3. manufacturability and other aspects of the design and analysis model processes, and
4. relating the analysis to other product and process design and production management information.

Much like work statements, a significant amount of analysis is performed that does not relate to a specific part or product component. As shown in *Figure 6-22A*, much analysis done at the early stages of conceptualization is so general in nature that it can only be related to the highest level of product geometry and specifications in an information envelope.

As the design proceeds, the analysis can become specific to a process for a particular component. This more directly related engineering analysis is shown in *Figure 6-22B*. At the end of the design process, engineering analysis can be related to a specific component, or even one detail element. This very close association is reflected in *Figure 6-22C*.

When it is possible to relate engineering information, analysis information and results more specifically, the configuration control process for geometry must also be used for engineering analysis information and track to version, level of detail of the WBS/product, etc. Also, engineering analysis information, like geometry, should be stored, managed, and related to other information as shown in *Figure 6-20*.

The software programs for this engineering analysis are often custom produced by the engineers themselves, and sometimes for specific problems. This can be dangerous if this software is not captured and kept under rigorous configuration control. If not, the ability to reproduce the same design parameters may be lost. This is particularly important as the geometry becomes the point for analysis. Keeping the geometry, its software, the product design, and the engineering analysis tests, data, and software all in synchronization is a significant task.

The net summary impact of CE Design operating in a fully concurrent mode using the collaborative-concurrent design approach can be summarized in

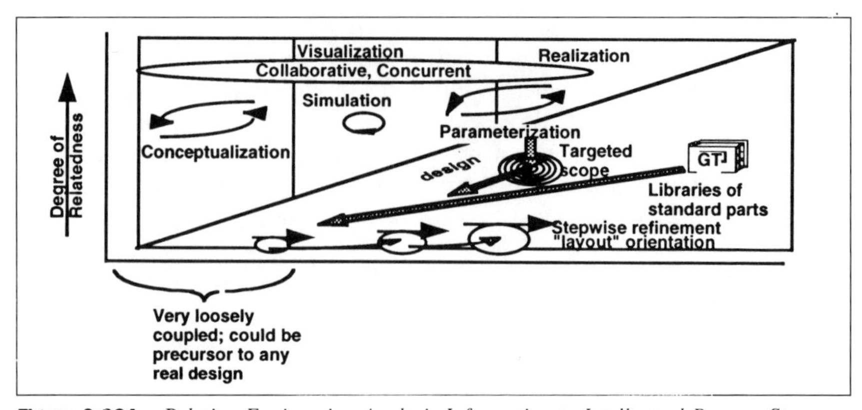

Figure 6-22A. *Relating Engineering Analysis Information to Intellectual Process Stage and to Design Approach.*

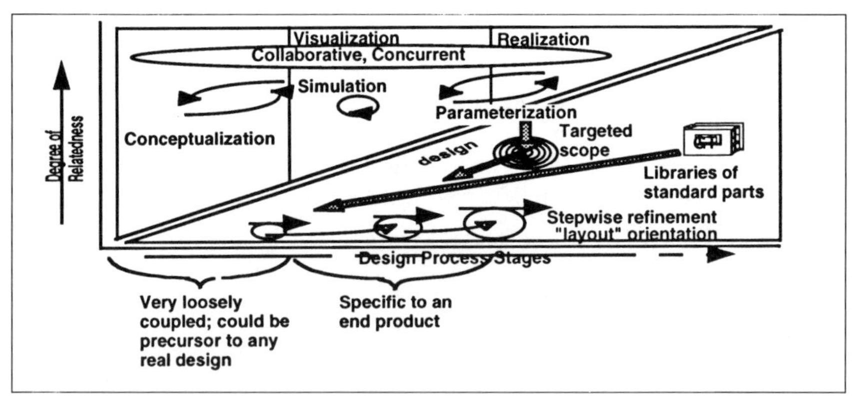

Figure 6-22B. *Relating Engineering Analysis Information to Intellectual Process Stage and to Design Approach.*

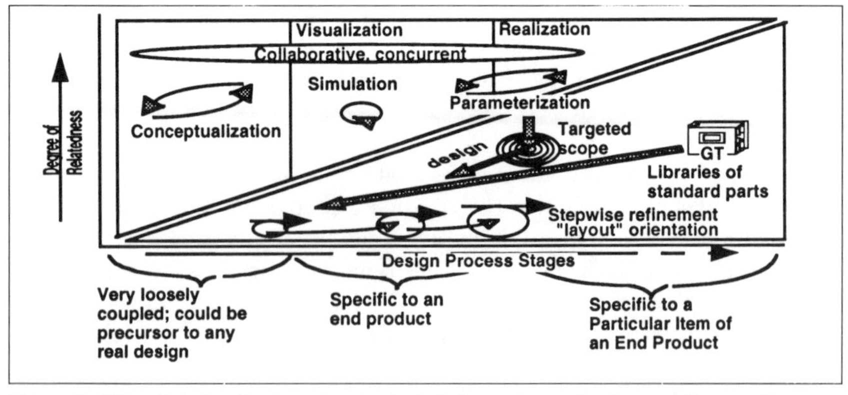

Figure 6-22C. *Relating Engineering Analysis Information to Intellectual Process Stage and to Design Approach.*

Figure 6-23. The levels of attainment and their definitions are contained in Chapter 9. The computing architecture required to support the various levels of attainment are contained in Chapter 8.

The design business process is one of the four integrated business processes which make up CE Design. This in-line, model-driven process uses and produces information artifacts in the form of work statements, geometry/models, and engineering analysis information. Moving from a step-wise refinement design process to one based on a collaborative-concurrent design process is a necessary component of the transition to full CE Design. As these information artifacts are being produced, Process Management is assisting in control over CE Design, and the manufacturability process is assisting in assuring that a usable design is being produced.

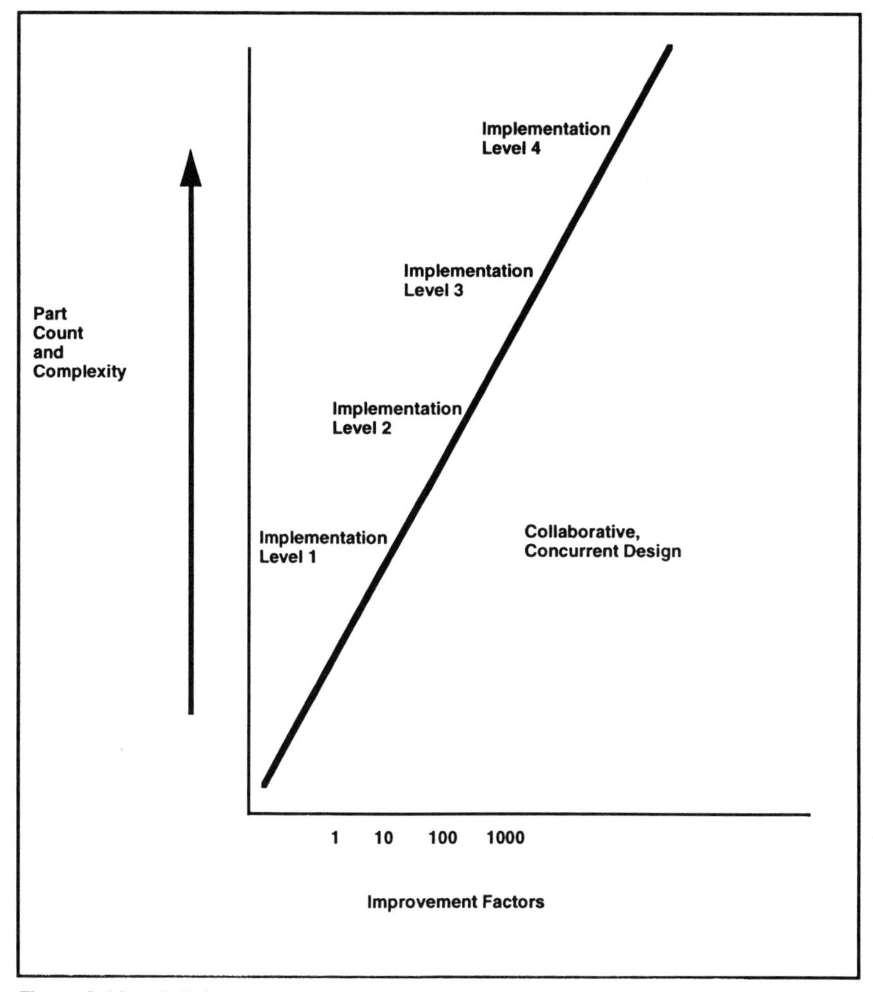

Figure 6-23. *Collaborative-Concurrent Design Implementation Level Benefits.*

7

CONCURRENT ENGINEERING DESIGN'S MANUFACTURABILITY PROCESS

Concurrent Engineering Design (CE Design) consists of four major processes: Process Management, discussed in Chapter 5, the Design Process, discussed in Chapter 6, the Manufacturability Process, which will be discussed here, and Automated Infrastructure Support, discussed in Chapter 8. Manufacturability is that process within CE Design which is focused on the transition of the product and process definition from a design perspective to a total production management perspective. It is this change in perspective which has created the "wall," shown in *Figure 7-1*. The wall is between design engineering and the rest of the organization in many complex manufacturing organizations today.

Design engineers have been releasing product and process definitions which they believe are complete. The design engineer considers the product as an integrated whole. Customer-originated operating characteristics are descriptive of the "whole" product. The rest of the organization must adapt these designs for production. As shown in *Figure 7-2*, the rest of the organization is concerned with "building." Because manufacturability represents and relates to the rest of the entire organization, its scope is large. It is clearly a cross-functional process. Its interests are many and varied.

Those involved in the manufacturability process take a "builder's" perspective. A builder's view begins with individual components as depicted in *Figure 7-2*. These individual components are combined into assemblies and then into products. Notice that if *Figure 7-2* were drawn for a design engineer, it might look like *Figure 7-3*. In *Figure 7-3* the design engineer considers the product as the focus; assemblies, components, and process plans, along with other "internal" elements, are considered along one plane of thought. Product

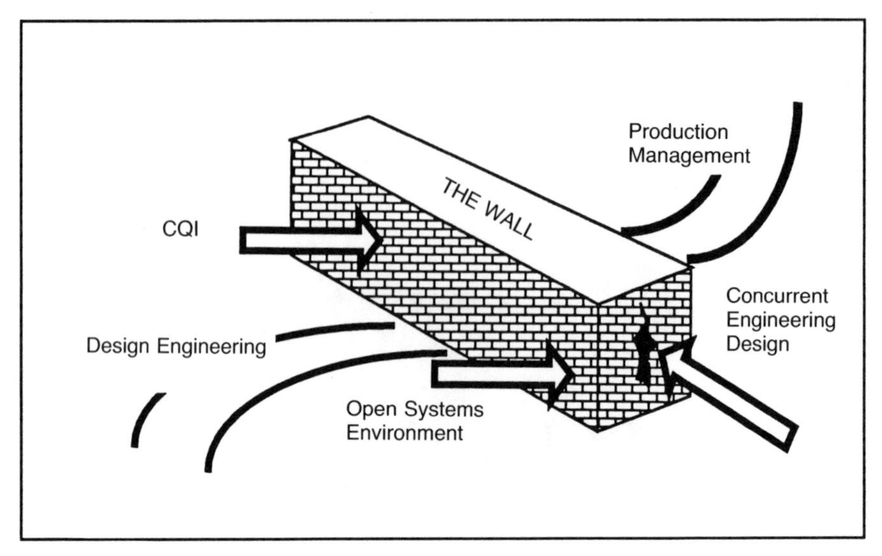

Figure 7-1. *CE Design is part of the Effort to Break Down the Wall.*

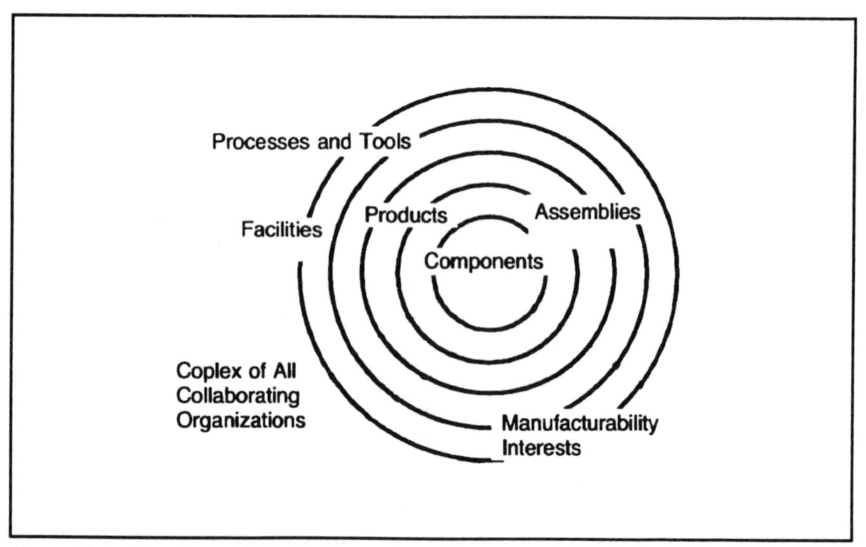

Figure 7-2. *Manufacturability Scope.*

packing is considered another area of thought.

For those involved in design, all these product and packing components are considered within the various design approaches discussed in Chapter 6. For those involved in manufacturability, all these product, packaging, and process matters are considered by starting with individual components, and then expanding them effectively into larger assemblies, etc. In addition to the product and its packaging, the manufacturability process considers the larger concerns of supporting processes, tools, facilities, servicing, and the organization as well.

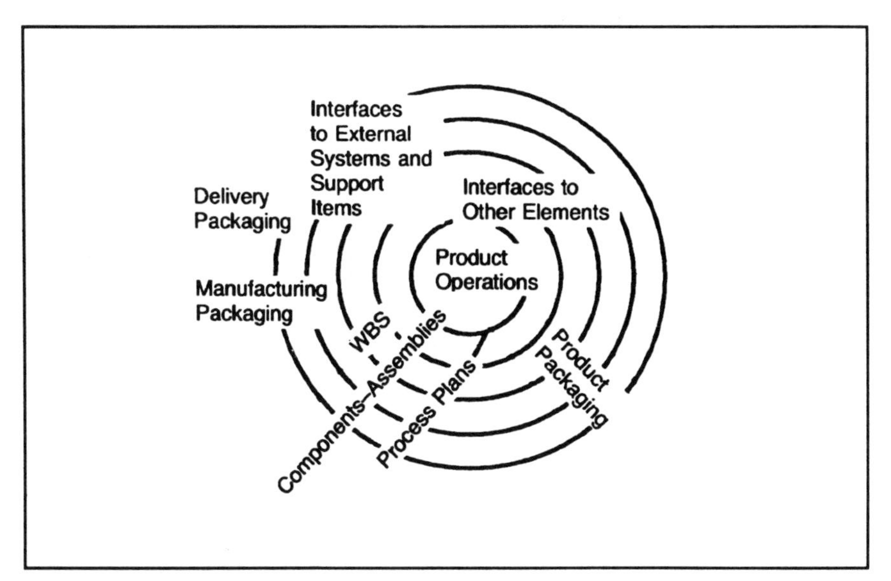

Figure 7-3. *Packaging.*

For complex manufacturers, suppliers and business partners are also within the scope of manufacturability.

Chapter 1 introduced the concept of CE Design teaming in *Figure 1-6*, which depicted a simultaneous emphasis from a design team perspective on both the design of the product and its ability to be manufactured effectively. Since that point, and throughout the book thus far, there has been constant reference to the teaming aspect of CE Design. Teaming is an important element of CE Design and becomes particularly important in discussions of manufacturability. Adapting *Figure 1-6* into *Figure 7-3* emphasizes the structure of the team and manufacturability's interest in building. This building interest is indicated by the two dimensions of the teaming arrangement. These two dimensions also vividly demonstrate the transition from design to production manufacturing. Along one dimension, multifunction product design teams are separated into personnel grouped by technical skills or major product components. The design groupings shown in *Figure 7-4* are focused on major product components. CE Design's manufacturability portion of the team effort focuses on "interests." The various areas of production manufacturing affected by different aspects of the design are indicated as members of the Concurrent Engineering (CE) dimension of the CE Design Team. The CE portion of the team can contain representatives of any or all of the various interest areas either actually or partially affected. It is this simultaneous interest of *design* and of *manufacturability* which makes up CE Design.

There are two concerns about the team's structures as depicted. The first concern is scalability. For smaller product designs, are all members still required? Design teams could become cumbersome and counterproductive if all these people had to be present all the time on each project. In fact, most of the

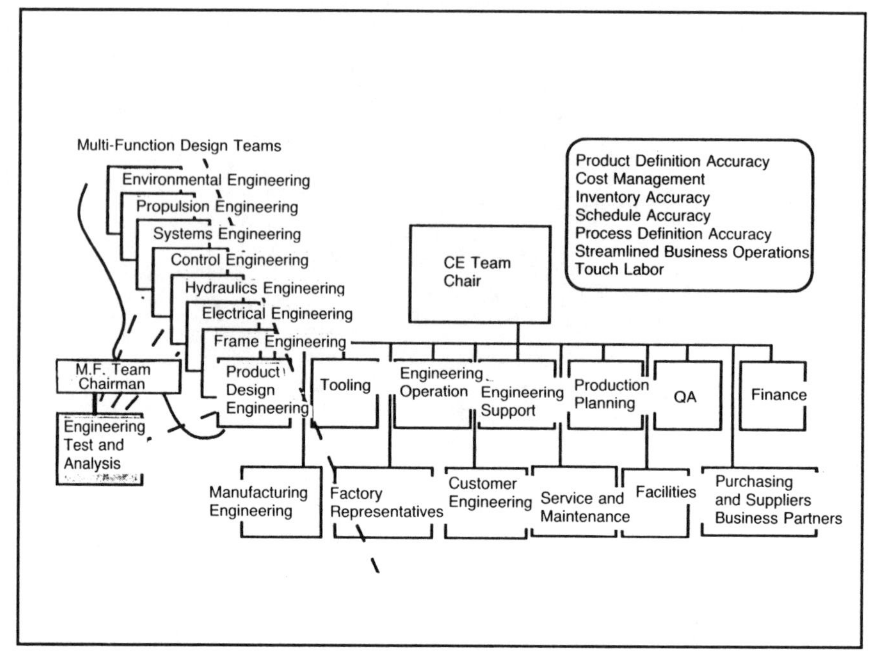

Figure 7-4. *Manufacturability's Focus.*

time they would be unproductive. This need for collaborative, concurrent activities, but only as needed, is at the core of electronic proximity, the virtual teaming arrangement of CE Design. Electronic proximity allows the various team members to be involved throughout the process.

Electronic proximity permits proper involvement if the second issue, knowing when a design issue of interest to one of the various specialists has been created, can be managed. The intent of the design process should be, for example, to note as each new addition or change is made, what has changed, why (design intent), as well as provide an index of areas within the overall team that might be affected. This allows the potentially affected or "interested" portions of the team to become involved as necessary.

Transition from design to manufacturability is also reflected in the different CE Design intellectual activities' interaction. *Figure 7-5* reflects these different involvements as the intellectual process proceeds. As the initial product's outlines, general specifications, and component strategies are developed during conceptualization, product production planning can begin. As more details occur in visualization, more detailed product and component production scheduling, material acquisition, and other details of planning are conducted. This greater detail can include facilities, personnel training, process details, and other manufacturing planning. Finally, as product realization recurs, and detailed design is completed, production of early detailed parts can begin. Of course, this intellectual process occurs several times as more detailed assemblies and components are designed and the design proceeds "down" the product WBS.

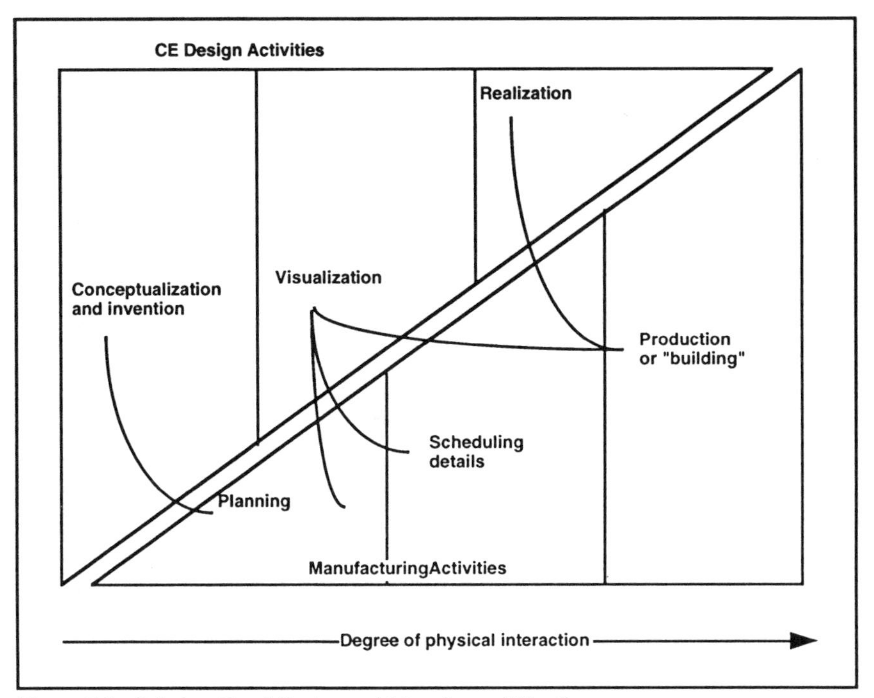

Figure 7-5. *Relationship of Design and Manufacturability Processes.*

This progress provides the potential for the parallel execution of many aspects of manufacturability during the execution of the design process.

In *Figure 7-6* the various design approaches described in Chapter 7 are added to the diagram. Placement of the approaches in the diagram shows the timing of these approaches. For example, group technology utilizes existing, fully realized and available parts, and so it originates far to the right in the diagram where ''libraries'' of existing part designs have already been collected.

Simulation can be used in all three general stages of the intellectual process, and it spans the stages. The relationship of the approach to manufacturability is also depicted in this diagram. Major manufacturability activities are shown along the bottom of the diagram. Simulation is also used by the manufacturability interests. For example, during conceptualization, as the initial design is begun at the entire product level, the facilities interest area can use these early product designs to analyze the current or required structural designs of the production manufacturing's actual building of the product to determine improvements or changes needed. Simulation can ''move'' the product through the facility's design model, looking for appropriate size, jigs, work center stations, and painting opportunities, for example.

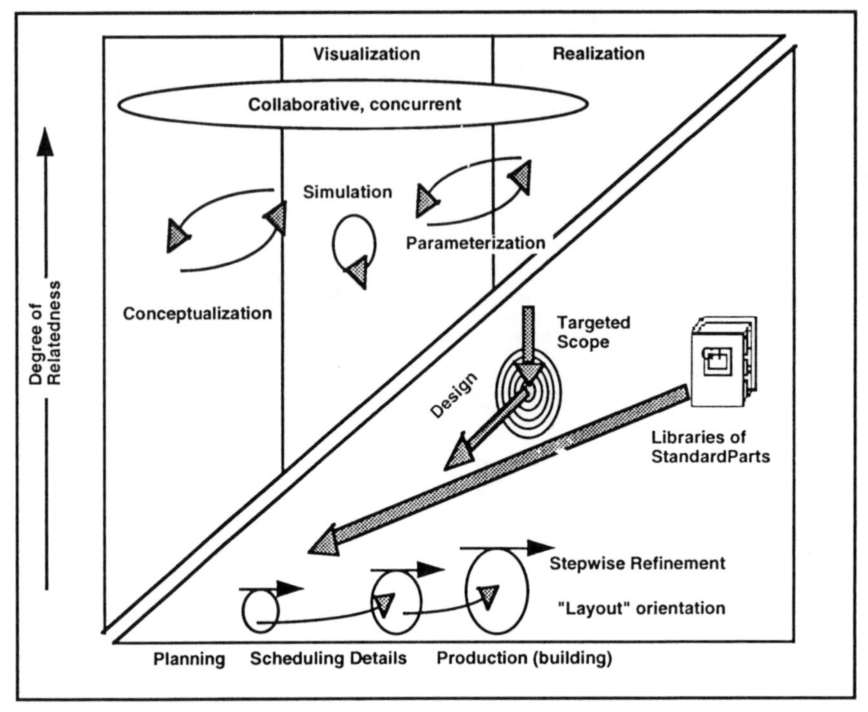

Figure 7-6. *Intellectual Process Stages and Various Design Approaches.*

THE ORGANIZATIONAL ENVIRONMENT FOR CE DESIGN

For crossfunctional processes, such as CE Design and its four major supporting crossfunctional processes (including manufacturability), to be effective, an understanding of the actual pattern of the organization's execution of its assigned task must be understood. From this understanding can come the knowledge to manage the effective execution of these cross-functional processes. *Figure 7-7* depicts the relationship between the actual, process-oriented organization and the traditional view of the organization's structure. On the right side of *Figure 7-7* is the traditional pyramid of organizational structure, with the primary focus assigned to each area. On the far left of *Figure 7-7* is a vertical table outlining the type of perspective each layer of the organization takes when performing assignments. For example, middle management relays schedule information to executive management, manages processes, and assigns activities and monitors their execution and successful completion. In the center of the diagram is a representative process map for this organization.

What is really happening in the organization relates more to the flow of information and materials in various ''chains'' or processes than to traditional organizational units and their functions. These processes are a representation of the old ''grapevine'' or informal organization discussed in an organizational theory context for years. Processes are composed of activities requiring one or more functions for the successful transformation of their input into desired

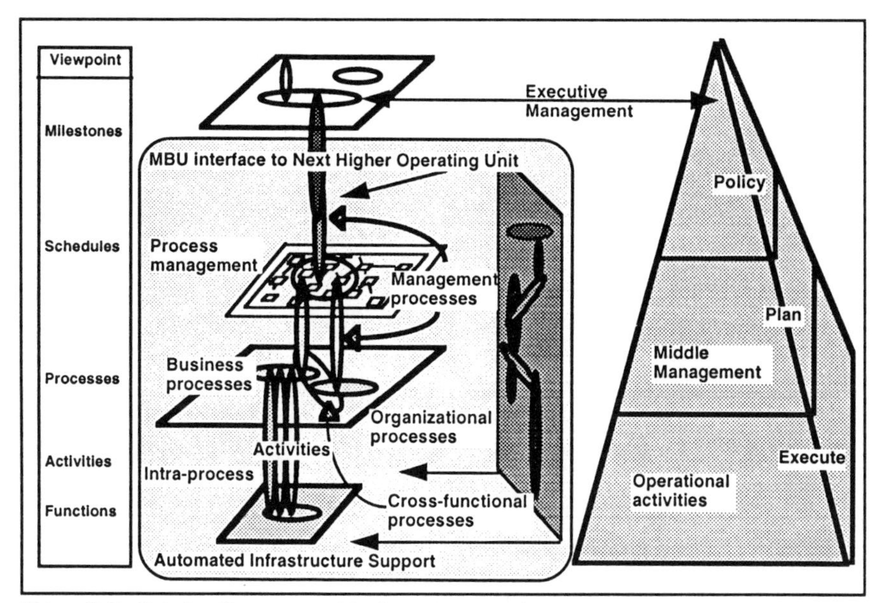

Figure 7-7. *Relating Processes to Traditional Organizational Structure.*

output. In *Figure 7-7*, process management, the subject of Chapter 5, is the bridge between executive and middle management. Business and technical processes comprise the interface between the group and individual and middle management. Notice how management processes are vertical in nature, and link the levels of processes. The organizational processes used to sustain the organization, but not adding value directly to the product and process value chain, are shown in a separate set of processes. When these activities are properly operating, they involve the people in the primary value chain, but operate orthagonally to product and process related processes. These primary value chain processes directly produce that which generates revenue for the organization.

Figure 7-8 depicts a representative process map for today's complex manufacturing organization. In this extension of the basic process model depicted in *Figure 7-7*, the *Figure 7-8* model depicts the multi-managed business unit (MBU) style organization, with its extended enterprise perspective. In this extended model, multiple sets of MBU-level process models are operating simultaneously, with extensions to outside operating suppliers and partners. In addition, the MBUs report into a parent which manages them.

The management style for a multi-MBU parent appears to be divided into two camps. Either the MBUs are managed as independent business units or as components of an interrelated operation where synergy between units is sought. Attempts to achieve synergy are more difficult than managing for individual goals. The complex product manufacturer, who is a part of an extended enterprise, as shown in *Figure 7-8*, is the subject of this book. The process of manufacturability is about making the broader aspects of the entire production

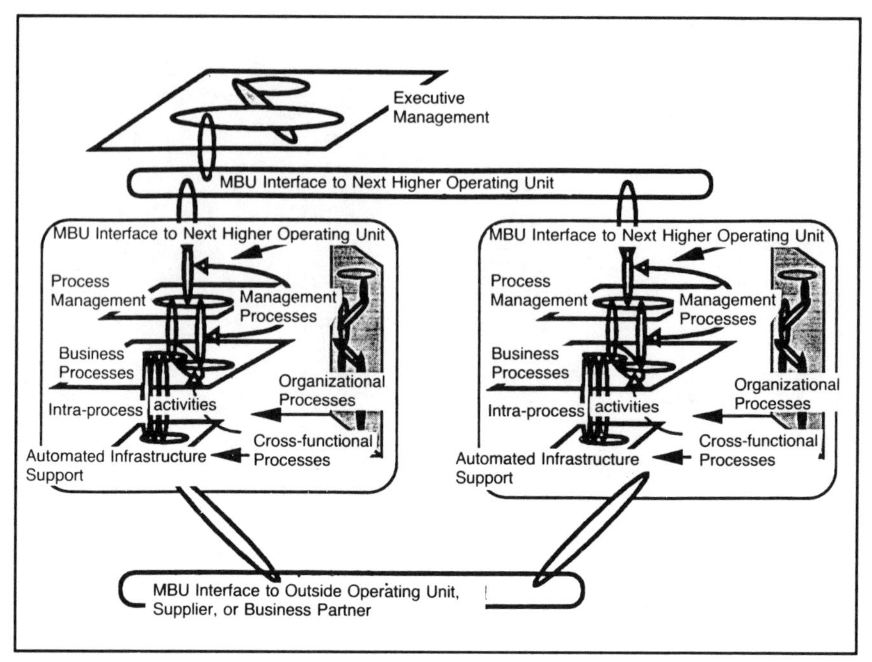

Figure 7-8. *The Multi-MBU, Extended Enterprise Model for CE Design.*

manufacturing operation work effectively through integrated product development (IPD), and integrated product management (IPM) after initial design. These processes usually comprise most of the primary value chain of this style of organization.

Manufacturability operates as a crossfunctional process to the design process in overall CE Design. As a CE Design's design process proceeds, some of its activities are devoted to manufacturability. *Figure 7-9* and *Figure 7-10* depict the two types of general design processes introduced in Chapter 6. *Figure 7-9* shows the step-wise refinement approach. It reflects how Chapter 6 described a step-wise refinement approach which has been enhanced to operate within overall CE Design. During each of the serial design process activities, manufacturability activities are included in the loops. As each iterative step-wise refinement in the direct design is completed, manufacturability of the iteration is considered. All of the potentially affected interests receive a copy of the current version of the design, as it exists at that point in the iteration cycle. The different areas are expected to analyze the design, and comment or suggest changes, modifications, or concerns regarding their area of interest. This is usually done in a sequence so minimum time is spent by each portion of the team until those "upstream" from their function have had their review opportunity.

For example, production engineering usually adds to the basic design for production engineering concerns first, because planning does not want to review the design until the production engineering portion of the review is complete. In stepwise refinement, it is believed that planning's review should be done later

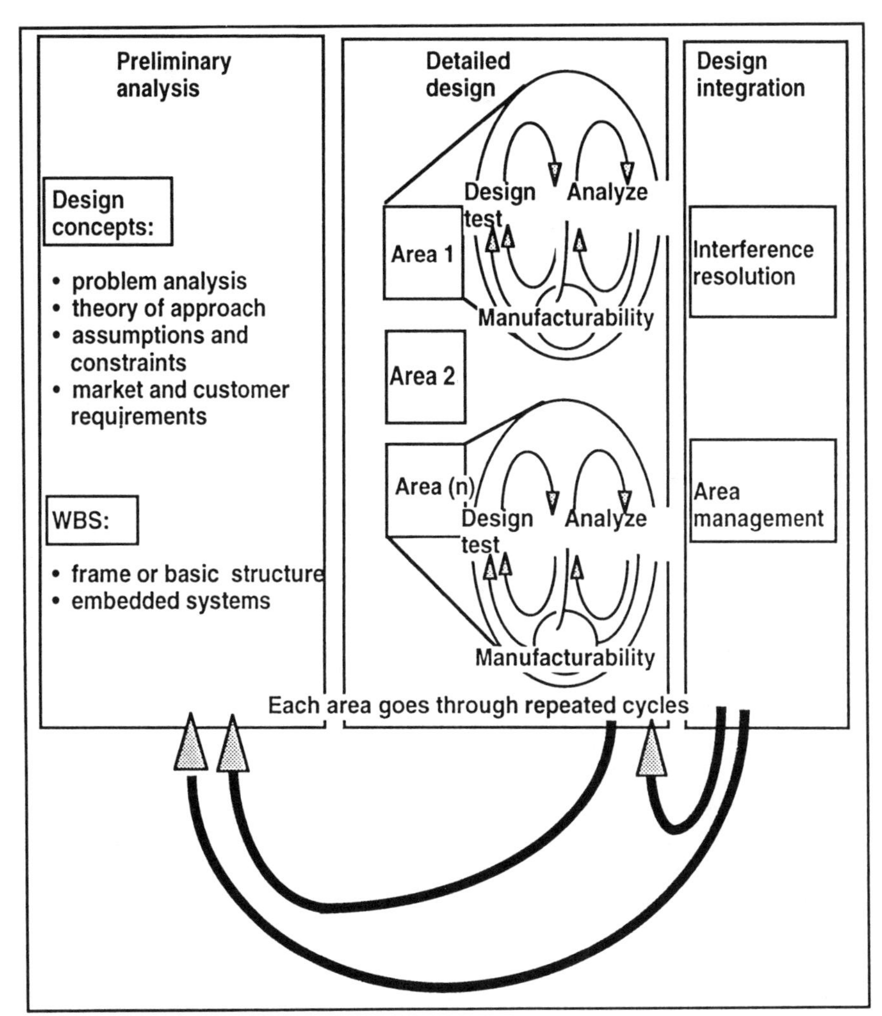

Figure 7-9. *Step-wise Refinement.*

because they need the additions of production engineering before performing their analyses. This sequence of activities continues to loop back on itself until the process's set of activities and the design are complete. This addition of manufacturability *inside the loop of design activities* is of course different—and arguably much better—from a traditional serial, iterative process, which delays manufacturability analysis until the product design is deemed completed by the engineering design part of the organization.

Manufacturability in Collaborative, Concurrent Design is inherently "in the loop." As the design process proceeds, the product and process design is considered by the manufacturability process and its activities. At each intellectual stage, the design is enhanced by manufacturability analysis. The sequence of activities in each process flow is arranged so added elements of the product

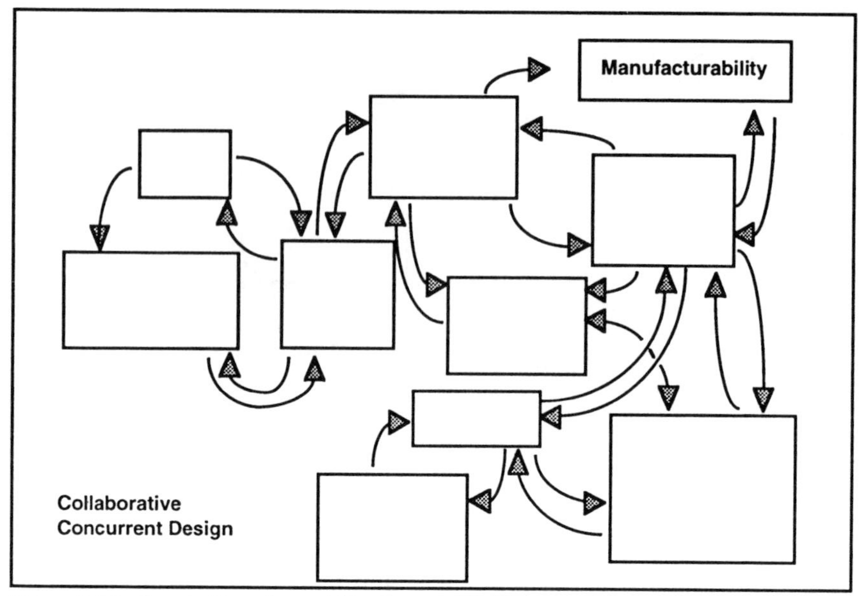

Figure 7-10. *Sample Manufacturability Process Model Approach.*

and process design are cumulatively productive and considered in the order most appropriate. As the manufacturability process itself proceeds, design intent in all its forms is added to the design information store. Process management's configuration control capability permits the collaborative, concurrent design process to proceed while several current copies of the same product or process model are examined simultaneously. During this review, individual changes are monitored and communicated as net changes to the central information store for the product or process design. Change types can be used to permit automatic notification of potentially interested areas of the manufacturability subteams as net changes occur. This automatic notification allows a greater concurrency, perhaps even complete concurrency, in design process activities.

To construct a set of process models to represent the manufacturability process itself, a generalized model of the overall manufacturability process must first be constructed. *Figure 7-11* depicts a general model of an example of the manufacturability process. Its most important component is its expression of formal subteams based on a Work Breakdown Structure (WBS) of the various interest areas for manufacturability. By using a WBS, appropriate coverage of all interested areas is ensured. Each organization should develop such a "checklist" of interests, develop a WBS to reflect these interests, and then assign personnel to each subteam group which is developing out the WBS. Each subteam will need to develop its own set of checklists and operating activities to ensure that its particular interest area is thoroughly covered during the design process. Also important is the establishment of authorization and approval procedures as each subteam adds to the design intent, and perhaps to the substance, of the product and process designs that they analyze. In addition, each subteam may develop

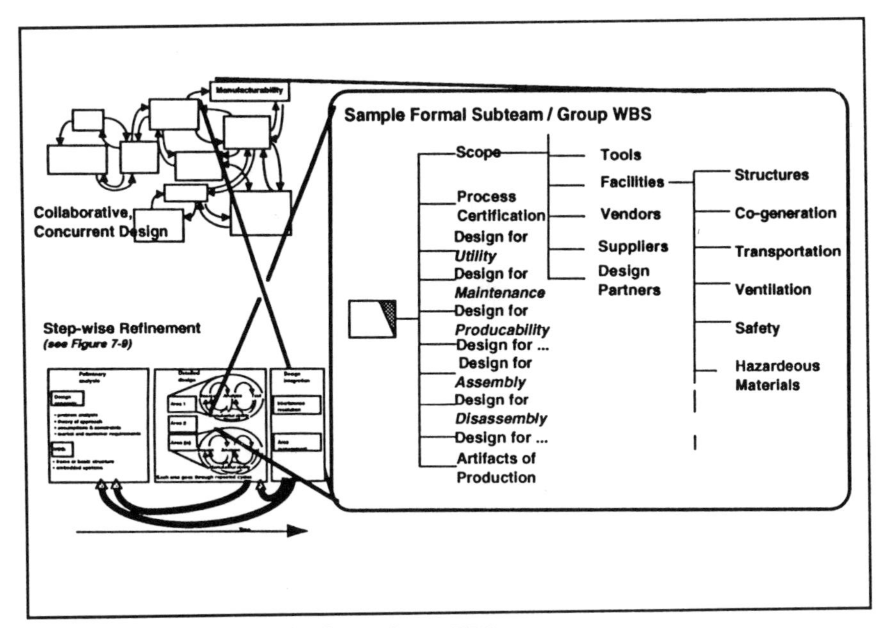

Figure 7-11. *Sample Formal Subteam/Group WBS.*

new or changed designs for affected areas of the production manufacturing process.

Figure 7-11 also indicates that these interest area WBS subteams can be part of either the collaborative, concurrent or step-wise refinement design process models. Finally, there should be a process plan interest subteam, and a subteam for interest area certifications. *Table 7-1* describes a portion of a sample product WBS for a "medium lifter" product.

Table 7-1
Product WBS of Drawings/Digital Models Table

WBS Number	Title	Description
1.0.0.0	Midsize Lifter	Whole Product Title
1.1.0.0	Frame	Framework for rest of product
1.1.1.0	Base Rods	Frame composed of these
1.1.1.1	Center Rod	Main Base Frame Rod
1.1.1.2	Bolts	Connectors for Rods
1.2.0.0	Rear Transmission Assembly	Rear wheels & assembly
1.2.1.0	Wheels	
1.2.1.1	Rims	(2)
1.2.1.2	Axle	
1.2.1.3	Tires	(2)
1.2.2.0	Rear Transmission	

The Manufacturability process is that part of CE Design bringing together different types of involvement and interests which production has in the design process. It provides an integrated and interrelated set of processes through which manufacturing and engineering can clearly see, cooperatively communicate, and act together to maximize the effectiveness of the overall complex product manufacturing organization. *Table 7-2* describes a portion of a product WBS from a manufacturability perspective, using the same ''medium lifter'' product. Notice how the product is grouped into slightly different perspectives, reflecting the detail or component ''up'' perspective of this area of interest.

Table 7-2
Manufacturability WBS

WBS Number	Title	Description
1.0.0.0	Midsize Lifter	Whole Product Title
1.1.0.0	Rod Assembly	Framework Assembly
1.1.1.0	Rod Kit	2 sets of side rods & front/back rods, plus center support rod
1.1.1.1	side rods	(2)
1.1.1.1	front rod	
1.1.1.1	back rod	
1.1.1.1	bolts	(24)
1.2.0.0	Rear Assembly	sets of wheels, transmission, axle, & assemblies
1.2.1.0	Rear Transmission Kit	gearing, casing, central drive shaft, connector shafts, bearings, miscellaneous O-rings and cotter pins
1.2.2.0	Rear Axle Kit	rear axle, bearings, O-rings, connector shafts

The scope of the manufacturability process, and its interest in product design can be quite large. In *Figure 7-11*, the issue is larger than improving the ease of manufacture of a component, although that is where it starts. Just as CE Design attempts—in the preferred collaborative, concurrent design process—to view the product holistically, the manufacturability process needs to address all potentially affected elements of manufacturing to include processes, tools, the facilities in which the product(s) are produced, and the entire complex of all collaborating manufacturing organizations. *Table 7-3* describes a portion of a specific application of the generalized model. It reflects the potential interest areas which may become involved in the analysis of the product's design.

Table 7-3
Process Teaming WBS

WBS Number	Title	Description
1.0	A single CE Design Team	Whole multifunction, manufacturability team for a portion of the overall product
1.1	Multifunction Design Subteam	Design Engineering subteam

1.1.1	Frame Group	Portion of subteam devoted to the design of the basic frame for the product
1.2	Manufacturability or Concurrent Engineering Design Subteam	Manufacturability Engineering Design Subteam
1.2.1	Design for Producibility Group	Portion of subteam devoted to ensuring producibility; should include planning, manufacturing engineering, production manufacturing, etc.
1.2.2	Design for Maintenance Group	Portion of subteam devoted to ensuring that the product is maintainable; should include customer service, field service, quality manufacturing engineering, etc.
2.0	Another Single CE Design Team	A team devoted to another portion of the same product
2.1	Multifunction Design Subteam	Design Engineering subteam
2.1.1	Power Group	Portion of subteam devoted to the design of the power train for the product
2.2	Manufacturability or Concurrent Engineering Design Subteam	Manufacturability Engineering Design subteam
2.2.1	Design for Producibility Group	Portion of subteam devoted to ensuring producibility; should include planning, manufacturing engineering, production manufacturing, etc.
2.2.2	Design for Maintenance Group	Portion of subteam devoted to ensuring that the product is maintainable; should include customer service, field service, quality manufacturing engineering, etc.

As mentioned in Chapter 6, complex products are increasingly being viewed as integrated systems, manufactured by integrated systems of people and tools within integrated processes. Just as the design team's preferred working model is collaborative between designers, there needs to be equal status and collaboration provided to all elements of the manufacturing complex as represented in *Table 7-3*, but only when necessary.

MANUFACTURABILITY PROCESS OVERVIEW

The design process has been described as a model-driven set of technical, business, and managerial activities monitored and controlled by a Process Management process. These processes can be accomplished manually, but they only become practical when appropriate automated support is provided. This is also true of manufacturability. In discussing *Figure 7-9*, the natural assumption and practice is to consider the manufacturing aspects of a product's design only after it has been designed. It is perceived to be an inefficient use of manufacturing engineering resources if they evaluate a design version not actually released.

Complex products are becoming increasingly complex, and increasingly taking on integrated systems characteristics. This complexity is driving the design process to consider involving production manufacturing. In some circles, this is called concurrent engineering. As discussed throughout this book, manufacturability becomes an integral part of the initial design process in the CE Design concept. This appears to be superior to just ''including'' production manufacturing. This immediate involvement yields such large benefits, as ''opportunity'' costs, which are so great that the organization must immediately involve production manufacturing through the design process. Quality considerations alone may account for up to 20% of the product's cost (source: KPMG Peat Marwick Survey).

A word of caution, however, is that the measurements of success for CE Design must be carefully described, and appropriate measures established, because overall product direct design cost may go up somewhat even as product costs and cycle times are reduced. Production manufacturing's cost to support CE Design may also cause its own budget to go up somewhat. Care must be taken to recover these additional costs through product and process improvements, and market share and product profitability improvements.

The total product life cycle cost will go down substantially when a good design is produced. This reduction occurs for the following reasons:

1. The constant involvement through a flexible manufacturability process brings experts to the design just in time to detect and correct concerns before they occur. This prevention type quality improvement and error reduction is from 100 to 1000 times less than the cost associated with a product failure in the field.
2. The cycle time of the IPD, and the entire manufacturing process, is completed quicker, though design takes longer, as the product will move quickly through planning and initial production activities because all or most concerns discovered at this point have already been detected, corrected, and prevented.
3. The quality, from a customer perspective, is substantially improved as the design is more mature and thoughtful because of its various interest area reviews.

The cost benefit harvesting process should include measures of improving product cost using Activity Based Costing, as well as other measures of success directly traceable to these improvements. These measures should include pre-established targets acceptable to management. This pre-establishment is done to ensure a steady commitment to this involvement.

Within that single box entitled "Manufacturability" in *Figure 7-9* and *Figure 7-10*, as well as the overall process detailed in *Figure 7-11*, there is a large and significant set of potential activities. Unlike the design process, when most activities of any process are similar over similar product component types, the manufacturability process may involve substantially dissimilar sets of activities for what seem to be similar parts. For example, a metal part with the same dimensions may be forged, machined, or cut into its shape, "blown" into a mold, or welded together from subcomponents. Each ensuing manufacturability analysis would be very different for each circumstance. This is the principal reason why the process model for the product is as important to manufacturability as the product description.

In *Figure 7-12*, a suggested overall manufacturability process model is

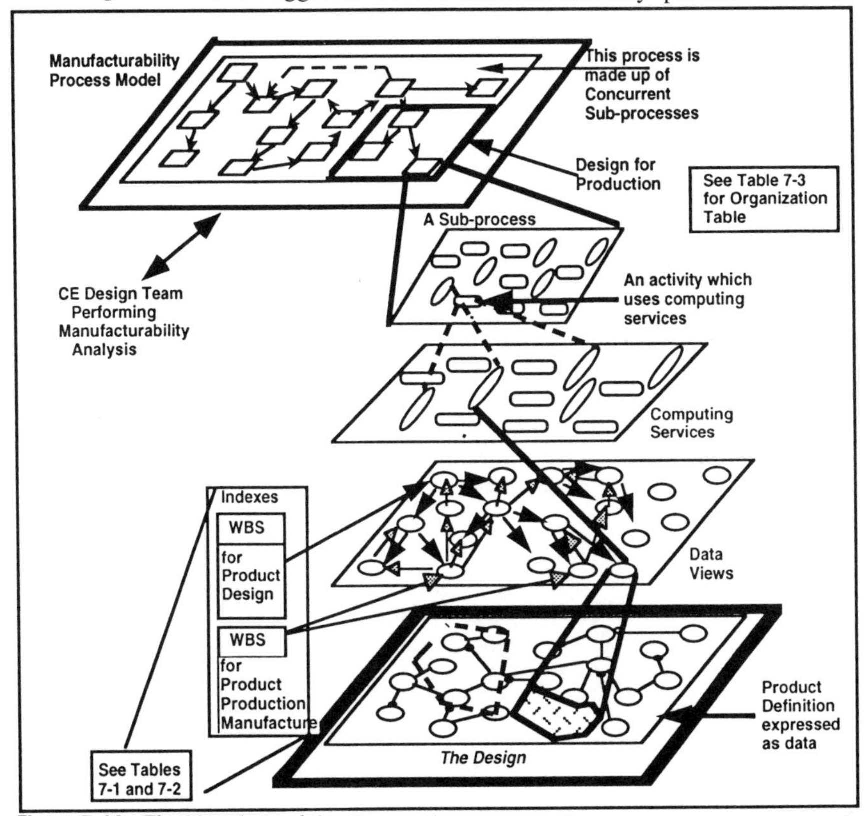

Figure 7-12. *The Manufacturability Process from a User's Perspective.*

depicted. This complex model of the manufacturability process outlines three suggested structures. First, a process model represents the process management level "in-line" model process; this model represents a particular instantiation of the generalized design process.

Second, for each activity within the process, computing services are used to accomplish the purpose of the activity. This support is of three categories:

1. Recording what has already occurred (bringing the process model into synchronization with reality);
2. Determining the status of various aspects of the model process or the product and its process; or
3. Actuating activities for information processing when the process's activity requires computing for its accomplishment.

Third, a capability which relates services to views of the overall design's information stores can keep product and process information consistent with process requirements.

In *Figure 7-13*, a sample high-level manufacturability process model, which would occupy the high-level model depicted, contains:

(a) An initial step in which the product/assembly/component ("item") design version is evaluated as to its current overall situation. Various analyses are made to determine what aspects of manufacturability are potentially involved.

(b) Several routing or manufacturing process-type decisions (required to manufacture the item) are made and various interest area subprocess design sets of activities, depending on the general type of item, are initiated. These subprocess design activities can support either a step-wise refinement, serial process, or a collaborative, concurrent process. These subprocesses can stand alone, or can be integrated into the larger design process of which manufacturability is part.

(c) All of the "design for ...s" are then executed with duplicate copies of the current design areas under interest area consideration. Recommendations are provided in net change format. This is not unlike providing hand sketches and notes on drawings in a colored pencil on some other overlay surface which allows changes by the designer or originator. As these changes are provided, the lead design team, as shown in dashed lines in *Figure 7-13*, determines who should review such changes, in addition to the designer, and then appropriately routes the designs.

Figure 7-13 depicts the different manufacturability process and its interest area, "design for ...s" relationships with the step-wise and collaborative design processes. For step-wise refinement, these various subprocess activities would be expected to be executed at various release points for the design. The typical approach is to have several incrementally complete releases at perhaps 25% complete, 50% complete, 80% complete, and final release of the emerging design detail. This permits periodic manufacturability analysis, interference analysis, area management, systems end-to-end validation, etc. This is a good, intermediate technique providing periodic concurrency, ensuring that the design

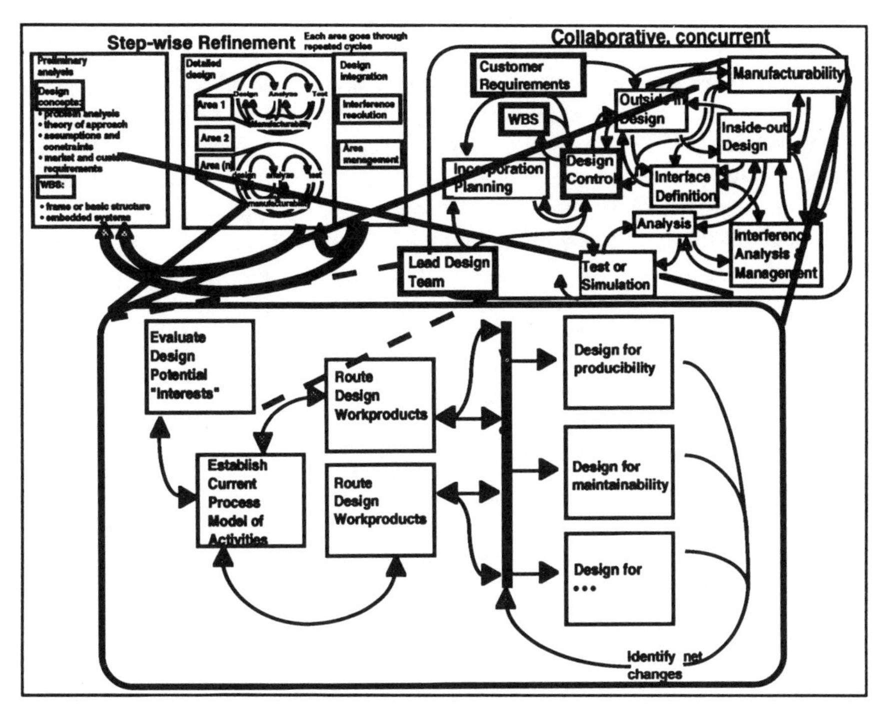

Figure 7-13. *Sample High-level Manufacturability Process Model.*

process and its result will exhibit the good characteristics of the CE Design paradigm.

In *Figure 7-13*, the collaborative, concurrent process can have its manufacturability major process executed in two completely different approaches. *Figure 7-13* depicts manufacturability as a separate process in the overall design process. In this approach to execution, manufacturability is invoked as a process each time it is presented with a design element to examine. The frequency of examination is determined by the lead design team. Periodicity is monitored and controlled by the design control group. The group, in conjunction with the lead design team, is constantly evaluating design versions and iterations for potential interest area impacts. It also is constantly asking multifunction design teams to estimate the probable interest areas which should look at their latest work. In this manner, the manufacturability process is implemented on a JIT basis.

A second approach is not directly depicted in *Figure 7-13* because it would be difficult to do so. This second approach is the constant interaction approach. In this approach, the heavily shaded boxes denote activities not performed by every design team. These activities are performed by a central group supporting the multifunction design teams. The remaining activities are performed by multi-function subteams who are each interfaced with Concurrent Engineering design subteams. The design versions which reach the manufacturability process and the interest area teams originate from each subteam, not from the central design control activity. This subtle but powerful difference permits more concurrency.

By initiating manufacturability involvement directly from the multifunction design teams, product designs need not reach an arbitrary point, such as a 25% complete design, before review, but can be evaluated at any point that the multifunction design subteam feels is appropriate. This permits a better flow of activities and produces better designs quicker than the more formally staged approach. It does require the multifunction design subteams to be desirous of feedback, yet sensitive to the problem of too much review too soon. This is best handled by experience. The Chapter 9 CE Design implementation recommendations contain a staged, formal, multiple release approach in its phasing plan.

MANUFACTURABILITY INTEREST AREAS

Many books, articles, and organizations, such as the Society of Manufacturing Engineers (SME), are devoted to the advancement of the manufacturability process. This book will not attempt to recreate the advice provided by such sources as the *Tool and Manufacturing Engineers Handbook: Design For Manufacturability*. Instead, the balance of this chapter focuses on aspects of those interest areas of particular importance to the manager who must ensure the success of the CE Design process, and the full integration of CE Design's output with a successful production manufacturing operation.

Process Planning

During the manufacturability process, two sets of production manufacturing planning activities occur. The first focuses on aiding the design team in examining aspects of the design version under evaluation which affect the manufacturing planning and process. An example might be looking at either generated (directly from the proposed geometry, if possible) or derivative (derived from components of similar types previously manufactured) processes to identify material, shape, performance, weight, and other issues considered during the design to reduce cost and time or increase item utility. The second process activity is to begin to design the manufacturing process itself for this item, and to consider these issues with the manufacturing process as well. This is commonly called process planning.

Process planning proceeds through three stages:

1. **Developing a nonspecific process plan**. At this stage, during early design activities, not enough design work has been done to determine how the component will be produced. This generalized, nonspecific process plan is the type usually produced by design engineers in today's serial, step-wise refinement design environment.
2. **Developing a specific process plan**. During the later stages of CE design, this process plan is produced to coincide with approximately the 80% release point. Enough is known about the product design to be able to accurately determine the exact production sequence. If possible,

simulation of the production process, and essentially a preproduction prototype of the component can be produced. This can be done in time to influence the design before final release. This is a significant improvement over step-wise refinement's approach.

3. **Developing the routing**. This process plan actually identifies people, equipment, and other details that cannot be known until the design has been released to be produced. This can only be done after formal release, but the release should be almost without change to the specific process plan. This also is a significant improvement over step-wise refinement design output.

There are several popular and evolving tools for process planning such as CAPP (Computer-Aided Process Planning), which facilitate the development of the process models. The CAPP area of production manufacturing planning is well described in many other media, including handbooks, texts, and articles. The specific process plan can also be developed in the CAPP tools, but routings are usually found in MRP and factory control systems. There is thus the need to provide for their translation and feedback. When the process plan becomes reality on the factory floor, there is a tendency to ignore feedback to the designers. However, this feedback is as important as the early feedback, which was conceptual in nature. The manufacturability process must include formal steps for the feedback and incorporation of these shop floor experiences.

Manufacturability Scope

It is important to note that the scope of the issues which the manufacturability elements of the team must consider, both from their own interest area perspective and for further consideration by the designer, must include two often overlooked issues: tools and facilities.

Tools include machines and various tools to create the item as well as the various jigs and other pieces of equipment used to assemble items, and gages and test equipment. This aspect of manufacturability is also well discussed, but in its own ''world'' of discussion. Tooling is, of course, supposedly developed only once, and then just used. It has been a widely held perception that tooling does not need formal manufacturing techniques, nor must it be included in other major management systems. This is not the case in a CE Design environment. *Tooling is important, constantly changing, and critical to the quality of the end-product*. Perhaps it is the key element of quality, and of cycle time reduction as well. The most important aspect of CE Design's tooling related manufacturability subprocess is the ability to utilize geometry and other product design related information to plan and design tooling, handling, and packaging, within the context of the design process itself.

Facilities, in the CIM approach, especially when much of the product and process definition is translated directly from the geometry, include the facility, its computer networks (WANs and LANs in the facility) and the wide variety of other automation and people issues facing the facility environment. These considerations are as much a part of the routing or manufacturing process design as the products themselves.

Environmental and product safety issues have become as important as the product's producibility. Producing the product in a safe and environmentally acceptable manner has become a "pacing" item for product development. However, the design team has not historically been concerned with these issues. In CE Design, the environment and safety are just another subteam interest area inherently "in-the-loop" of design, a most important feature of the collaborative, concurrent design process.

The experience of many complex manufacturers is that these two issues (tooling and facilities) are beginning to dominate the design process. Infrastructure, capital, safety, environmental concerns, and the organization's built-up capability are becoming serious constraints on the design process itself. There is a constant attempt to develop products which extend present capabilities (save time and capital, and are safer) and do not stray far from the current expertise of the organization. This is the essence of the currently popular trend toward "core capabilities." Do only what you can to be world class. The problem, of course, is that this relatively narrow product focus, going forward in time, can be dangerous. Each organization must include this concern as a part of its present decision-making process. At the same time, designers need to be aware of the limitations which these two issues can create. They can save themselves time and energy by considering them, through planned subprocesses in manufacturability, from the beginning of the IPD and throughout the IPM life cycle.

At different levels of the product design process, a great number of different manufacturability issues in these two areas, especially facilities, might be considered.

First, consider the product level. If the large complex product—or a major portion of it—must be moved, some sort of lift and move device is required. The ability to move the completely assembled product out of a facility is unfortunately ignored in some cases. The design process might be affected by this manufacturing process support equipment concern in three ways:

1. The need for designed attachment points, and the need for special stress analysis for the lifting process itself on the product at that point. A sample of the attachment point concern is shown in *Figure 7-14*. In this diagram, a complex shape is being designed. It is heavy, and therefore it must contain an attachment point for a temporary jig to hold this portion of the product at an appropriate height for further assembly.
2. The assembly level. The layout of the assembly process and the use of a zero mass insertion technique for placing a circuit into its packaging may be required. For the design process, this means designing holes in the packaging which match the capabilities of the machines and the needed diameter of the holes. It creates the need for attachment grooves in the packaging to firmly and accurately hold the packaging during the insertion process.
3. The component level, where the manufacturing process requires facilities

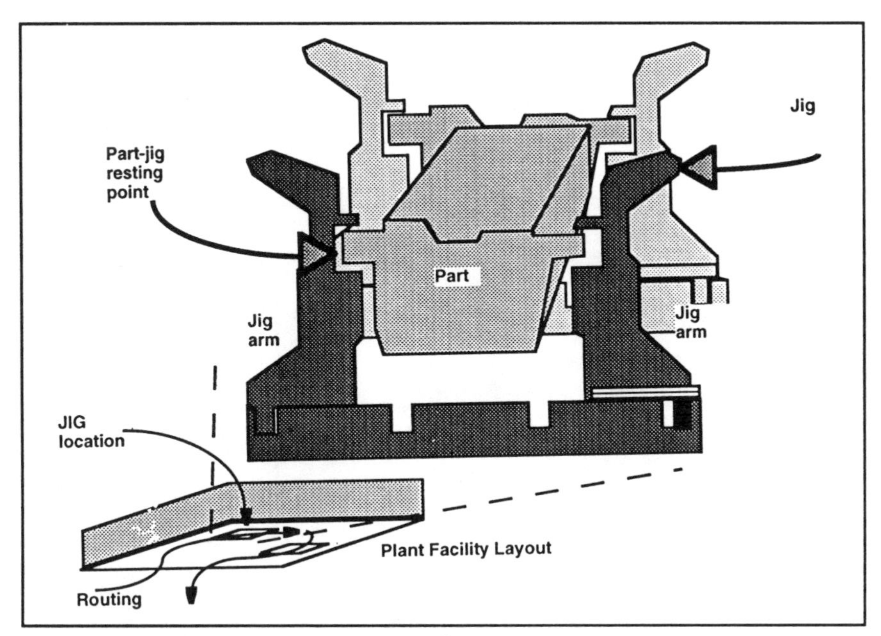

Figure 7-14. *Relating Designed Parts to Facilities.*

to prepare a wiring bundle to connect different electronic circuit assemblies. These might include long tables, predesign forms, and connectors. Connectors supplied by another manufacturer might be utilized. The design process must consider the pre-existing requirements of the connectors, the specifications of the supplied wire, and the performance characteristics and requirements of these elements of the component bundle. Other design concerns might include determining the shape of the bundle to be placed in the end-product, and translating that shape back to the table layout form. Another example of component level concerns can be seen in *Figure 7-15*. This diagram is intended to depict the dependent relationships between product, assembly, and component design intent, its nesting opportunities, and the relationships to equipment, drill bits and their composition, raw materials, scrap materials, and the many other factors to be considered during design for production as an interest area of manufacturability.

4. Second, consider business partners, vendors, and suppliers. The concern of manufacturability, when components and subassemblies from other manufacturing organizations are to be used, includes several design process-related issues: design control, shared liability, and shared customer support.
   1. *Design control*. In a complex product design it is important to be clear about who has design control over which portions of the product. It is relatively easy to manage when smaller components are involved. The outside group, which has perhaps already designed a product, provides

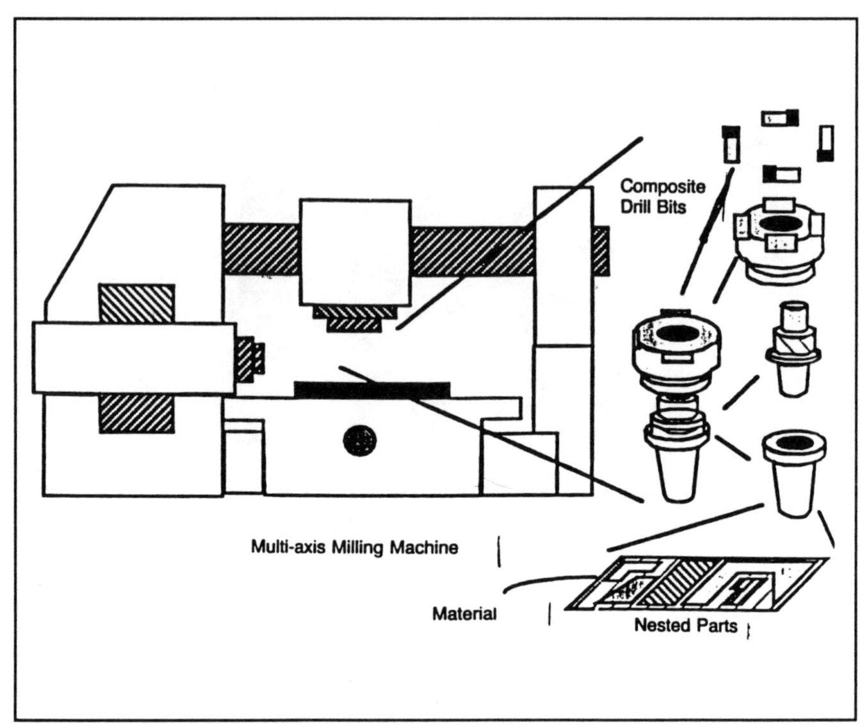

Figure 7-15. *Matching Machine to Bit to Part to Material to Design Intent to Performance.*

performance specifications. Design control over the smaller component's *use* in the end-product, however, remains with the lead complex manufacturer. Maintaining design control is not easy when significant subassemblies are involved, and care must be taken when it is a system supplied by another; usually the design responsibility is shared. Responsibility for interfaces, relationships, and other components, and inclusion in the overall physical packages stays with the lead complex manufacturer who takes end-product responsibility. The supplier provides predesigned and manufactured product which meets pre-agreed subassembly or system performance specifications. The design teams must coordinate these interfaces and performance specifications, and keep them and their associated information envelopes under configuration control. The constant communications in each direction can expose the lead manufacturing organization to the loss of design intent, information security concerns, and loss of specification control through repeated informal communications. If the system flows through multiple areas of the product, not only is there a need for shared product design responsibility, but shared manufacturability process analysis also must be jointly pursued. For example, in the case of a simple addition of another set of connectors, or the removal of a set of connectors, either may simplify assembly, but both parties must again agree on design intent, performance specifications, and use in the end-product.

2. *Shared liabilities*. Both parties may share liabilities for errors and failures, especially in the case of shared design responsibility.
3. *Shared customer support*. Designing for customer support, discussed in the following paragraphs, is also a shared responsibility.

Final end-product configuration control is the responsibility of the lead complex manufacturer under all these circumstances.

Manufacturability Interest Area Types

In addition to ensuring that the scope of manufacturability analysis includes tools and facilities, several other types of manufacturability interest areas deserve mention. These types include designs for producibility, maintenance, and assembly.

1. *Design for producibility* is an analysis process type that focuses on simplification, greater tolerance consistency, lower cost, and lower failure rate production manufacturing processes. Product simplification (part count reduction), was one of the early approaches. A sample of this simplification, produced by NCR Corporation in the redesign of a keyboard, which reduced the part count from 91 to 3, is shown in *Figure 7-16*.
2. *Design for maintenance* is an analysis process type that focuses on ease of product maintenance. An example might be placing the fuse box high up in the engine compartment of a utility vehicle so that it will not be "drowned out" as it goes across streams and through high water. This high placement keeps the fuse box dry, and makes it accessible for quick and easy replacement. Another example can be seen in *Figure 7-16*, where encapsulation reduces part count and changes the concept of the product design and the operations concept from mechanical slicing to laser cutting. These changes mean maintenance is reduced from complicated mechanical part replacement, including disassembly and reassembly, to electronic circuit element and laser tube replacement.
3. *Design for assembly* is an analysis process that focuses on the ease of assembly of various components and assemblies. An example might be maintaining very tight specification performance on certain key design points to assure a consistent fit together all the time (see *Figure 4-8*). An example of tolerance management shown in *Figure 7-16* is the move from three sigma to six sigma on the rollers in the moving box. The design concept in this case is to reduce the cumulative tolerance variance to a cumulative minimum. Such an improvement in cumulative tolerance control means that the fit will be correct each time; i.e., no returns, or during the assembly "adjustments." SPC (Statistical Process Control, a measurement process important to design for assembly,)

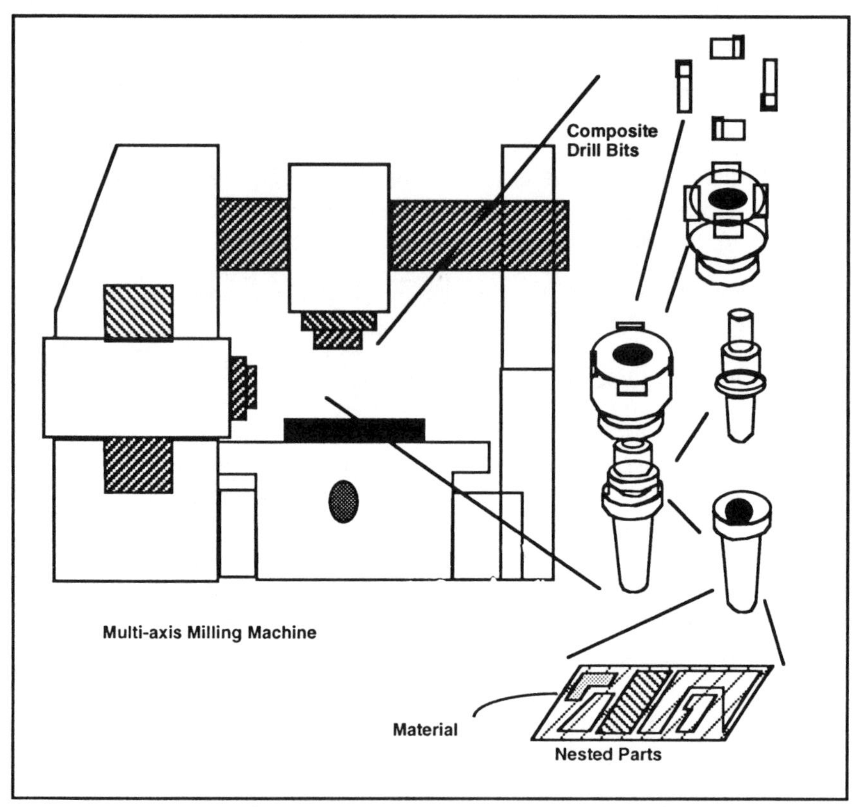

Figure 7-16. *Various designs for ....*

is the technique useful in measuring and monitoring tolerances and their cumulative variance. There are several excellent books and articles on SPC which should be of interest, including *Quality Engineering Using Robust Design* by Phadke, which describes the use of the Taguchi Method as applied to design.

Each of the "design for ..." subprocesses should contain two additional process activities:

(a) robustness analysis and

(b) cumulative tolerance analysis.

In either of the design processes, process management uses SPC to focus on process quality. Defining key features and product robustness and tolerance characteristics should be *specifically detailed* activities. In either design process type, attention to these activities focus on those aspects of the product element's design reducing sensitivity to new material variation, manufacturing process variation, and operating environment variation. It also promotes reusability and extended product life. SPC should assist in the application of these analyses. By applying these analyses in either design process type, as well as in the manufacturability subprocess, product quality will increase.

Design for disassembly. This manufacturability analysis process type focuses on the ease of disassembling the product. The sequence of assembly may be easy, but not duplicitious in the field because welding or special assembly equipment is utilized. In this analysis-type process, the focus is on full disassembly/assembly. An example is a personal computer held together with four snap-together plastic and metal panels and needing only a regular screw-driver to disassemble/assembe.

Design for utility. This analysis process type focuses on the useful life and range of use or substitutability of the product or component. Can it be designed to last long, be usable across many other products, and be easy to manufacture, all at once? This "design for ..." considers *core competencies* and related products and substitutes or the ability to substitute the design into other products, as well as for substitute manufacturing processes. GT is also part of this manufacturability analysis process type. Two examples of this approach are the Swiss Army knife and small engines. The same small reciprocating engine can be used in many different types of products, including hedge trimmers, children's toy automobiles, go-karts, lawn mowers, small boats, power generators, air pumps, etc. Being world class at small engine manufacturing means the most important element of all these products is world class also. In addition, the product strategies of variety and complexity are supported by such a core competency.

Each of the "design for ...s" can be the basis of a core competency or world class product. This is how design is having a profound impact on business success as whole.

Enhancing Product Design Information

During manufacturability, the process must produce additional information or design information artifacts for use by the production portion of the complex product manufacturing organization. These artifacts must continue to be produced until such time as all products are manufactured directly from the geometry.

These artifacts include:

1. Features, attributes, and standard and particular design notes;
2. Product, process and activity costing, and
3. Bills of materials, nonspecific and specific process plans, perhaps actual manufacturing routings, and incorporation plans and schedules.

The most important concept about information systems support for CE Design is the need to separate the information on the various pieces of paper now associated with design from the physical medium itself. Essentially, the piece of paper (either letter or drawing size) is acting as a storage and retrieval device for design information. It is not a piece of paper, but the information on which a designs resides. This concept holds true whether it is text, test data, or the actual geometry associated with the design. Once the design is separated from the physical medium, all types of important things can happen. The computing architecture and information system support described in Chapter 8 are based on this separation of information from its physical medium.

Features and Attributes

Key features and important attributes are elements supplemental to the design on a separate piece of paper, whether added as a note, or shown as an additional piece of information in the geometry.

If engineering organizes the CE subteams according to the WBS outlined in *Table 7-1* through *Table 7-3*, this provides for substantially improved support for any of the design approaches selected by the CE Design management team. Each of the "design for ...s" subteams must coordinate changes in a net change fashion. That is, each change to product designs going through the design process must be annotated with rationale, and indicators of which other interest areas should review such net changes.

Managing key features and their attributes is critical to this coordination effort. Key features should be maintained. Changes to key features should result in immediate notification of potentially affected subteams: for example, changes in tolerance might result in completely different process plans, or changes in dimensions might result in the need for interference analysis. Changes in attributes, such as material choice or finish, may be incidental or important. If specific attributes are considered high priority, they should have the same high visibility afforded key features.

In an information system supported environment, such as the one described in Chapter 8, these features and attributes are added as additional folders, packets, or objects of information linked to the basic design through indices, or physical pointers. This information becomes physically separated from the paper, but is still logically linked by the computing systems to the geometry, and to the rest of the design information necessary to completely communicate design. This approach permits the separate, but coordinated capture and updating of this information. Changes to the geometry portion of the design could, for example, result in automatic notification of changes to the key features and attributes.

Notes and Standards

The total set of design information includes a variety of additional information beyond geometry, product specifications, key features and attributes, and the nonspecific process plans for each of the components and assemblies described in the design. Instructions for those who will do the manufacturing, and information related to the design are also assembled and should be included in this information set. Of particular interest are design notes that describe key features. These are of particular interest to the manufacturability process, as they drive aspects of the different "design for ..." analyses previously described. Finally, design notes about key features can illuminate those aspects of the design changed or enhanced for "design for ..." reasons. Examples include adding the jig holding attachment points as shown in *Figure 7-14*.

Standards, in such areas as materials, processes, performance criteria, and schedule information, are also important to the manufacturability process's execution. There are number of types of standards in overall manufacturing that affect product design, including design documentation standards, material standards, drawing, or model content standards, standard notes (such as the company's copyright and ownership statements) and process standards. The standards of interest in manufacturability as an element of CE Design include those that affect the effective execution of CE Design, and their notation as an element of the design information set, which accumulates during the design process.

From a note perspective, the development of standards used in the product design should include an information system to maintain standard notes for quick access, uniformity and universal accessibility, and cross-reference. Any information system containing the design artifacts should also be indexable by the developing set of design information through geometry notes. These notes are added to the geometry information in the information store as they are developed throughout the design and manufacturability processes. Notes can be contained in the geometry itself, or added as cross-referenced items in the information store. The difficulty is that when they are part of the image in the geometry, the CAD system may not be able to tell a note part of the image from the part image itself.

It is difficult in the abstract to determine what the impact of standards in an individual complex product manufacturing organization will be on the design process. However, there are several key issues to analyze during a CE Design implementation process, including design documentation and process plans.

1. *Design documentation*. This documentation is usually prepared by specified groups and individuals. During the implementation process, these assignments should be reviewed to determine if they conflict with specific process models developed for the conversion. An example of such a conflict is the assignment of a draftsperson to clean up the model after its initial preparation by a more experienced design engineer during the step-wise refinement design approach. Such a task is good training, and utilizes less skilled personnel for less demanding tasks, but it would be important for the design, in that state, to be first reviewed by the manufacturability teams and the design engineer. Suggestions would then come quicker and involve less rework by the design groups.
2. *Process plans*. Usually, nonspecific process plans are prepared by the design engineer. Subsequent detail is provided to the nonspecific plan as more understanding is developed by the manufacturing engineering and planning groups. Over time, many sets of nonspecific process plans can be developed. Defining and using standard process plans (indexed and retrievable by area of interest and other keys), and encouraging reuse, are important elements of the benefits of CE Design.

Product, Process and Activity Costing

Costing information is an important part of the design information being developed. As discussed in Chapter 4, the costing process for CE Design utilizes the Activity Based Costing technique, or ABC. This is convenient because activities are the executable elements of the design and manufacturability process in-line models and take advantage of the extended capabilities of Routing and Queuing Process Management as discussed in Chapter 5.

There are several categories of costing which should be considered during manufacturability. These include:

- CE Design Process Costing.
- Production Manufacturing Process Costing.
- Trade Studies.
- Product and Material Costing.
- Measuring and Demonstrating Success.

CE Design Process Costing. This costing category considers the cost of conducting design itself and includes all the costs associated with the execution of the design process in-line model, including people, machines, management, and other resources consumed. As the activities of these processes are executed, the costs associated with their execution are captured via their activation and conclusion.

Production Manufacturing Process Costing. As the process plans to produce the product are developed, the costs of executing these processes are also captured and calculated. This permits the manufacturability portion of the matrix design team to consider process costs, cost, quality, speed, safety, and other issues in their comments to the balance of the team, as well as communicate their resolution to the production teams.

Trade Studies. These cost analyses are performed by various subgroups within the design team and sometimes between teams and/or groups of teams. Trade studies are comparative cost analyses. These are especially effective during the collaborative, concurrent design process, because they can permit the quick comparison of various design options before they become "too finalized" and difficult to reverse. These studies take advantage of other aspects of costing improved by the ABC approach, including revised process cost, material and product costing, and marketing and finance costing information from a customer perspective.

Product and Material Costing. Comparative product costing is the type of costing for which ABC was initiated. There are a variety of computerized tools and associated manuals, seminars, and tutorials on how to conduct these product costing analyses.

These studies, performed on existing product families, many times produce results which indicate that the products, costed and based on "burdened labor hours," are substantially incorrect in their costing. Many complex manufacturers with large nontouch operations, designed to handle increasing complexity, believe that they cannot manufacture simple components on a cost-effective basis as compared to small, focused manufacturers. When ABC methods are used, the results typically look like *Figure 7-17.*

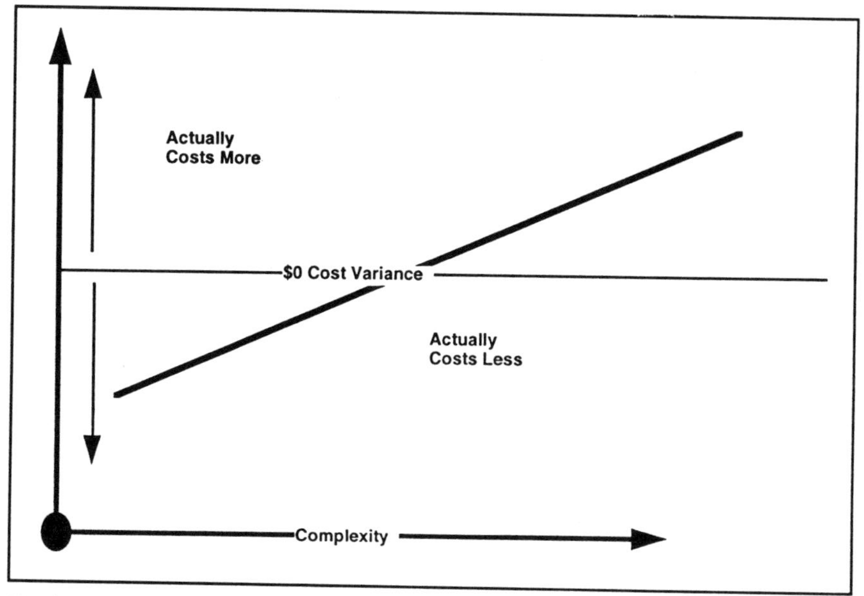

Figure 7-17. *Costing Distortions.*

After the overhead allocated to each hour is dissected and costs are reallocated on the basis of activities using the ABC methodology, as shown in *Figure 7-17*, most organizations find that their simple components are overcosted, and most complex components are undercosted. The organization is typically outsourcing components on which it can make money and not addressing cost improvement opportunities on more complex components for lack of priority.

Once this relationship is understood, the elements of the more complex component's cost can be addressed and reduced. Advantages of The ABC technique include:

- Deals with cost issues before inappropriate costs are incurred;
- Provides more accurate information for trade studies, design and process alternative evaluations, and potential market/competitive product comparisons;
- Fits into the CQI improvement scheme by supporting the small, incremental improvement methodology with a more reliable method of forecasting improvement impacts.

One concern about conducting this type of analysis for complex products is the ramification of these studies if the complex product is, or could be, part of a government procurement, or is performed in a facility using similar processes for which a government product is produced under contract.

The problem is one of disclosure. If the organization knows that its costing techniques are incorrect, it must disclose these to the Defense Contract Administration Agency (DCAA). For new products, this is not a problem. It is when existing products, already under contract, are considered that a problem might occur. New allocation approaches, cost goals, and product costs and profitability analyses may also be required. Yet, it is this aspect of costing which is the most important to CE Design.

Measuring and Demonstrating Success. One of the challenges of CE Design is demonstrating convincingly that, as the organization proceeds through the transition from today's environment to the vision of CE Design, it is constantly worth the substantial effort and change required. One of the advantages of the in-line model-driven approach is the ability to collect activity-based costs, as described previously. Once the raw information is collected at an adequate level of detail, a set of previously researched and agreed-to performance measures can be used to gage progress and demonstrate success.

The basic performance measure for CE Design, and the manufacturing organization as a whole, is profitability. The issue is how CE Design can be an integral part of taking the organization to profitability. *Figure 7-18* depicts a

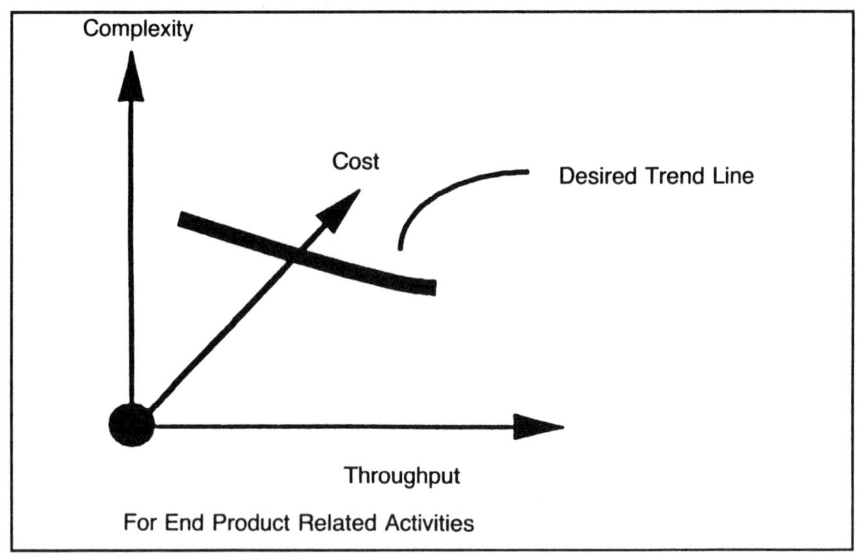

Figure 7-18. *Basic Performance Measurement.*

basic CE Design performance measurement graphic. In this performance metric approach, the objective is to increase or maintain throughput, while lowering or maintaining costs, in the face of increasing complexity. Reducing complexity should result in lower costs and higher throughput, if practical. The basic composite measure between complexity's increase and the composite vector of these three factors is the Δ. Each of these three measurement vectors should have its own measurements, goals, and objectives.

Each organization should establish overall organizational and process goals. Departmental goals which support the overall goals and do not reinforce intra-organizational competition should be developed and coordinated. Individual unit goals within processes which support overall organizational processes, and departmental goals are permissible. Some example measurements at each level include:

- Organizational Level
  1. Active part number count.
  2. Active end products.
  3. System versus nonsystem component ratio.
  4. Order to delivery time period.
  5. Time to change incorporation.
  6. Aggregate product cost and revenue.
- Process Level
  1. Aggregate process cost.
  2. Time to release versus plan versus model experience.
  3. Incorporated change rate.
  4. System component count.
- Unit Level
  1. Aggregate product revenue contribution.
  2. Product requirements satisfaction level.
  3. Average queue size and clearing time by type of workstatement or product area of responsibility.
  4. Variety of change handled index.

These measurements, goals, and objectives encourage change and solidify support for CE Design because when they are integrated into the management processes and routinely describe improvements and progress, they create enthusiasm for more.

Bills, Routing and Schedules

At some future time, all production manufacturing will be directly derived from geometry plus instructions using some type of intelligent software. Until that time, the manufacturability process, as the design process, must produce information artifacts which communicate to manufacturing what needs to be built (BOM), with what process (routing), and to what schedule (JIT, MRP, TPPS, CDS, OPT).

There is an active and aggressive industry focused on providing information systems in this area of manufacturing. Many use Joseph Orlicky's book *Manufacturing Resource Planning* as their basic starting point for scheduling and routing functions. There are now a set of differing philosophies based on the relative success of each of these competing ideas (MRP, JIT, OPT, and TPPS).

These techniques are depicted in *Figure 7-19* and discussed in the following paragraphs. Each of these has good capabilities to offer to a complex manufacturer. The general situation is that different elements of the complex manufac-

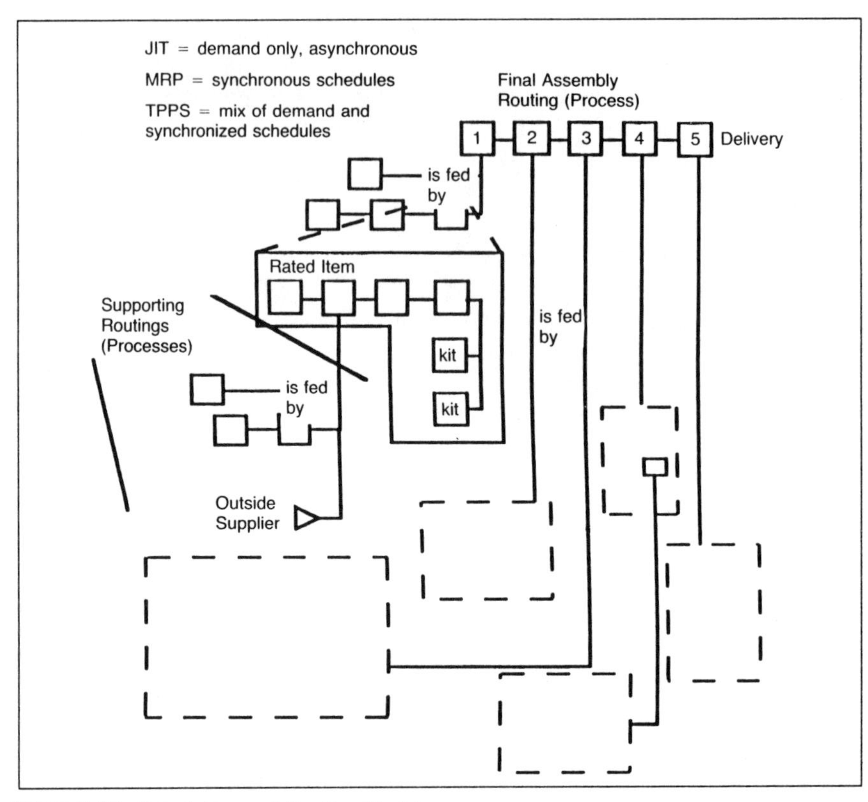

Figure 7-19. *Scheduling and Routing Models.*

turer are best served by each of these manufacturing management and scheduling approaches. The discussion which follows only focuses on the interrelationships of these different management approaches and CE Design.

Manufacturing Resource Planning (MRP) requires high discipline in all related activities. Highly effective MRP also is dependent on a very accurate bill. This accuracy must substantially exceed 99.997% for the complex product manufacturing organization to reach a high achievement status. As mentioned in Chapter 2, in a complex manufacturer the target composite error rate is now 0.0010% or 10 errors of some type in every million opportunities. Because of the need for a highly accurate bill, good MRP discipline is critical.

MRP is also batch oriented. This batch orientation can be deceiving because batches can be one or more. The fact that MRP is described as focusing on batches is really more about the "push" which comes from MRP's issuance of work orders, schedules, and the expectation that these schedules will be adhered to. These "push" control activities are similar to work statements, in-line process models, and the like. Thus, there is considerable affinity between CE Design and MRP, although CE Design actually better fits other techniques.

The intent of CE Design is to produce highly accurate, more timely product definition driven manufacturing information. MRP and CE Design operate off schedules and model-driven processes, and are integrated with other elements. MRP permits the final assembly process to drive procurement and subcomponent manufacturing and assembly in an overlapping fashion. MRP, like CE Design, is scheduling based on forecasts, experiences, and expectations.

Just-In-Time (JIT) utilizes a "pull" to the final assembly processes' rate requirements philosophy. Components and subassemblies are consumed during final assembly. This consumption pulls on feeder processes to replace the items consumed so production can continue. As the demand moves backward through the various feeding processes, more of each item required to fill the next upstream process is completed. There is no paperwork to be filled out, or schedules to be kept.

JIT fits well within certain elements of the complex manufacturing process. For example, subcomponent assemblies built at a particular rate, with change added incrementally (using the "as of a date" or "as of a serial number" approaches to configuration control) are potential opportunities. JIT has problems with high change rate items, long lead time items, or items better manufactured in batches due to unusual circumstances, such as "mix and use or lose" chemicals, processes with high setup or other costs, or when imaginative solutions to the need for batching cannot be found.

The CE Design process is creating something new, i.e., a "push" to a scheduled process of manufacturing design and manufacturability information, as opposed to drawing on something already created. One might think that there is a difference in philosophies between JIT and CE Design. It is not clear that there is a difference, or if it matters. The basic orientation of change management in the CE Design environment is very similar to JIT. Resources, authority reviews, and design changes are "pulled" by production manufacturing as an element of the communication between production manufacturing and design. When changes are required, based on an improvement or problem encountered on the shop floor, engineering responds in a pull fashion using pull oriented scheduling.

JIT pull processes can also be adopted to manage the customer specials and special bills per final assembly end item, via incorporation points for products promised to customers. These "specials" pull the generation of the needed design and production manufacturing information to be produced by the CE Design process and must accommodate overlapping schedules. TPPS (Time Phased Procurement Scheduling), is, in reality, a "modified" JIT or "pull" scheduling technique.

TPPS often appears in complex product manufacturing operations as "waterfall" schedules. The schedule looks like a waterfall because each end product's final assembly sequence is fed by a series of supporting assembly sequences which tie to product areas and their associated bill of materials levels. The rate of final assembly is set, and the supporting number of processes are established. Final assembly demand pulls from the support assembly sequences, which pull to suppliers and other internal manufacturing. To achieve flexibility in customer ordering, many components are planned and orders for these pushed to suppliers and internal operations in anticipation of the customer order. TPPS is a mix of JIT and MRP as the situation dictates.

The TPPS final assembly rate is based on a combination of backlog and anticipated demand. The rate is published as a date driven schedule, and CE Design's activities, including incorporation (via the CDS as in Chapter 5) are then scheduled into this rate. Customer specials are set up by using end item or block-oriented effectivity to pull standard and unique requirements through the manufacturing processes. TPPS uses schedules to force changes to meet up with the ''rated'' items at the correct time. TPPS works well with CE Design. TPPS seems to offer the best of MRP and JIT at the same time for the complex manufacturer.

Optimized Production Technology (OPT), developed by Eli yahu Goldratt, is based on a set of nine principles which focus on scheduling. One of the key points made by Goldratt was that a ''balanced'' factory—where all resources are being utilized optimally, at full capacity, and to the shortest combination of elapsed duration—is impossible (see *The Goal* and *Theory of Constraints*). This is impossible because a single bottleneck can throw the rest of the balance off, and it can never be recovered because it is a bottleneck in the first place. Instead, the focus should be on finding where there are bottlenecks, and on priorities and improvements in production in those bottlenecks. The use of Load Centers (control points for work centers which stage and release until completion for ''push'' jobs) to constantly improve throughput through optimized flow at the bottleneck, or constraint points in the process, is a technique which organizationally facilitates the use of OPT techniques.

In addition, OPT principles can be applied to the routing and queuing in-line model scheduling of design and manufacturability processes. Since step-wise refinement is a serial process model, OPT scheduling rules can apply. Bottlenecks include vertical authority processes and various resource constrained activities within the serial process.

The collaborative, concurrent design process is also affected by OPT principles. If there is a problem with collaborative, concurrent CE Design, it is the highly interdependent collaborative approach with which it approaches the development of the design and its manufacturing information artifacts. Various circumstances such as encountering serial processes, resource availability, and error detection and recovery all mean that dynamic scheduling adjustments are required. OPT-oriented scheduling, which attempts to resolve all these issues, is a good tool in this situation. Adding OPT-type scheduling algorithms to the schedule analysis portion of the in-line routing and queuing model management automated infrastructure support capability is preferable to the backward scheduling, critical path method (CPM) currently used, and is of significant benefit to the collaborative, concurrent design model.

Bills of Materials

The primary means of communication through design's manufacturing information artifacts, is the bill of material. The "hand off" between the matrix design team and production manufacturing has been hampered by two significant phenomena.

The first is the different view each takes of a product. As depicted in *Figure 7-20*, design takes an "outside-in," "whole-to-part" or WBS, top-to-bottom

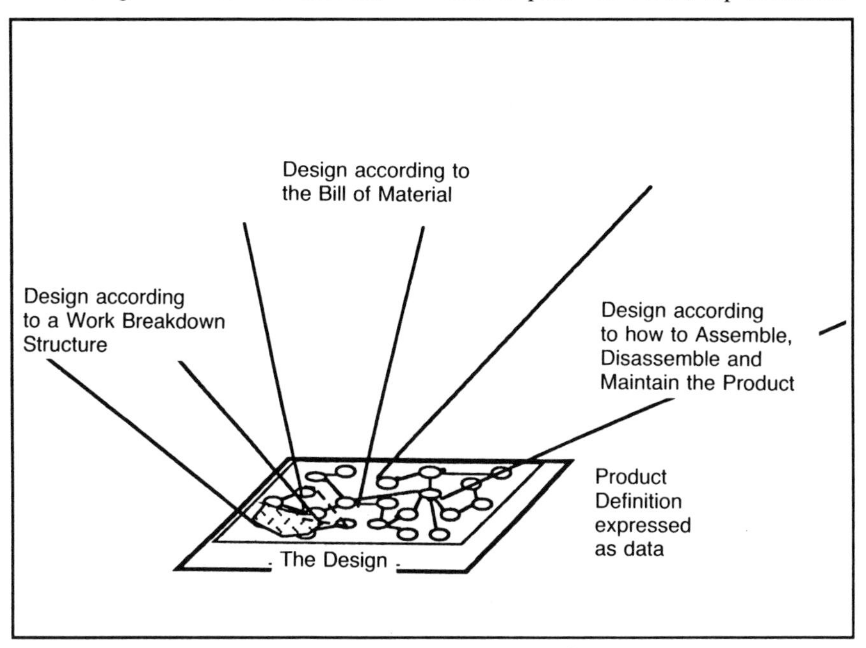

Figure 7-20. *The Design Has Many Different "View" Requirements.*

view. When the product is conceived, it is thought of in a holistic fashion first. This is why finally enabling the collaborative, concurrent design approach with CE Design is so powerful. It is natural.

Production manufacturing, on the other hand, is bottom-up oriented (what part numbers are assembled in what sequence to "roll up" into this particular product). To get to the end product, one must first get the smallest, detailed parts together, and then build larger and larger assemblies until the end product is produced.

In addition, marketing, customer service, and maintenance support take different views of the product. These views are oriented toward their interest in the product's characteristics of operations, assembly and disassembly, and servicing.

Traditionally, there are several bills of material in a complex manufacturing environment:

1. "As designed" bill, which the engineer/designer (or more likely an

assistant) prepared after the design was complete. The part numbers are assigned once the design is stabilized, except in those cases when existing parts are utilized.

2. "As planned" bill, which reflects how manufacturing engineering and production planning see the product being actually manufactured. Typically, there are "phantoms" (nondesign subassemblies created for ease of manufacturing) and requests for temporary parts which all result in this bill being different from that envisioned by the designer. Many times, the CAD data set of the design is enhanced to include this additional information.
3. "As built" bill, reflects all the minor changes made as the product was manufactured, and shows accumulated substitutions, minor "fit adjustments," incorporations.
4. "As supported" bill, may (if good control with the customer is maintained) reflect all the field maintenance, enhancements, upgrades, and modifications made to the end product after it is first delivered to the customer.

The second phenomenon affecting this communication is part numbers. Designers, in the abstract, do not worry about part numbers until after the design is completed. They are an incidental byproduct of the design process. Manufacturers, on the other hand, want part number-oriented information, in detail, as early as possible. How many? Are some of them already being manufactured? What is the make/buy ratio? The answers to these questions may affect not just schedules and responsibilities, but jobs and careers. Eliminating parts may also eliminate jobs.

CE Design addresses these issues aggressively through the manufacturability process. The end result of the CE Design process is an as planned bill of materials with part numbers firmly understood. More attention has already been given to the use of existing parts, tools, facilities and the like because all of considerations are intrinsic to the manufacturability process. Additionally, the automated infrastructure support necessary to support CE Design facilitates the improved operation of the as built and as supported views of the product.

The ideal situation is to have a composite bill of material for *each* individual end product. In the complex manufacturing environment (capital ships, commercial airplanes, large buildings), each end product is unique in some way. From a practical perspective, it is only in the last few years that computing power has grown to the point that it can contain such a large composite bill. The PDES/STEP standards effort intends to provide a database view of the complex product which supports these various "views" simultaneously through the use of "contexts of use" descriptions.

One of the important issues affecting the bill of material in a complex product environment is how to construct the bill to reflect options and customer specials and to facilitate their identification with block effectivity. Block effectivity, a Marker type, was discussed in Chapter 5 as it affected the design and its information artifacts. Once the design and its part numbers are established, the Bill of Materials is created. As changes occur, however, both the design, its WBS, and the associated bills must be modified accordingly.

Referring to *Figure 7-19* and *Figure 7-21*, one of the other principal

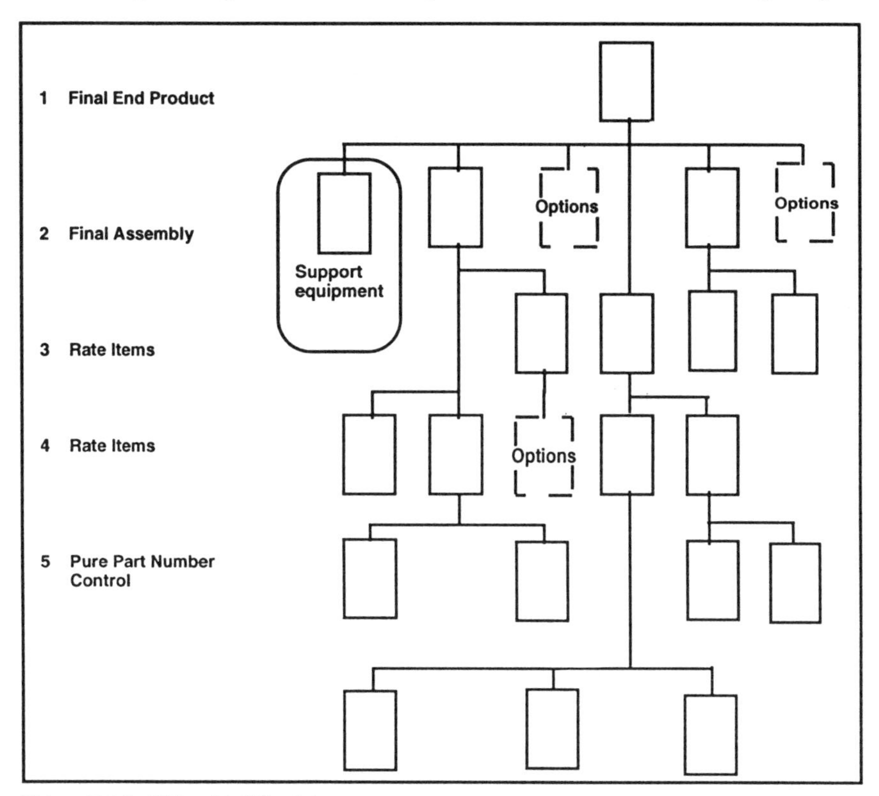

Figure 7-21. *Bills with Effectivity.*

objectives of the manufacturability process must be to organize the product, from a manufacturability perspective, during the design stage, to break down the product to match the TPPS view of the product's manufacturing activities, and to organize the bills to support this view. This organization prevents very complex ambiguous activities, such as phantoms, from occurring in production planning.

In this approach, all final products, final assembly subcomponents, and rate item subcomponents are controlled by effectivity. The relationship between more detailed elements of the product are associated with the product by effectivity. The cross-reference boxes in *Figure 7-21* show this effectivity controlled relationship. All options are designed to be effectivity controlled as well. However, all subordinate elements below the rate item level are not affected by effectivity. A change in these items causes a new part number to be assigned.

This approach minimizes the confusion that would ensue if effectivity was not used. Every change would be a new part number and the end item part number would be constantly changing, or changes would require effectivity at every level, which would create substantial complexity throughout the organization instead of in those few controlled, manufacturing areas associated with final assembly. This approach substantially reduces the amount of computing resource required to do explosions (accumulating production requirements by part number) and implosions (identifying where in the product each part number is used) of the bills as well.

The manufacturability process brings production manufacturing into the design so they and the designer operate on a co-equal, collaborative basis. This is one of the most important aspects of CE Design. Manufacturing and engineering develop a ''win-win'' working relationship.

SECTION III

Concurrent Engineering Design Architectural and Implementation Framework

In this third and final section of the book, the Automated Infrastructure Support necessary to fully realize the potential of CE Design, and an Implementation Strategy and Plan for CE Design are discussed.

Chapter 8 discusses the functions and operating characteristics of computerized support for CE Design. As shown in *Figure III-1*, Process to Infrastructure Relationships, these computerized support requirements affect all levels of the CE Design business processes. *Figure III-1* is intended to show how process architectures drive, yet reflect the computing architectures which support their execution. This is particularly important today and in the future as the computer continues to become more and more critical to process execution and cost effectiveness.

The major issue facing complex product manufacturers is how to evolve into CE Design while maintaining and improving today's products and activities. In Chapter 9, an implementation strategy and sample executive level plan which intends to enable this evolution are described The *Figure III-2* series points out that CE Design, in its fully concurrent and finally implementable form, provides 100 or more times return on investment and is thus a highly desirable goal for the complex product manufacturing organization. This is because CE Design is made up of important, but lesser value processes.

In *Figure III-2A,* the functional systems, which provide computing support to a function within a process, return is significant, but requires at least several years to recover its investment because it is so dependent on other processes for its overall impact on the cost and profitability of the product and the organization.

In *Figure III-2B,* various functional systems are combined into various complete processes such as purchasing(representative functional systems for purchasing include buying, accounts payable, *n* .

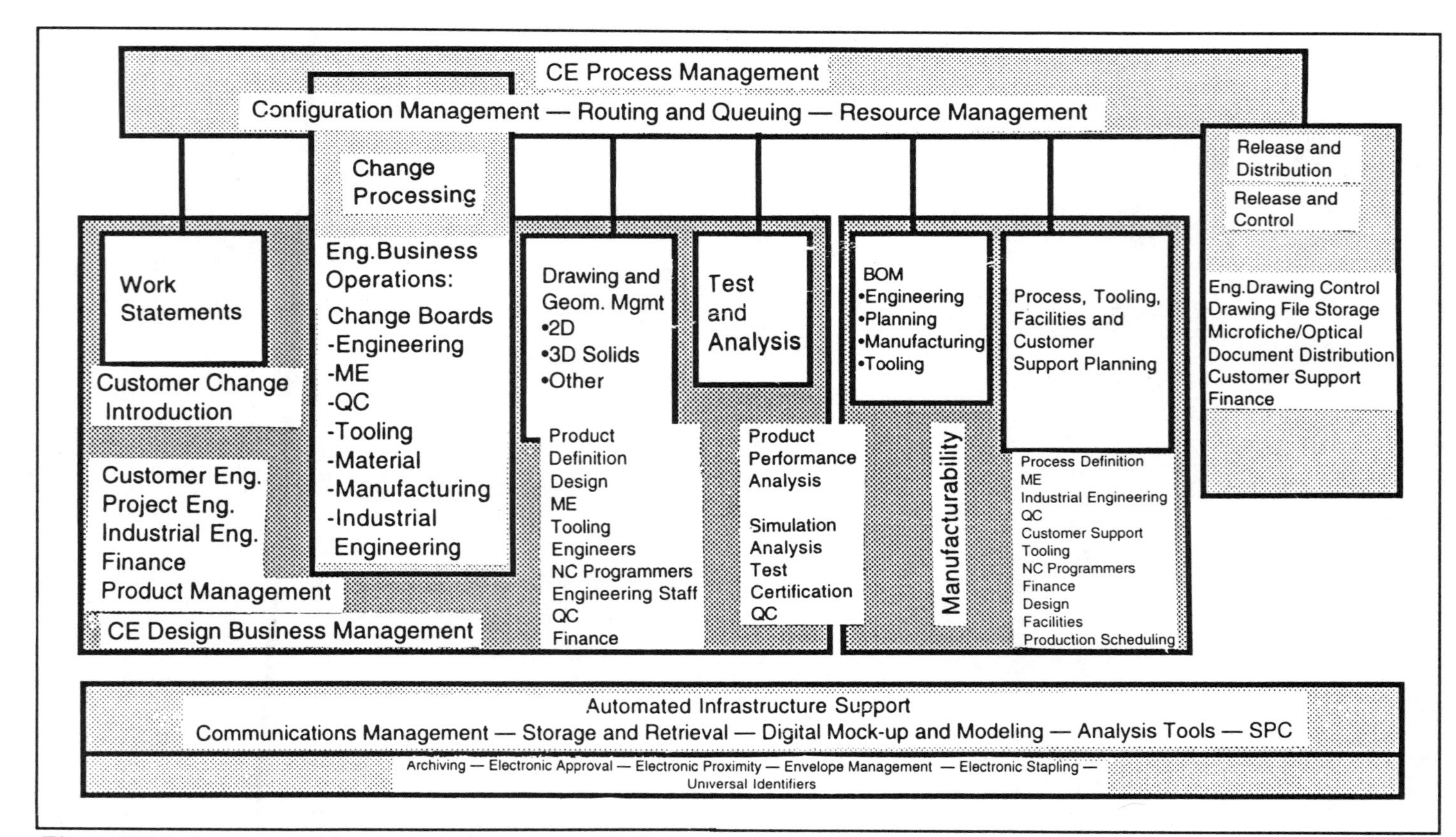

***Figure III-1**. Automated Infrastructure Support for CE Design.*

In *Figure III-2C*, several of the major process are combined into a process management level cross-functional process such as CE Design. It is at this point that tremendous leverage can be achieved if the implementation can be successful. the issue has been that integration at this level has rarely been achieved using traditional architectures. This section of the book is about the architectures needed to support CE Design, and the implementation approach and plan necessary for successful CE Design introduction.

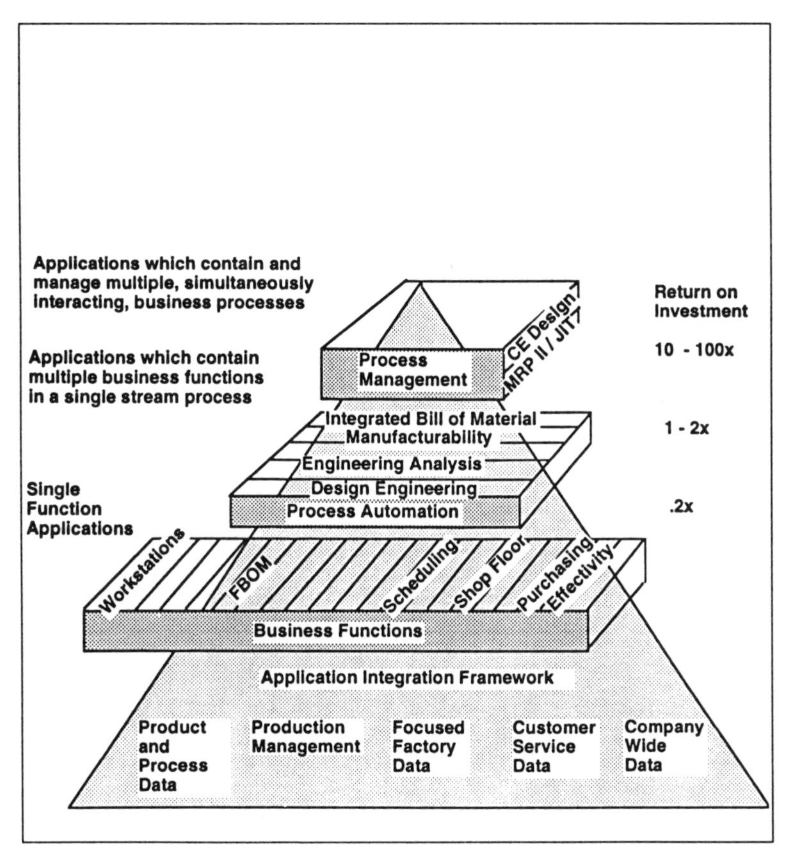

Figure III-2C. *Application Framework and Relative Return On Investment by Level.*

receiving, claims, cash flow management). This level of integration yields significant benefits if this implementation can be achieved. This represents the typical "best case" of integration in most firms because the architectural approach taken is inappropriate for the next level of integration and process management.

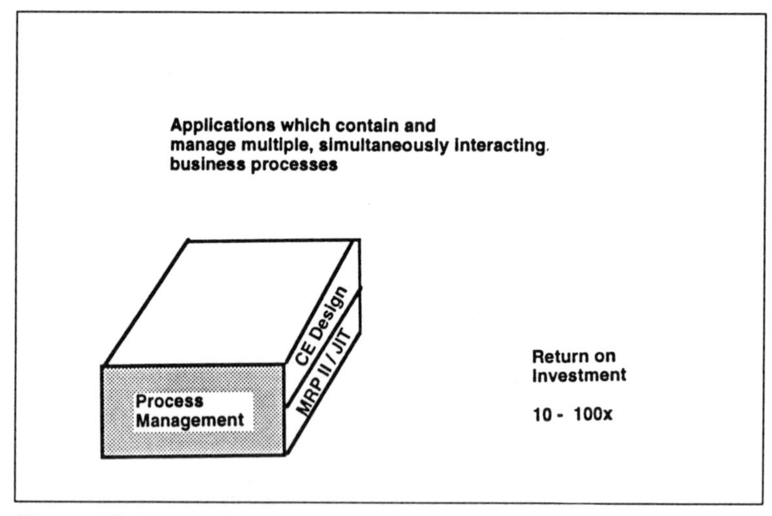

Figure III-2A. Application Framework and Relative Return On Investment by Level.

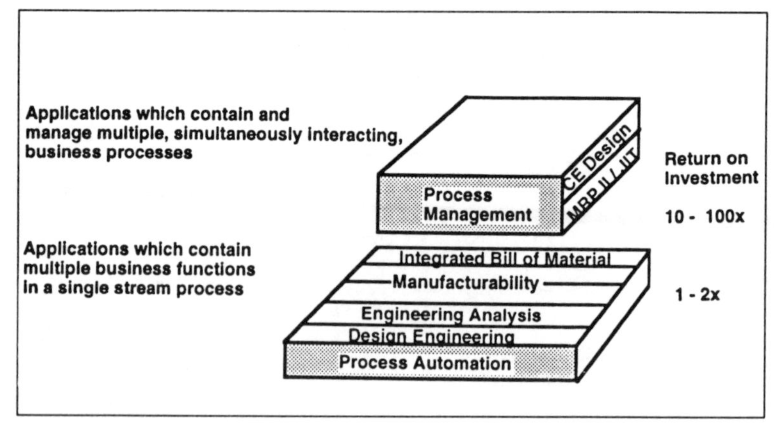

Figure III-2B. Application Framework and Relative Return On Investment by Level.

8

CONCURRENT ENGINEERING DESIGN'S AUTOMATED INFRASTRUCTURE SUPPORT

The CE Design process has several objectives:

1. To facilitate completion of product design;
2. To preserve knowledge and experience gained with each design;
3. To produce information artifacts necessary to communicate the design to manufacturing, and
4. To achieve these objectives at lower overall cost, faster, with greater accuracy, and with resulting products of higher quality and profitability.

Reaching these objectives prepares the complex manufacturing organization for world-class manufacturer status and moves the organization to a new, higher plane of competitiveness.

However, these objectives cannot be achieved, nor can the processes described in Chapters 5 through 7 be implemented without automated infrastructure support.

AUTOMATED INFRASTRUCTURE SUPPORT

Automated Infrastructure Support for the other CE Design processes is shown in *Figure III-1*. Most of the CE Design processes described in Chapters 5 through 7 are implementable on a manual basis. However, for portions of each process, the cost would be impractical. For example, 3D solid CAD is a computerized design tool which replicates the solid model mockup. This model usage would be impractical without automation. A number of aspects of these processes *require* automation. An example is the execution of the collaborative concurrent design process across multiple, matrixed design teams discussed in Chapter 6.

The business, technical, and managerial processes described in previous chapters operate most effectively when the information utilized and produced is:

1. Captured once;
2. Captured as an integral part of the process;
3. Captured "in the loop," or as it is generated, and not later as a recording of prior events;
4. Part of a single integrated view of information, at least from a user perspective;
5. Available concurrently to the number of users necessary for that problem set; and
6. Reusable across the various processes.

The *Figure 8-1* series shows a single flow of information ranging from the initial capture of customer requirements, through release of product and process definition, to manufacturing and its subsequent use. Information is stored in the product definition information envelope, which grows with additional input.

Figure 8-1A represents activities occurring during traditional design engineering. The processes which originate this information have been described in Chapter 6. Customer engineering information leads to the generation of work statements that authorize and document the processing and management of either initial product development (IPD) or changes to products. They also document the schedules, resources, and activities required to complete the design-related work. These might typically include design analysis and testing. All of these

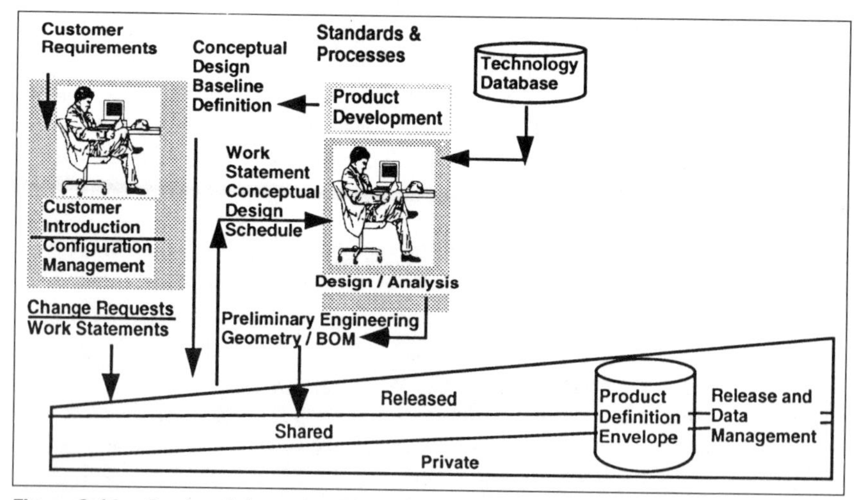

Figure 8-1A. *Product Information Flow; Design Engineering.*

process activities should constantly add to the product information being accumulated as design proceeds. If a change is involved, adds, deletions, and changes of information are occurring.

Figure 8-1B adds the manufacturing engineering type of activities to the process. These activities also add to product information. The additions are usually in the form of enhancements to previously generated information originated in engineering design, but also include new information. Because of the process, this information results in changes in originated information from engineering design. The activities of Chapter 7 occur during this portion of the information buildup.

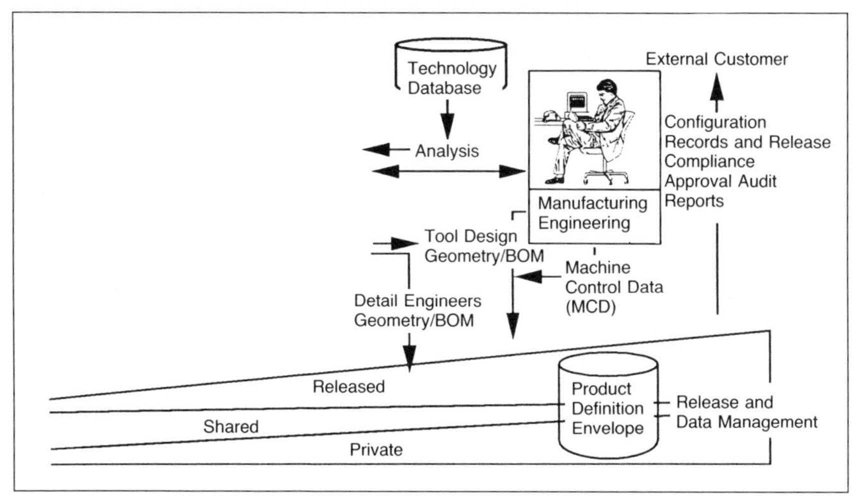

Figure 8-1B. *Product Information Flow; Manufacturing Engineering.*

Figure 8-1C adds many of the management activities needed for production. During the activities of *Figure 8-1A* and *Figure 8-1B*, the originating information is mainly product-related. Several of today's Product Definition Management Systems (PDMS) are focused on storage and retrieval of product definition geometry and some of its associated data. This information is originated in design engineering. The management activities and information are not directly product related. They utilize portions of the information generated for different reasons. For example, the material hardness called for by the designer causes the manufacturing engineer to analyze available materials to find an acceptable hardness. Engineering management would consider the hardness and the material call-out during a product cost review to determine if cost savings and quality product performance could be achieved.

Process Management activities, described in Chapter 6, utilize the information generated in all areas according to the processes identified. In addition, the information used to manage the processes also is used to enable local management to manage itself against predefined objectives.

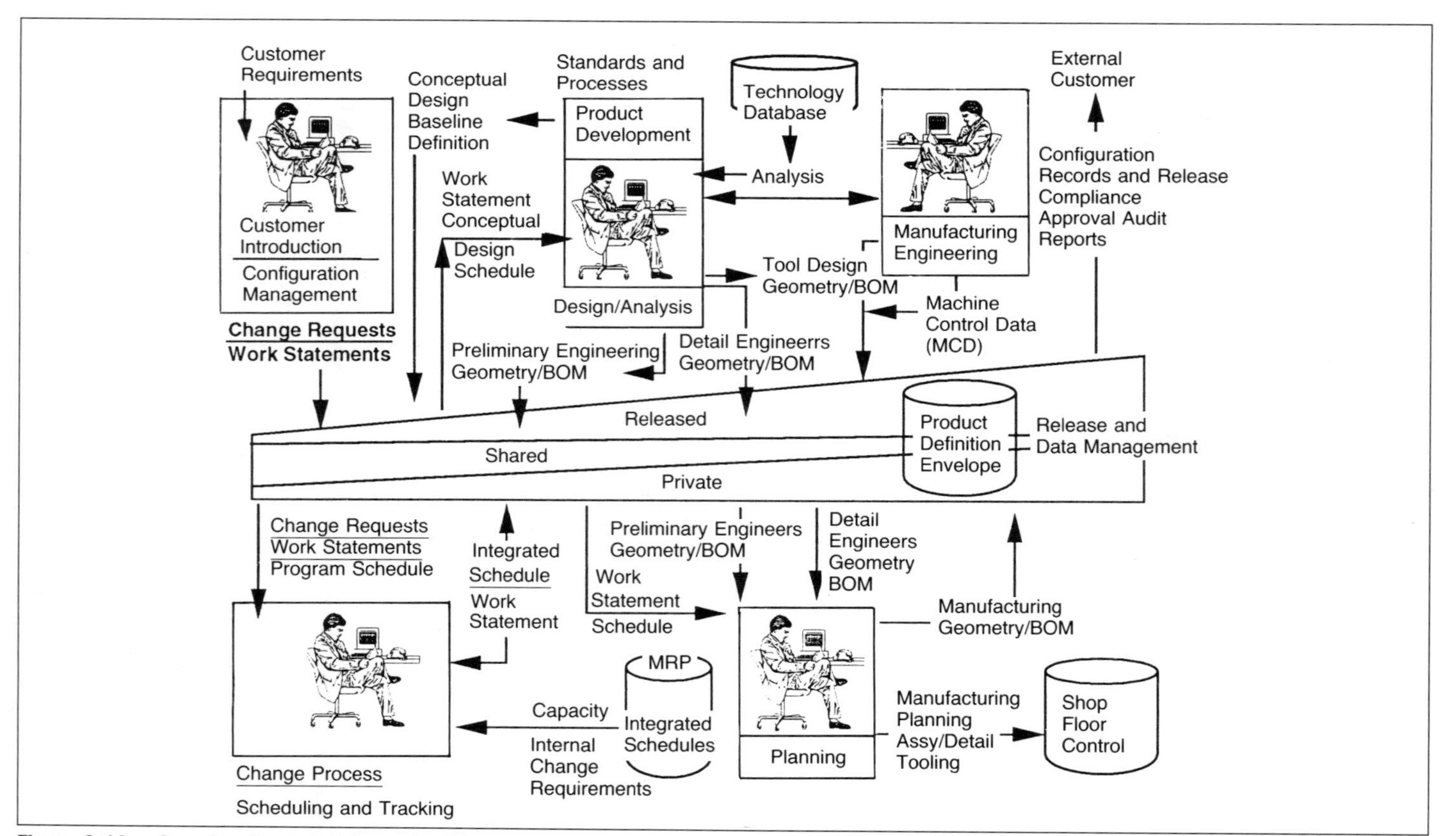

Figure 8-1C. *Complete Product Information Flow.*

DIFFERENTIATING FUNCTIONAL CHARACTERISTICS OF CE DESIGN

From an Automated Infrastructure Support perspective, CE Design has three fundamental operating characteristics which separate it from production manufacturing processes, such as MRP II, and those supported by accounting or finance systems, which are highly proceduralized and structured. Basic characteristics of CE Design are different and reflect the nature of its underlying processes. These characteristics include *Dynamic, Unstructured Work*; *Complex Information*; and *Loosely Coupled, but Interdependent Processes*.

In *Dynamic, Unstructured Work*, various functions within each process are structured, but the sequence and mix of the activities vary in actual use during each iteration of the process.

In *Complex Information*, the processes take in information (multimedia text, data, voice, graphics, CAD, etc.) used in schedules, resource allocations, budgets and analyses, engineering analysis, conceptualization, and impact analysis.

A third characteristic reflects the ability to work in an environment which assists in managing activities. In loosely coupled, but interdependent, processes, each process is chained to another. All share or use information, but each is highly interdependent across multiple design teams across multiple physical locations.

These three characteristics vary so from traditional computing architecture that CE Design is considered new territory.

CE DESIGN'S COMPUTING ARCHITECTURE AND AUTOMATED INFRASTRUCTURE SUPPORT

It is important not to confuse what appears to be a single flow of information with a single database. The basic architecture required to support CE Design can be described as a *distributed, integrated architecture* which may not have even a traditional mainframe involved, as the decision is up to each individual organization. The nature of the basic CE Design process characteristics and the process activities they represent is impeded by hierarchically driven systems.

The balance of this chapter describes, for CE Design, the general requirements and design of the application architecture, information (data) architecture and its computing hardware, system software, and communications/networking environment from a user/manager perspective. This support environment in time provides an increasingly smooth environment in which the "single view" of information and its flow, as depicted in the *Figure 8-1* series, can eventually be simulated operationally.

INFORMATION CATEGORIZATION

In the complex manufacturing organization, the activities which comprise CE Design capture, manipulate, and present different types and uses of information. *Figure 8-2* depicts a sample of the information variety in CE Design.

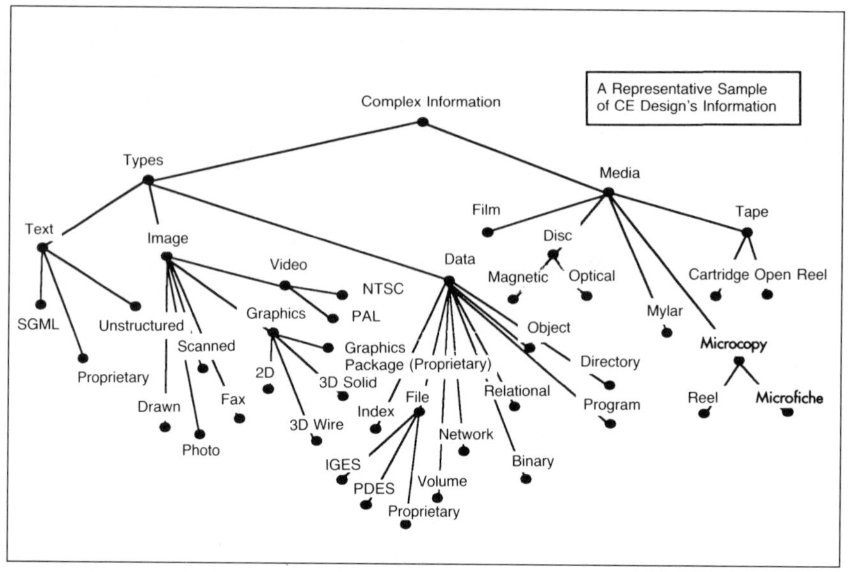

Figure 8-2. *Complex Information Categorizations.*

Information technology has made great strides. However, there is still tremendous diversity in types of information. The differing methods of capture, manipulation, and presentation reflect many factors, including inventions of each new type of information. Film, fax, text, handwriting, drawing, 3D CAD, etc., and the information media have their own sets of necessarily specific procedures for creation and use. Systems built to manipulate only one type of information are shown in *Figure III-2* as function-level systems. These divergent, single media, or information-type single-function systems reflect the continuing presence of special handling requirements. Using and understanding levels of abstraction, standards, and computing architectures are approaches being used to better address the integration of these capabilities. The use of in-line process models to manage handling of process activities also bridges or chains functional systems.

ARCHITECTURAL CONSIDERATIONS

The complex information sharing processes supporting the CE Design process must be managed within an integrally planned computing architecture. *Figure 8-3* depicts the relationships between the business processes described in Chapters 5 through 7 and underlying computing architectural layers, described in this chapter. These underlying layers are Application Systems, Information (data types and structures), and Automated Infrastructure Support. Application Systems provide tools to perform work. The information architecture acts as an information management guide for these tools, and is supported by an infrastructure of computers, networks, databases, and systems software.

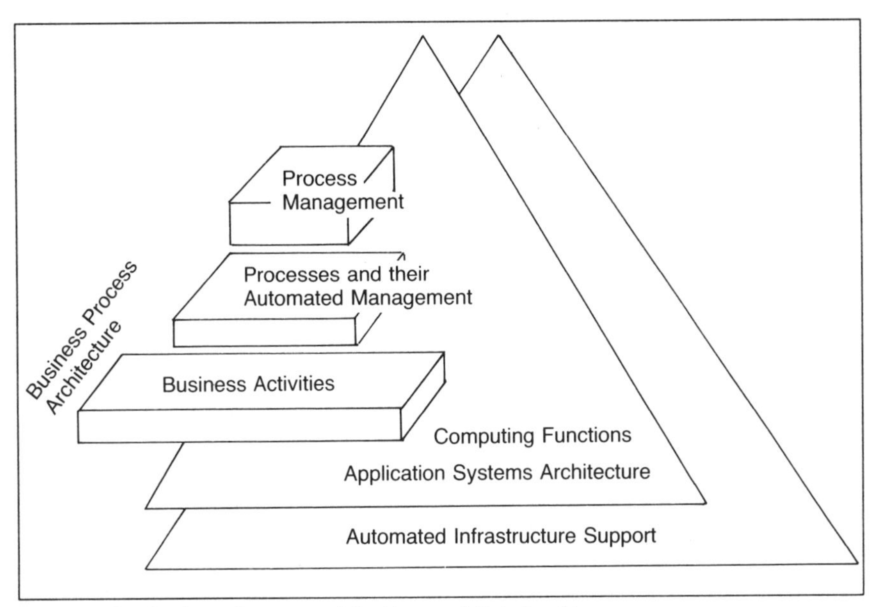

Figure 8-3. *Business Process and Architectural Relationships.*

The application systems architecture must take on almost the same structure as the organization which it supports. *Figure 4-9* depicts a much flatter, or network-type organization for the 1990s, with autonomous units responsible for themselves within the complex product manufacturing organization, including business partners, suppliers, and vendors. The three-tiered business process architecture of the CE Design (Process Management, Business Processes, and Business Functions), depicts the three layers of processes which a 1990s organization requires. This architecture also directly reflects the overall process structure introduced as a part of Chapter 7. (A very high level view of the CE Design Process architectural components is shown in *Figure 8-4*).

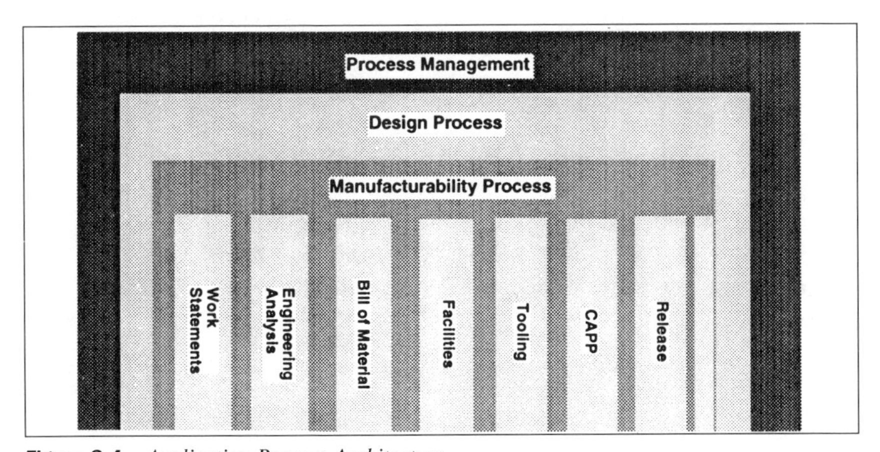

Figure 8-4. *Application Process Architecture.*

Because this is an overall CE Design and Business Process-oriented book, engineering analysis has not been its primary focus; however, it is very important. Engineering analysis affects and drives the design. It is part of the design and manufacturability process descriptions (Chapters 6 and 7) as a process activity. Today, engineering analysis is computer-oriented. Within the CE Design set of architectures (application, information, support), there are two considerations for engineering analyses which should be mentioned.

First, engineering analyses are conducted before, during, and after design. These analyses and simulations lead to, or even generate, elements of the product's design. The analyses assist in the development of the design, and are conducted after the design to aid in problem analysis and problem solving.

Second, engineering analyses must be configuration controlled even though the product is several layers of analysis away. As discussed in Chapter 6, and depicted in *Figure 6-22*, engineering analysis has a place in all portions of the intellectual processes, the design processes, and within different design methodologies. *Figure 8-5* shows that analysis activity includes R&D preliminary

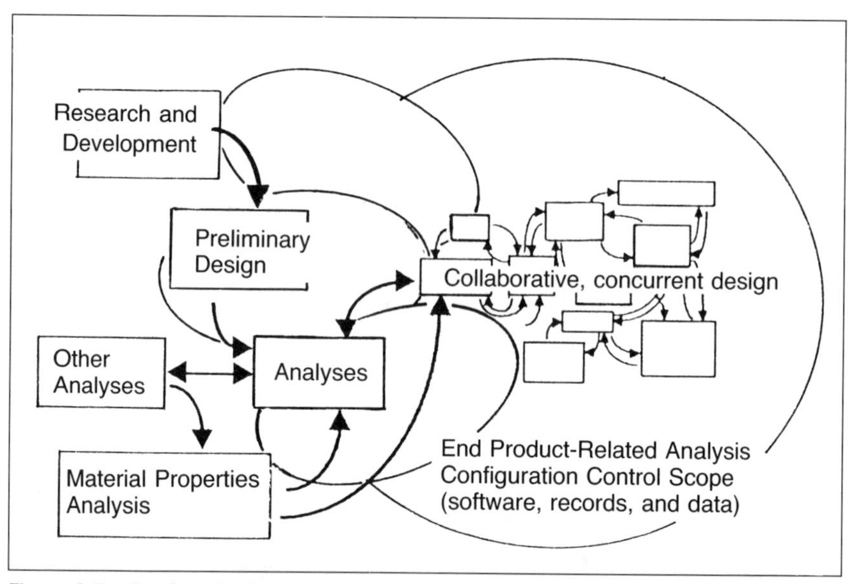

Figure 8-5. *Product Configuration Management and Engineering Analysis.*

design before product design begins. After design begins, there are a number of analyses which are not product-specific (e.g., material properties analyses) in that they affect design material call-outs, are part of design intent and product performance characteristics, but are not part of the design. These analyses, therefore, should be configured as a part of the product's fully populated information envelope. The shaded areas of *Figure 8-5* are intended to indicate that judgment is required when selecting those analyses that are important to the description of design intent and other aspects of the product definition.

In addition to illustrating engineering analyses, *Figure 8-4* points out that process management pervades the application architectures. Both *Figure 8-1* and *Figure 8-4* also depict the close relationship between the design and manufacturability business and technical processes. Of course, the application software for CE Design is not built as simplistically depicted in *Figure 8-4*, but the control relationships as built in *Figure 8-4* must be maintained.

Application software rarely provides for application systems that manage other systems. Additionally, most application systems are designed and built using a single-entity view of systems. This single-entity view is easy to understand, and thus easy to implement within organizations which have no previous computing experiences. *Figure 8-6A* is one way of representing such a

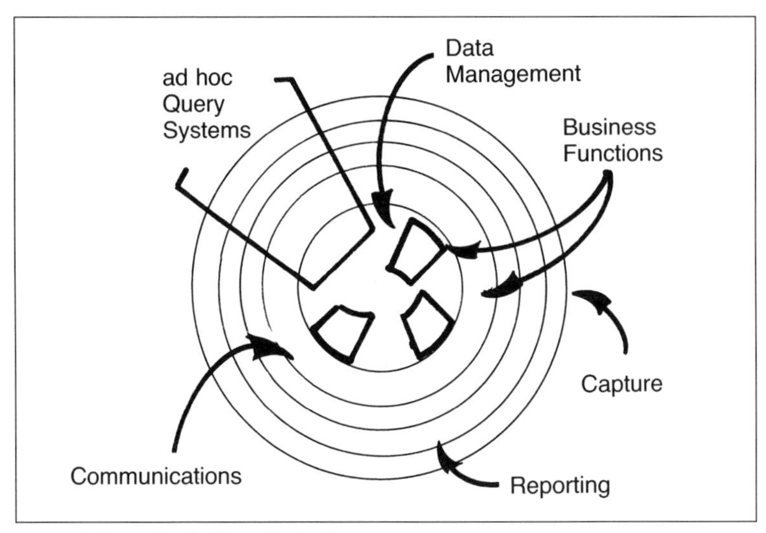

Figure 8-6A. *Single-View Entity Systems.*

concept. Here a database is at its core, with layers of software to accomplish increasingly general functions. Ad hoc query going to the core of the structure was added to permit access to the database without going through all the layers. These systems are increasingly integrated, meaning that an increasing number of business functions operate off the same database. The database is extended to add more data elements as their support becomes necessary. Using this technique, capable business management systems have been developed. These systems do have limitations, which are discussed in the following section.

System Limitations

Data-only systems for short business transactions do not attempt to store large amounts of complex information as depicted in *Figure 8-2*. Rather, they manage data (numbers and text about business transactions) intended to be stored for a short period of time. *Figure 8-6B* depicts some of the differences between short

Simple Business Transaction Systems	Systems Which Manage Complex Information
* short messages * short lived value * single point of control * geographically integrated * simple update rules	* large, "lumpy" datasets of complex information * long life requirements * must tolerate diversity and concurrency in a net change environment * geographically dispersed * complex, multistaged, update based on configuration control rules

Figure 8-6B. *Relationships Between Complex Information and Transaction Information.*

business transactions and CE Design's information-oriented environment. To respond to these information type differences, the CE Design environment needs an architecture which is *substantially different* from today's short transaction-oriented architecture.

A classic example of the short transaction environment is an inventory system with a data element for inventory-on-hand for each different item. This data element's contents change as soon as the next issuance or receipt of that particular item occurs. No configuration management of this type of data is generally necessary.

Information has value to the organization. *Figure 8-7* depicts the three axes

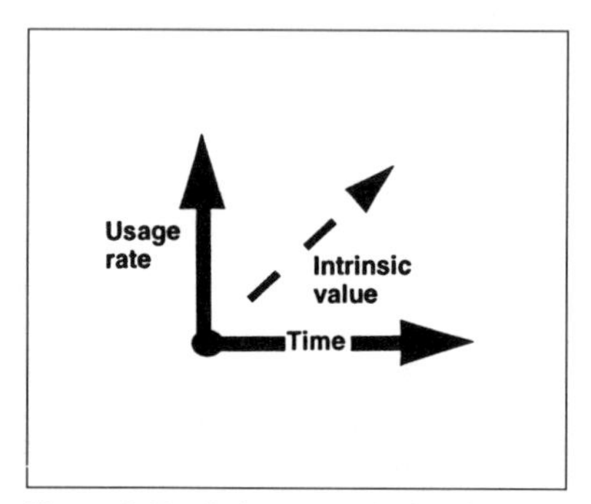

Figure 8-7. *Information Value Over Time.*

of value—*time*, *usage rate*, and *intrinsic value*. *Figure 8-8A* shows values for different types of information. Transactions of the short business-transaction type, produced by single-entity systems, have a high immediate value (the inventory count must be accurate until the next increment/decrement), but value trails off quickly. So quickly, many systems summarize these individual transactions and do not allow for multiple occurrences of the same item of data

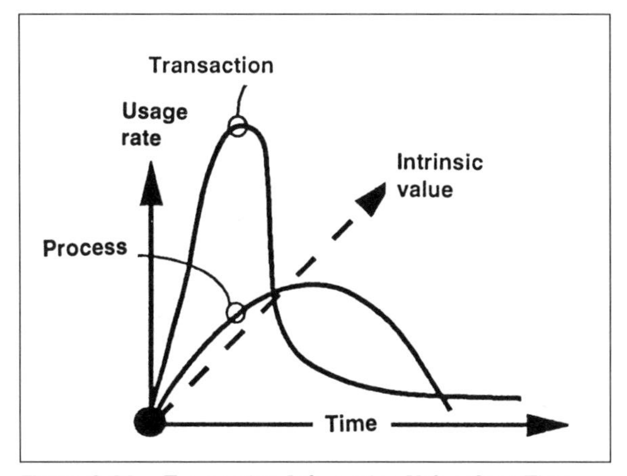

Figure 8-8A. *Transaction Information Value Over Time.*

in the system simultaneously. Instead, these types of systems relegate the noncurrent data to audit trails, summary fields, etc. There is no need for configuration control.

Process and product information is not short field, short message length, transaction-oriented data. It is complex information associated with analyses, each of which have longer lives. *Figure 8-8B* illustrates the value of process and product definition information. Product analyses and product and process data must be maintained for the life of the product. Such life cycles may be as short as three years, or as long as 50. These analyses have little value unless there is a product-related problem. The need for configuration control of this type information is created by: (1) the requirement for long-term storage and availability; (2) the need for stability; and (3) where information once stored as accurate must be kept in that state and any change must be carefully noted.

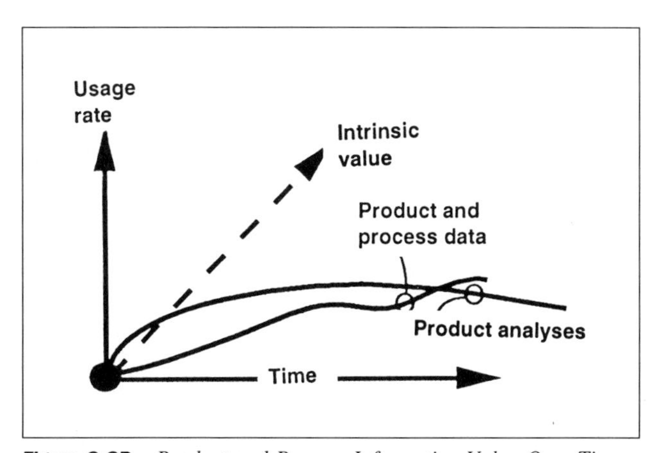

Figure 8-8B. *Product and Process Information Value Over Time.*

A carefully executed conversion from this data-oriented approach (as represented by today's PDM systems) to object-oriented (as represented by complex information management) is essential to avoid the hidden high cost of constantly updating hardware, software, and the data itself. In most of today's systems, data is used to recompose the images of the products and their associated information. This low level of data abstraction, and the systems which produce the images and information, are heavy cost burdens when substantial complex product data for large products must be retained for a significant period of time.

When the complex types of product and process information are included, with meaning attached to them, in a storage scheme appropriate to the requirements of CE Design, their values continue to increase for an extended period of time. *Figure 8-8C* depicts the various types of information value

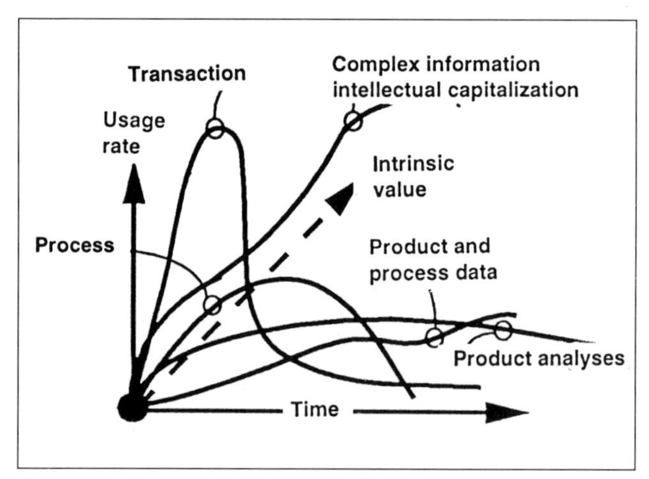

Figure 8-8C. *Complex Information Value Over Time.*

associated with complex information. Intellectual capitalization (which results from information increasing in value over time) is made possible because information can be used to describe reasons for decisions or can act as guidance for the next occurrence of that type of issue. Configuration control over the updating activity is essential to its intellectual capitalization value. The information-oriented architecture recommended here permits product and process-related information to be accumulated and combined at higher levels of abstraction. Support for such indexes as change order number, process number and product number can be used to provide key word access, context searches, abstracts, and other information retrieval techniques. It is this roll-up into many diverse envelopes through a variety of indexes and access methods that provides the *knowledge power*.

Single-entity systems were not developed to operate on multiple (i.e., different manufacturers) platforms. For single-entity systems, when needed for storage of data in more than one location, a synchronization process is established, leaving a single database as the control or master copy of data. Application software and data are typically carefully controlled to maintain current data accuracy. The relationship between these systems and their computer platform (hardware configuration, operating systems, and database management system) has been tight. Slight variations in operating systems release or other systems changes make the system inoperable on another platform because of the great number of different software elements in the many layers of today's complex computing environments.

The single-entity system encompasses and contains all architectural elements (application, information, platforms), mostly by default. This single-entity systems architecture has proven to be difficult to extend to manage CE Design's varieties of complex information. Other technological advances and the need for better management of costs in computer technology's use also have contributed to the emergence of a different architectural approach based on the concept of openness and networks or integration via messaging.

"Openness" generally refers to the ability to operate the same application software on multiple hardware/operating systems platforms through standards of operation in underlying computing elements. These concepts mirror the changes in organization structure and function. They appear more likely to support increasingly rapid changes in the technical, business, and managerial processes that are supported in a CE Design environment. In addition, the emergence of these technical capabilities has led to the recognition that a more object-oriented approach to architecture is needed, instead of just layering software.

The various architectural components (application, information, and automated infrastructure support) have their own further architectural breakdown. They have their own sets of issues, emerging and existing technologies, competing companies, and opportunities.

The emergence of choice in every aspect of the application, information, and automated infrastructure support of architectures has created the need for a separate, identifiable process relating to computing architectural planning and its harmonization with business process improvement planning. Chapter 9 discusses such an appropriate planning process, with a focus on CE Design's implementation strategy. *Figure 8-9* depicts the architectural portion of the planning process described in Chapter 9. The general sequence is shown by the arrows, emphasizing business processes and their impact on computing architectures. Note that technology "pushes" on user requirements and that business, technical, and managerial processes need both "push" and "pull" architecture.

While the CE Design environment does have a need for transaction processing, the CE Design computing environment needs to be different from the transaction-focused, single-entity-type environment. In addition to the technical computing developments and the types of complex information which CE Design's computing must manage, there is also the nature of the work which it must support.

The transaction system usually handles structured work. Each work item presented to the system has only limited actions which must be performed. For

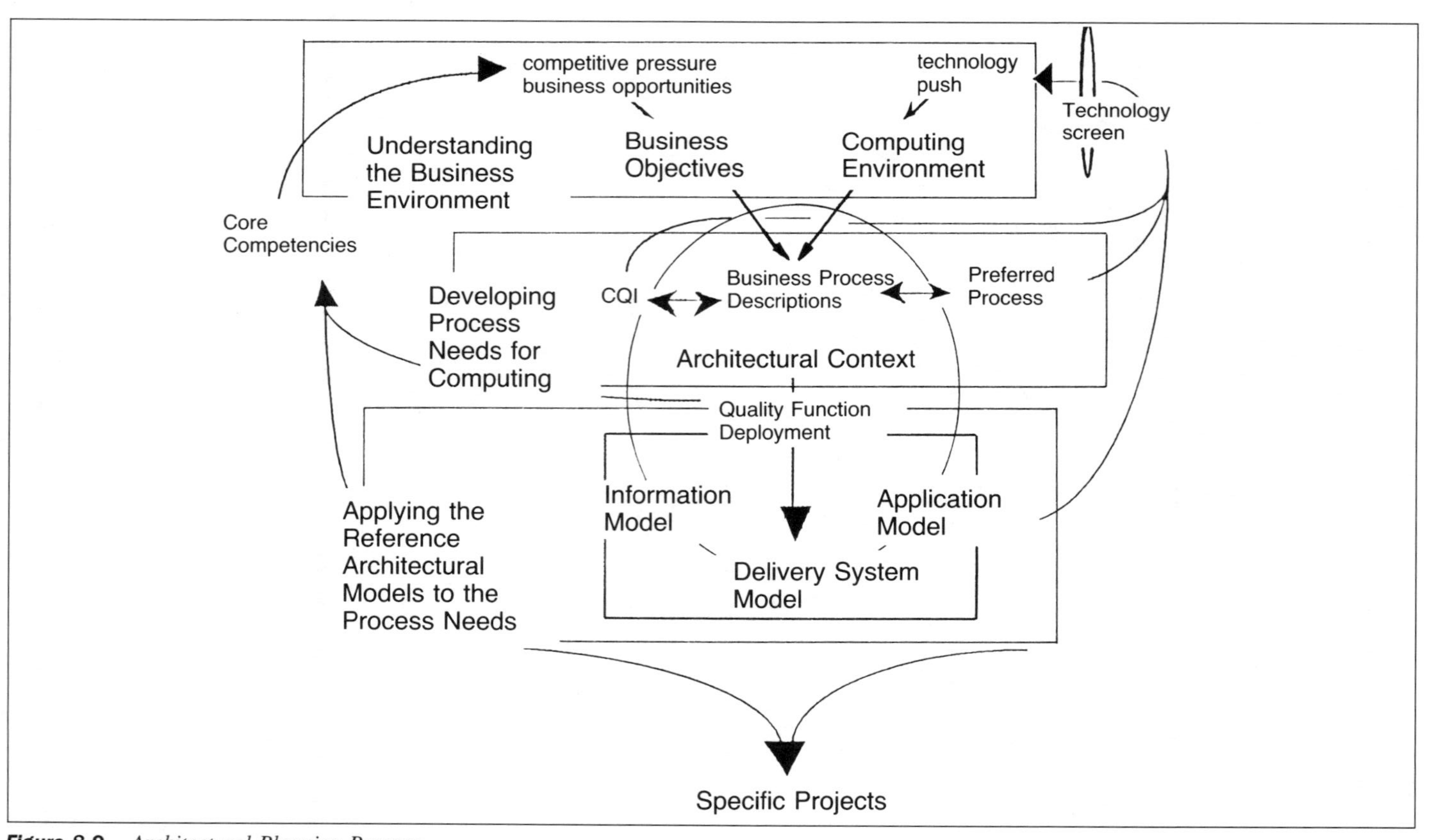

Figure 8-9. *Architectural Planning Process.*

example, in the inventory system, the transactions for the item inventory-on-hand can either add, query, or subtract. This type of system usually has one database because there can only be one correct occurrence of each piece of data at any one time.

Because the CE Design work environment is *unstructured*, the complex product manufacturing organization never goes through a new product design cycle, or even a major change to a product cycle, the exact same way twice. Trying to do it would be a grave, system-killing error in the typical transaction system computing environment. In the CE Design work environment, there can be many correct and different occurrences of the same data item at the same time. This is why effectivity, net change, and other elements of complex information configuration management are important to CE Design.

KEY CE DESIGN AUTOMATION CHARACTERISTICS

The technical, functional, and systems development environments for use in an unstructured business, technical and managerial environment must reflect the unstructured nature of their supported processes. The CE Design computing environment must, therefore, be constructed of a series of horizontal assemblies of flexible tools interconnected in a network of capabilities that operate on complex information.

This statement may appear to contradict the description and intent of Chapters 5 through 7, which described structured, disciplined, and self-inspected processes for the same premanufacturing environment. There is no conflict because the disciplined structured process only lasts as long as the set of activities for that product-related process. In the case of manufacturability, the process is designed to be constantly changing as its impact on the design evolves. A new structured and disciplined process, usually derived from a previous process but tailored to suit the slightly varying current situation, is then pursued. The structure and discipline of CE Design comes from an adherence to sets of rules governing which activities should be performed and how decisions are made, and not how individual activities are conducted.

What then are the key characteristics of each architectural element supporting such an application environment? They are summarized in *Figure 8-10*. CE Design computing application's key technical characteristics include being model driven.

CE Design's computing applications are expected to be "cross-procedural code" free. They are composed of a series of tools interconnected by the in-line process model available in a window on the workstation of each engineer or support person. Each computing tool may have its own set of operating procedures, but the rules on managing its information should be established by CE Design's process management and its configuration control capabilities.

The model process flow of interconnected tools has an embedded configuration control process. These tools are provided through process model support computing services so that it appears to be a single system. Substantial

Architectural Element	Key Technical Characteristics
Application	Model Driven Based on Tools Integrated Configuration Control Single View Systems Concurrency
Information	Object-Oriented Multitiered Storage Information and not "dataset" Management
Automated Infrastructure Support	Heterogeneous Hardware Network Based Operating System Geographically Dispersed Cooperative and Parallel Processing
Total Architecture	the "..ilities" .interoperability .consistency .reliability .stability .scalability .extensibility .security .degradability
Standards	Application Portability Profile "Confederacy" Approach to Control

Figure 8-10. *Flexible Architectures.*

concurrent work is permitted through multiple access to the same information and process models simultaneously. The information is stored in various degrees of granularity in an object-oriented manner. It is stored in a multitiered hardware structure based on cost, access frequency, and inherent value basis.

Figure 8-11 depicts the difference before and after the concept of in-line models is applied. The model-driven approach uses "late binding." This means the process to be used is not preset in the application, but is enabled with computing tools as it is actually executed. The software and the information then used are also not set until execution. Late binding is important because change is permitted until actual use. This is how unstructured work naturally occurs.

The Automated Infrastructure Support is composed of heterogeneous hardware and software, utilizing a network-based, object management system. This hardware is geographically dispersed, as are its many users. *They are interconnected to appear to be a single view or entity, but are not physically constructed that way.* Transition from single-entity architecture, where the physical is the

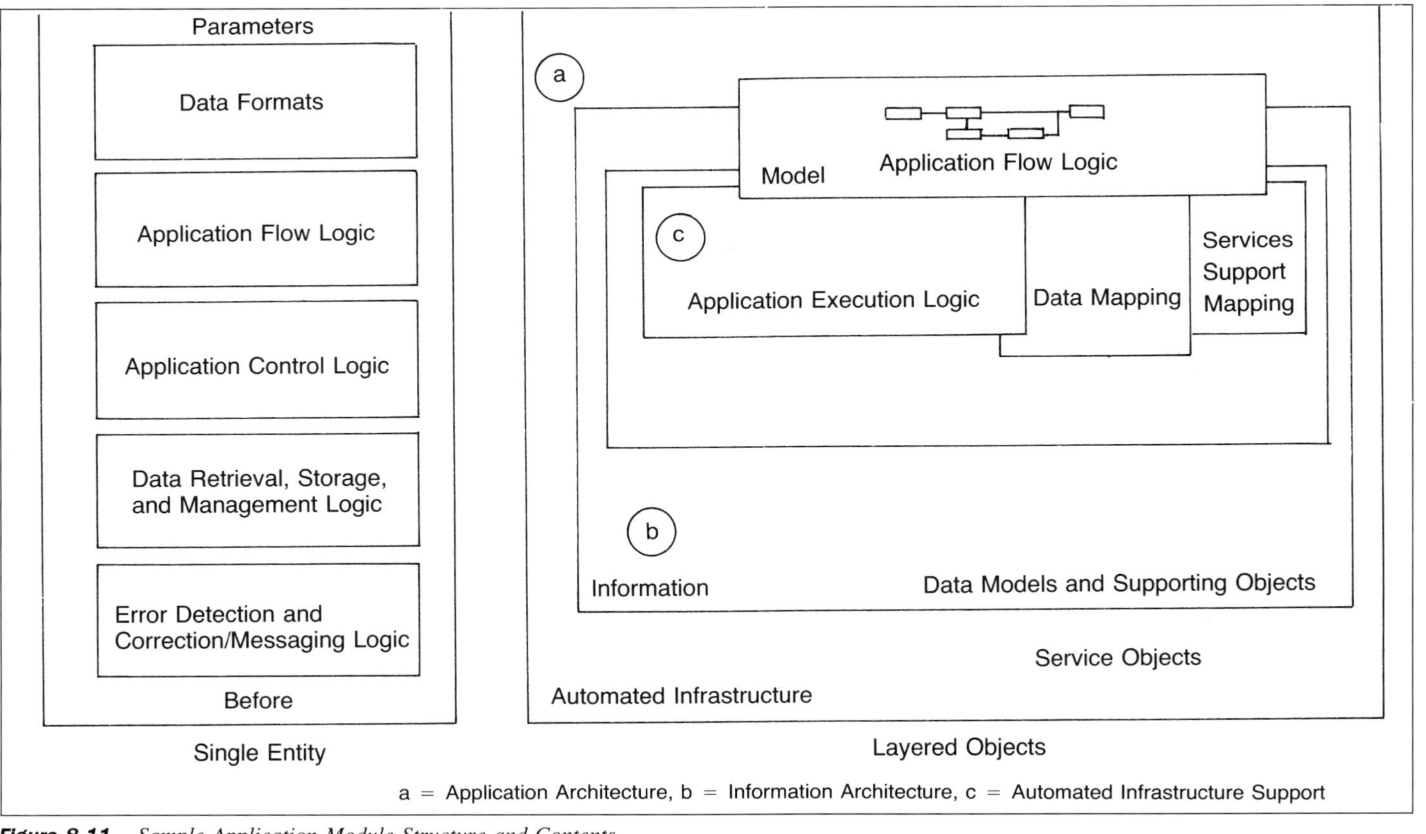

Figure 8-11. *Sample Application Module Structure and Contents.*

same as the view from a user perspective, to a logical, or virtual view, when its heterogeneous nature is masked from the user, is at the heart of the CE Design Automated Infrastructure Support concept.

The units of hardware are also multitiered, so processing can be moved to different units on the basis of performance, cost, and availability for both single, stream, and concurrent processing purposes. For example, the same computing tool may operate on a UNIX technical workstation and on other, more powerful computers, such as a Cray. The execution of the tool on this choice of platforms can change based on cost, priorities, and availability. The emergence of personal computers has had a profound impact on computing strategy and tactics, and on how work is conducted. A workstation is a personal computer attached to a network so that cooperative work across multiple types of computer platforms can be accomplished.

The evolution of cooperative workstation activity is depicted in *Figure 8-12*. The cooperative work concept develops through different stages. First, workstations share printers and storage, then processing chores, then data or information upon which processing is performed, then all resources, and then finally the work becomes unparticularized to any workstation or computer. This migration is part of the CE Design implementation strategy.

When combined into one object-oriented, message-integrated, computing architecture, the total architecture assumes certain performance characteristics. These characteristics include:

1. *Interoperability*. For applications, the ability to accomplish work operating on the architecture, in spite of the presence of heterogeneous hardware, without the need for a change to the software.
2. *Consistency*. All portions of the architecture operate in a similar, intuitive manner.
3. *Reliability*. Overall architecture is constant in its accuracy of operation.
4. *Stability*. Overall architecture is constant in its appearance of availability.
5. *Scalability*. The architecture can handle a wide range of volumes of activities and amounts of information without change or significant degradation in response and performance.
6. *Extensibility*. Enhancements of the architecture, to include new features and functions, can continue to be made with reasonable cost and in reasonable time.
7. *Security*. Control is maintained over the access to and manipulation of data throughout the architecture, yet the flow of work activity is undisturbed and information is protected.
8. *Degradability*. Overall architecture has complementary elements operationally unaffected by losses of individual communication circuits or processing elements. Performance may be reduced, or portions of applications and data may not be available, but operations can still continue.
9. *Standards*. These predefined profiles of functions and systems, usually set by industry and/or government-sponsored groups, provide an envi-

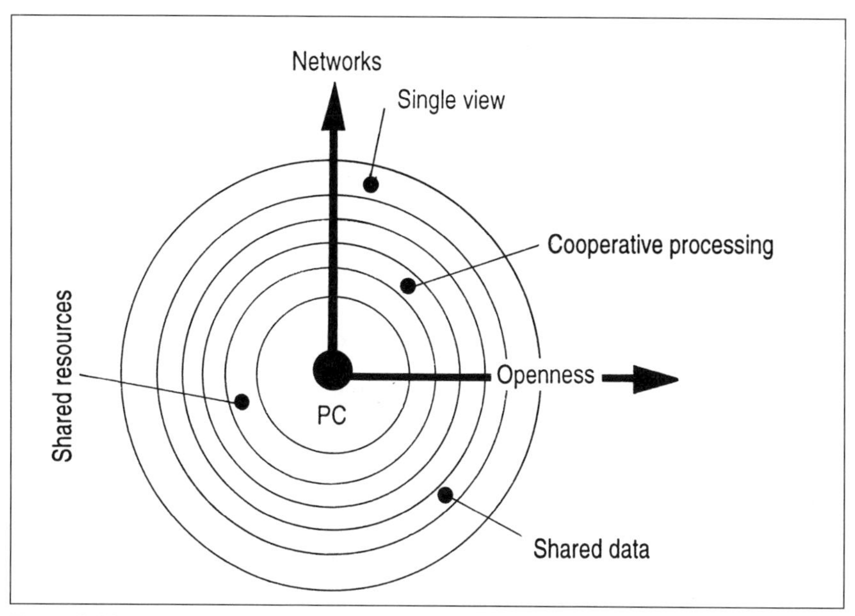

Figure 8-12. *Evolution of the Workstation Heterogeneous Architecture.*

ronment in which architectures can operate with these other "..ilities." The Application Portability Profile, a suggested profile to promote these "..ilities," is detailed in Federal Information Processing Standards Publication 151. While these standards do not yet provide for the architecture described in this book for CE Design, they provide an excellent start for a migration to this architecture.

Plans for a transition to this architecture from the present environment of complex product manufacturing organizations are described in Chapter 9.

In addition to these key technical characteristics and architectural performance characteristics, the architecture has certain other business goals and objectives.

First, the architecture must help lower the overall cost of computing relative to the growth in product, process, and organizational complexity. As shown in *Figure 8-13A*, the measurement of computing must take these three elements into consideration at the same time. Because complexity may be increasing, computing costs may be rising in actual cost but may be declining relative to these factors in combination.

Second, the architecture must help lower the cost of sustaining computing systems. As shown in *Figure 8-13B*, Composite Cost of Product Production support must go down in the face of increasing complexity.

While legacy systems sustaining costs have continued to rise, this pattern cannot be extended into the CE Design environment. Composite costs must decline in real terms. As shown in *Figure 8-13C*, the computive cost of the CE Design environment must decline in relative terms.

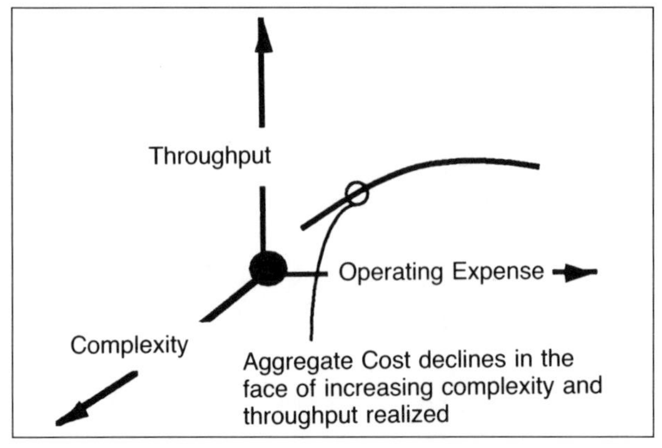

Figure 8-13A. *Performance Metrics.*

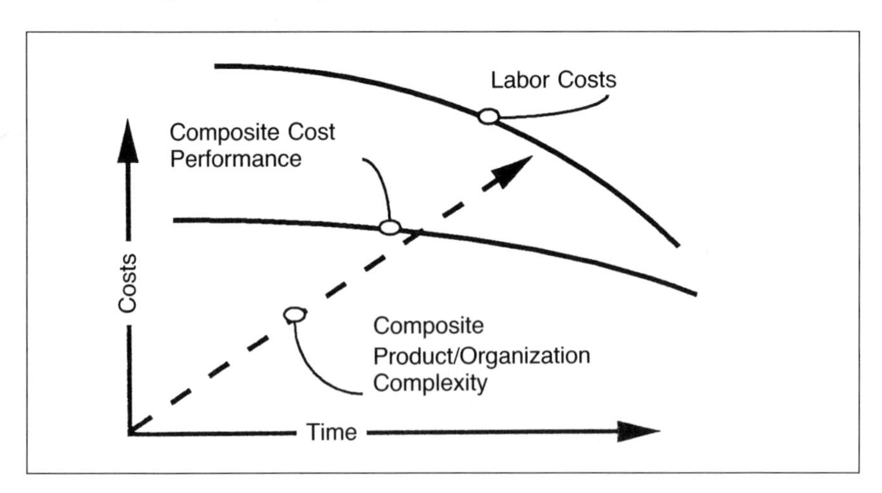

Figure 8-13B. *Overall Cost Performance Objectives.*

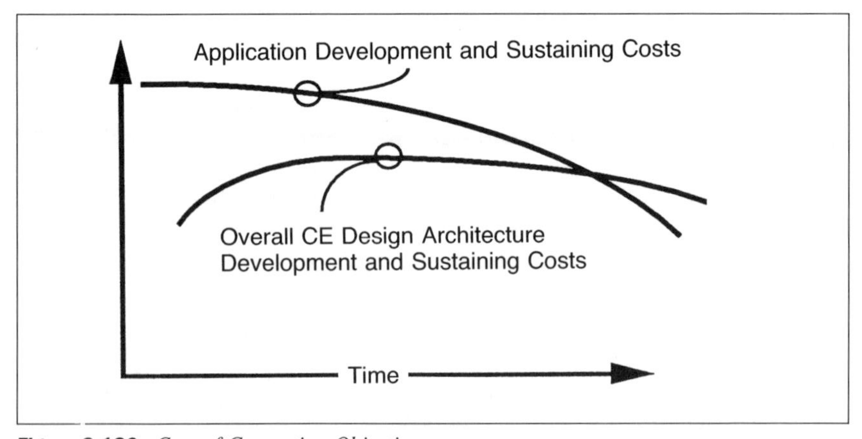

Figure 8-13C. *Cost of Computing Objectives.*

Figure 8-14 shows legacy systems compose a relatively small portion of the overall potential computing capability of the complex product manufacturing organization. Yet, these same legacy systems typically impede improvements in processes because:

1. They contain the procedures of the process, and the change cycle for these procedures is too long, too costly, and too cumbersome.
2. They consume too much of the talent and other resources available, given the rather small improvements. As mentioned before, these systems are usually function-oriented and do not yield great leverage.

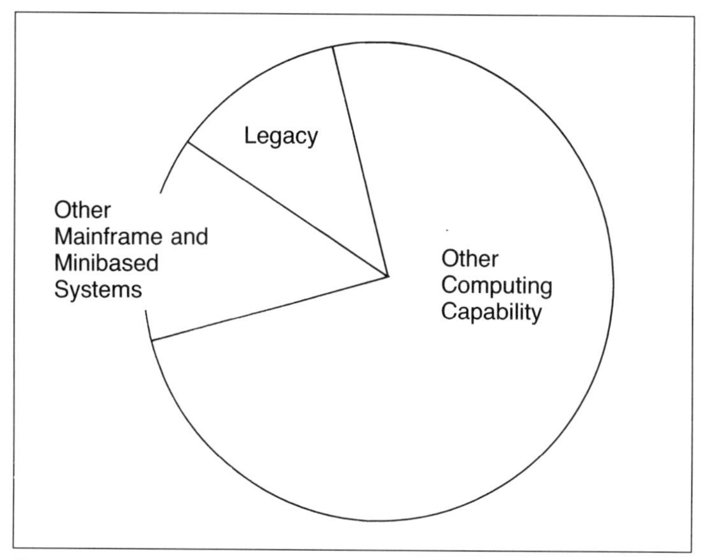

Figure 8-14. *Computing Penetration.*

A substantial portion of the "other computing capability" is personal computers of various capabilities and technical workstations. These will be replaced by increasingly powerful technical workstations. Without changes in the architecture, CE Design might flounder from its own complexity, resulting in the migration pattern shown in *Figure 8-12*. At the same time, it is important to remember that *the reason PCs are so popular is their relative freedom of use*. Their other appeal is the relative ease with which functional work can be accomplished through the use of tool-type software (spreadsheets, text processors, database application generators, CAD, CAM, CAE, and Windows™ software), without procedural impediments and the myriad set of evolving application development environments.

The architecture must be self-generating. When CE Design processes are used in producing architectural components supporting the execution of its technical, business, and managerial processes, the same benefits should be derived once the requirements are established.

The principal management concern regards the comparison of make, buy, cost, and time. As shown in *Figure 8-15*, the issue is how far ahead of the currently available hardware and software should an individual organization get? The cost of getting too far ahead is the composite impact of lower-cost industry software, lower cost of maintenance, lower risk, and lower cost of installation. This additional cost and resource utilization should be compared to the cost of time lost, benefits not realized, and competitive position lost. Examples of the consequences of an early decision to develop in-house what is provided by the marketplace include higher development costs, higher sustaining costs, higher costs to communicate with other organizations, and higher personnel costs.

However, the cost of waiting could mean a competitor getting a permanent competitive advantage by first developing the technology and using it to set standards in the industry. These standards can then be used as a weapon that forces competitors to adapt at higher costs relative to their already developed strategy.

For this reason, including the systems development process inside the application architecture is very important to the appropriate overall architecture for CE Design. As shown in *Figure 8-16A*, the current situation is one where the database, the software, and all components are designed outside the actual operation environment. They are combined at the point where actual operation occurs. Binding must occur very early in this process, but whether this binding, developed piecemeal, actually works isn't determined until it is tested. *Figure 8-16B* shows the future architecture concept, with its model-driven application structure and object-oriented infrastructure.

The appropriate overall architecture provides for a model-driven systems development process, model-driven applications, and an information modeling, object-oriented computing architecture and infrastructure. Within this type of *inclusive* architecture, the key technical characteristics can now occur. Notice in

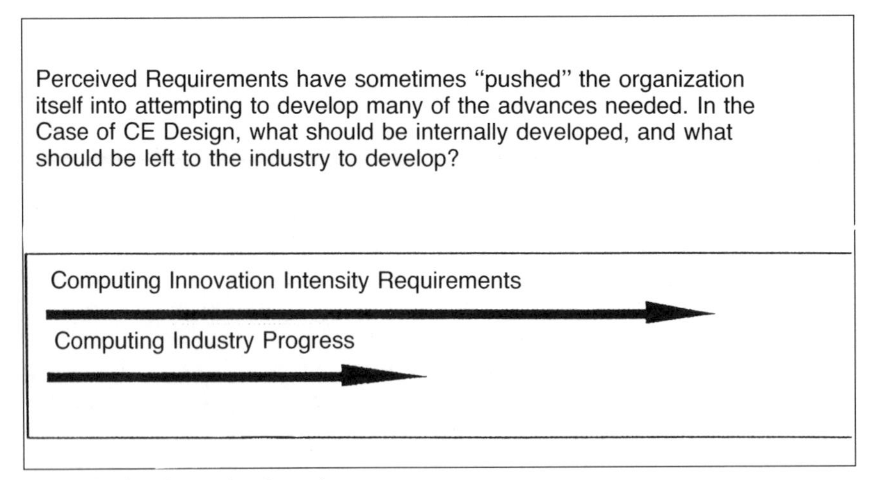

Figure 8-15. *Computing Intensity.*

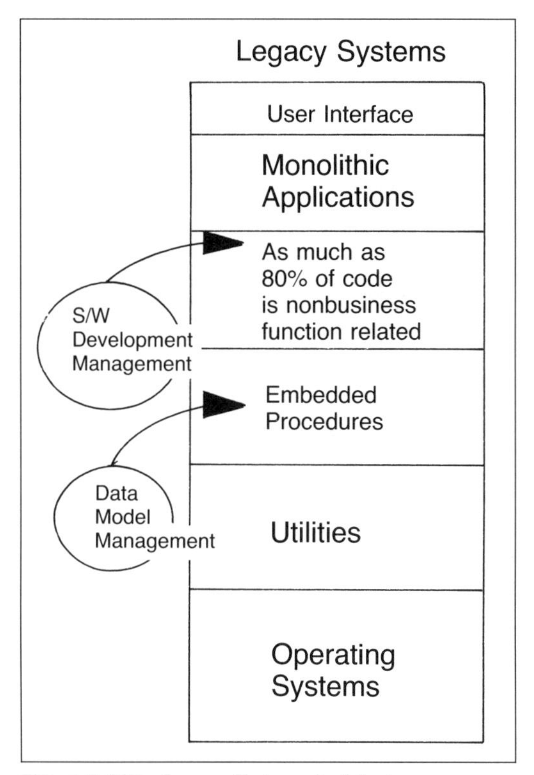

Figure 8-16A. *Legacy Systems Architecture.*

Figure 8-16B that there are not many layers, in contrast to current hierarchically architected systems. Increasing functionality, from an applications perspective, emerges from the base capabilities. The entire environment is object-oriented. All applications are combinations of tools, driven by models of the business processes to be executed.

CE DESIGN'S KEY TECHNICAL CHARACTERISTICS

Key technical characteristics result from the implementation of significant and fundamental computer system technical approaches for each architectural element. These fundamental technical approaches include:

- Application architecture—models and tools;
- Information architecture—objects; and
- Automated infrastructure support—open networks and platforms.

Each complex product manufacturing organization develops its own CE Design technical approach and implementation. The fundamental technical approaches appear to transcend any particular implementation. The transcendent nature of these fundamental approaches arises out of the complex nature of the managed information. Each of these fundamental technical approaches is discussed further.

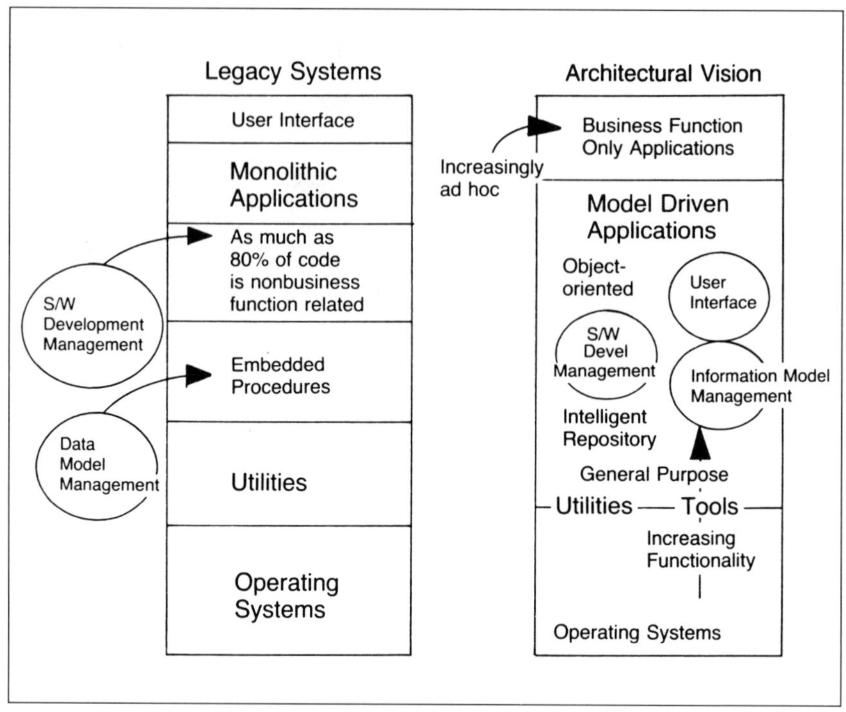

Figure 8-16B. *Legacy Systems Compared to Design Architecture Vision.*

MODELS AND TOOLS

The concept of models has been discussed throughout Chapters 5 through 7. From a computerized application system perspective, the use of models creates a fundamental change in how systems are designed and developed. However, this change is not as difficult to make as imagined. Organizations have been modeling their business processes since before the computer appeared. Flow charting and methods analysis are modeling techniques. Bubble charts, Work Breakdown Structures (WBSs), and Data Flow Analyses were extensions of these modeling approaches. For several years, software engineering has been attempting to develop tools which translate models that represent processes and data directly into programs operating as modeled. All these modeling processes have improved the quality of software. However, in all these cases, the model is used to generate a static, embedded, procedural set of programming, which must be regenerated if the conditions and actions it represents change.

Because of the unstructured nature of the CE Design environment, every time a variant on the previous design or manufacturability process is needed, the model needs to be changed and the computerized process support must be regenerated. Since this occurs essentially every time, it is impractical.

The important technical change in the case of the CE Design application environment is the move to in-line models, using a representation technique called semantic networks. In-line models are executed as if they were a script, *instead of being used to generate the procedural code*. These model activities can reference a subordinated set of model activities. Finally, the activity at its lowest level operates as if it were calling a program. It is referencing an object which operates the same no matter who invokes it.

The general concept is shown in *Figure 8-16C*. Process models are used to describe what has to be done. Computing services support execution of the process. Each computing service represents certain functions accomplished

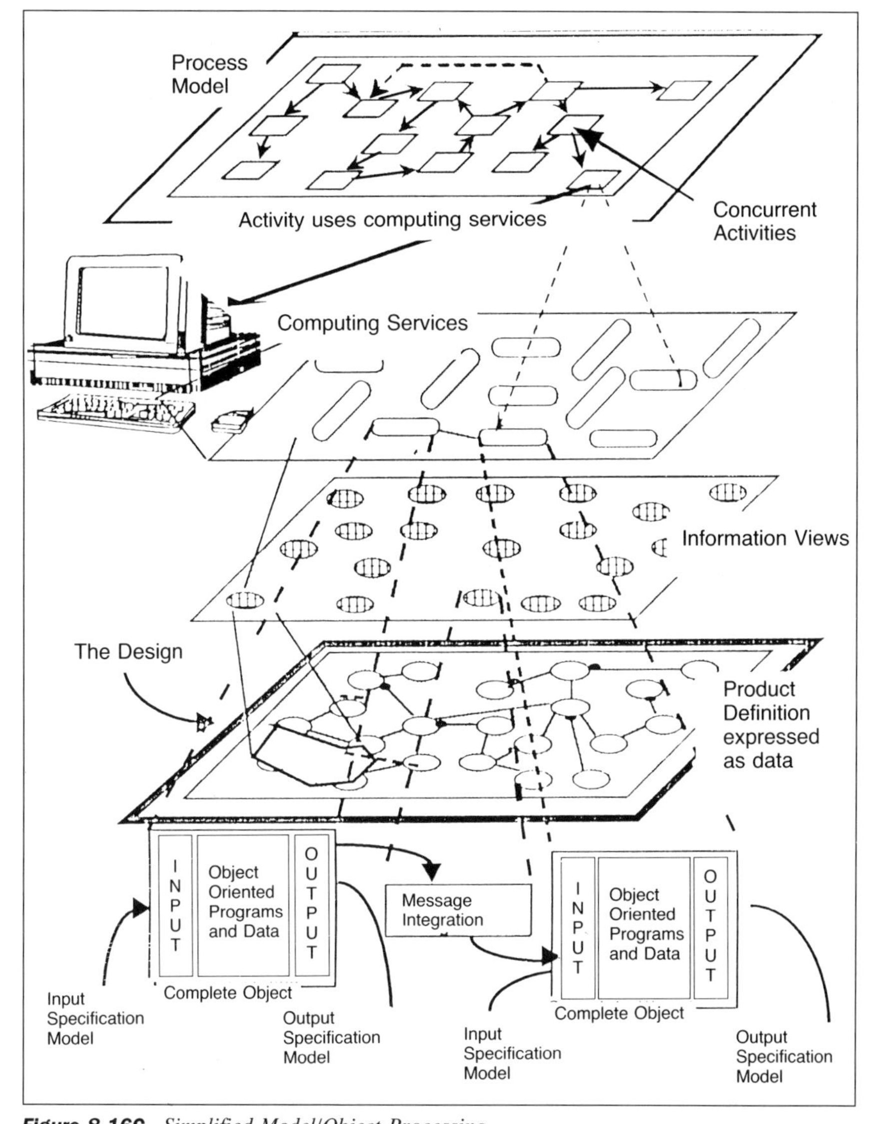

Figure 8-16C. *Simplified Model/Object Processing.*

using information models expressed, in the case of CE Design, as Product Definition information. In addition, the services also are represented by object-oriented classes of capability. These classes may represent traditional computing programming or actual object-oriented programming. In either case, messages serve as integration mechanisms. Instead of direct interfaces between programs, objects communicate with each other via messages. This permits more complicated relationships, since the message can be an independent, stable point in the relationship between the two objects. If the object has "enveloped," or acts as a surrogate for the traditional program, it captures the inputs and outputs via input and output specification models, and simulates messaging on behalf of the traditional program. This enveloping permits legacy systems and object-oriented systems to coexist as they are transitioned.

One of the many strengths of this approach to the CE Design overall architecture is its generality. For example, one computing service could be part of the common user interface for CRTs; others could be a print function, a database update function, or another part of the same user interface for output. Combinations of these general purpose computing services can be collected into *tools*, which are also important and useful. Like the PC, they do not impose a rigid structure on the user, yet they can provide a highly structured environment for accomplishing work. Examples of tool-type software in today's CE Design computing environment include CAD, CAPP (Computer Aided Process Planning), Routing and Queuing Modelers, financial analysis spreadsheets, ABC cost accounting models, and FEA (Finite Element Analysis) software. Just started up, these tools are quite powerful, yet do nothing until used to do something constructive. By using the object-oriented "enveloping" technique, these tools can be linked together to form integrated computing services. These linkages can be changed quickly using the process model as the integrating tool. This combination provides cycle times for computing tasks that take minutes instead of days or weeks using traditional batch computing techniques. These tools become more powerful when linked because of the user's ability to use the tools to iterate options. Evaluating options in near real-time permits concurrency in design.

OBJECTS AND THE INFORMATION ENVELOPE

There is continuing interest in Object-Oriented Programming (OOP). One of the principal concepts of OOP environments, provided by such open operating systems as UNIX, is the treatment of programs, data, and directories as abstract objects. This means that the same features of the operating system can operate on both kinds of information, since programming is stored as a different type of file. The evolution of object-oriented programming and architecture, summarized in *Figure 8-17*, is more than just combining data and programs into the same object. It also means changing the level of system granularity. *Figure 8-18A* shows that Object Orientation may enable the granularity level to go down

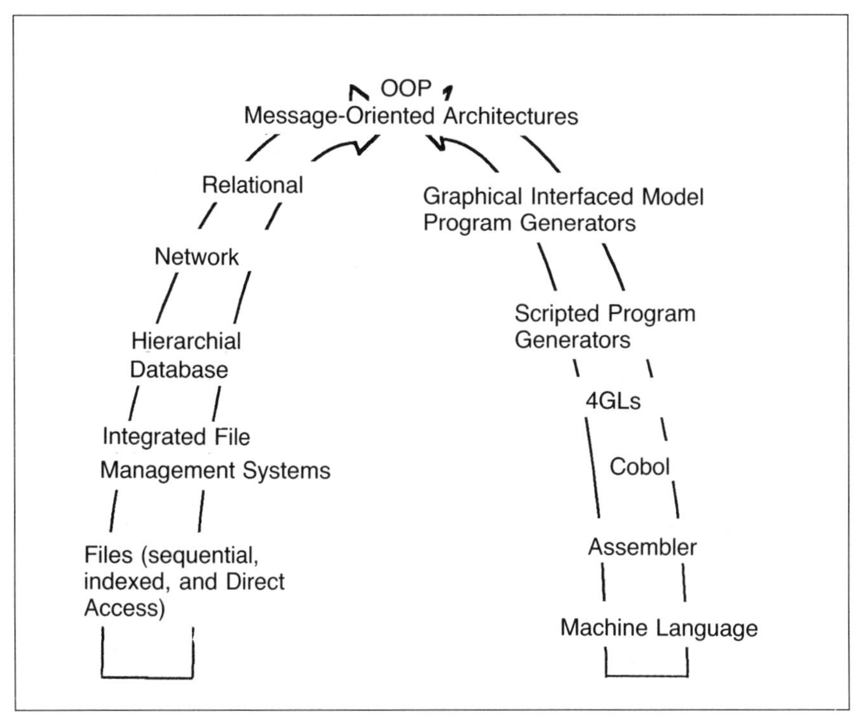

Figure 8-17. *Evolution of Objected Oriented Environments.*

to the individual piece of information, permitting the rules of processing associated with each piece of information to be its logical extension.

In *Figure 8-17*, the current approach of process models as program generators is shown as the next to last step in converging the technical and business process worlds. OO represents the logical conclusion of these evolving steps in achieving

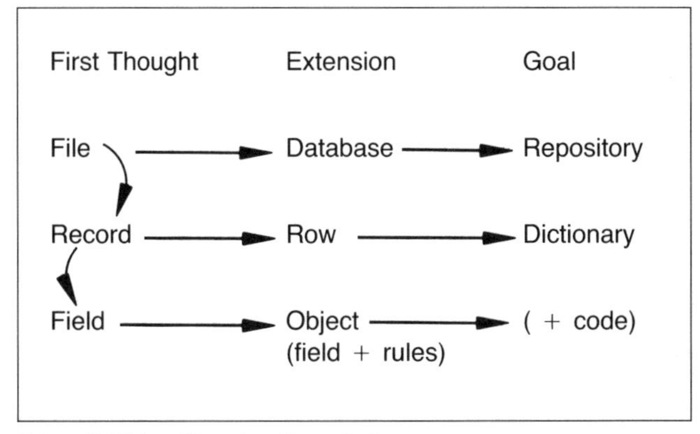

Figure 8-18A. *Changing Granularity.*

this integration. Remember, it extends process models to represent processes and their computerized tools. OO enables the process model to place application flow logic in a more manageable and usable fashion.

The concept of objects is powerful enough to permit simultaneous consideration of an entire database as an object, which could then pass information to another low level of granularity information object. This powerful concept is particularly important as the organization transitions itself through various intermediate stages to full CE Design, as discussed in Chapter 9.

The concept of object-oriented information, or object encapsulation, is called "envelope" in other sections of the book because envelope is more self-explanatory and because an envelope is created by an object. To the OO environment, envelopes are just another alternate combination of objects with a name. Envelope is also used because it is more descriptive of the activities likely to be encountered in the CE Design process. The term "object" is abstract, but envelope is not. It has a label, rules of use, fits in a filing cabinet, can contain anything, can be stored and retrieved out of other envelopes, and describes objects without using the current technical terminology of objects to do so. Some basic characteristics of objects are depicted in *Figure 8-18B*. Objects have attributes and elements, or features. For objects to be useful in CE Design, they also need to be stored in libraries with configuration control and with expanded information about their history, composition, and current status. This control is very important if objects are to be useful across multiple platforms and vendor offerings. They also must be built, or at least integrated, into an environment built on a message-oriented architecture.

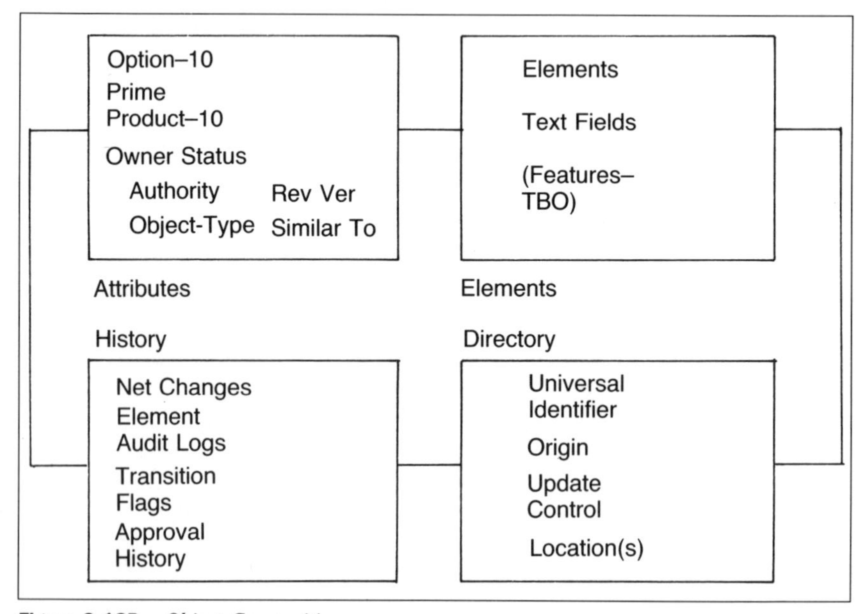

Figure 8-18B. *Object Composition.*

The message-oriented architecture resolves the problem of how all these objects will communicate with each other over this heterogeneous network of computing. By establishing the message exchange architecture, objects can be used, and reused over a broad range of computing services. This integration also permits wider use of dynamic binding and assumed correctness of objects. These two areas are significant problems in an OO implementation if the architecture of messaging has not been established.

OPEN NETWORKS AND PLATFORMS

For computing to continue expanding its capabilities and acting as an automated architecture for CE Design, the automated infrastructure portion of the overall architecture must promote technical perspectives. These may include:

1. ''Plug and play.'' Each component of each level of the architecture can be added to the system and utilized without the need to shut the system down to recognize it.
2. ''Substitution without notification.'' Each component of each level of the architecture can be changed-out without the need for shutdown or change in the using environment.
3. ''Retrieval without regard to storage media.'' Each piece of data is associated with its process-sensitive programming so its retrieval can be usefully accomplished whenever desired and from whatever archival media it was placed on. *Figure 8-18C* depicts one facet of this perspective. In this case envelopes have been built using objects as surrogates for underlying data, or other objects. Their essential characteristics are contained in a table of contents (TOC), which acts as an object abstracting service. The abstracting service contains various pieces of information, which further describe the object for use in its archival and retrieval. The archival portion of this storage scheme is important because persistent storage over long periods of time may be required in the case of CE Design.

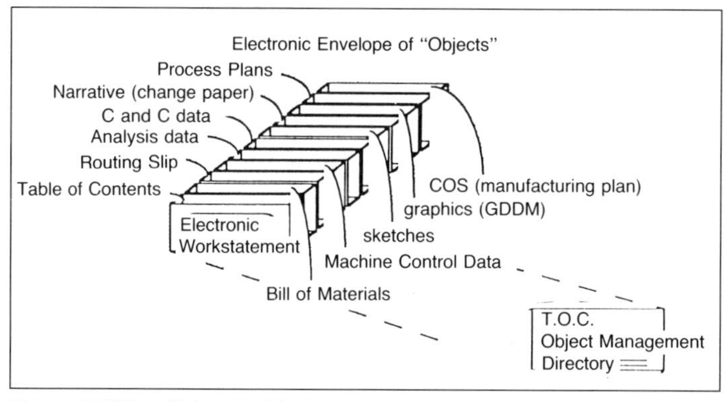

Figure 8-18C. *Object Archival.*

These preceding perspectives represent open networks and platforms. Just as concurrent design is collaborative and focuses on interfaces first, design of the automated infrastructure element of the overall architecture should focus on integrated open networks and platforms, distributed geographically. As shown in *Figure 8-19A*, the architecture of "Yesterday" was dominated by proprietary single-entity systems which had their own intrinsic and undifferentiated total architecture. In "Today's" evolving environment, shown in *Figure 8-19B*, these different types of systems have developed into three general tiers: 1–mainframes and supercomputers; 2–minicomputers and LAN servers; 3–workstations. Each tier still has its own set of proprietary systems, knowledge, and capability barriers, preventing their easy interconnection and interoperability.

Fully operating CE Design requires a total open architecture, depicted as "Tomorrow" in *Figure 8-19C*; its open networks and platforms facilitate simultaneous collaboration. This heterogeneous set of platforms provides a single-system image, made possible by any process on any element of the network able to find information and/or involve other processes on any other network element.

Figure 8-20 provides a detailed look at a representative CE Design architecture for a complex manufacturing firm which must retain large mainframe computing for its vast storage management or computing capabilities. The complex of central computing and the distributed, integrated environment of CE Design's primary process-driven activities are shown interconnected by communication facilities.

Concurrency is made possible in the complex CE Design computing example shown in *Figure 8-20* by involving a cooperative in-line process model (with planned concurrency built in), which accesses the same data while providing what appears to be simultaneous change. This apparent concurrency also is made possible by high-speed, high-bandwidth (able to move large amounts of data) network communications between all elements. Having all this data available is accomplished by tiers of different types of information storage devices and media tapes. For example, magnetic tape, optical disks, magnetic disk, and large RAM all might be storage media. Each would be used as appropriate for cost, speed, and longevity reasons–storing of compressed video, document images, and voice, for example.

PRESENT TO FUTURE ARCHITECTURE

The overall architecture depicted in this chapter is probably not fully implemented in today's complex product manufacturing organizations. This does not mean that CE Design and its implementation should not begin or continue. In fact, establishing a CE Design implementation process promotes and assists in the implementation of its architecture because of the power of CE Design's benefits and its cost justification. Also, CE Design's use promotes the use of its tools and processes, which benefit not just CE Design, but the entire organization's computing architecture.

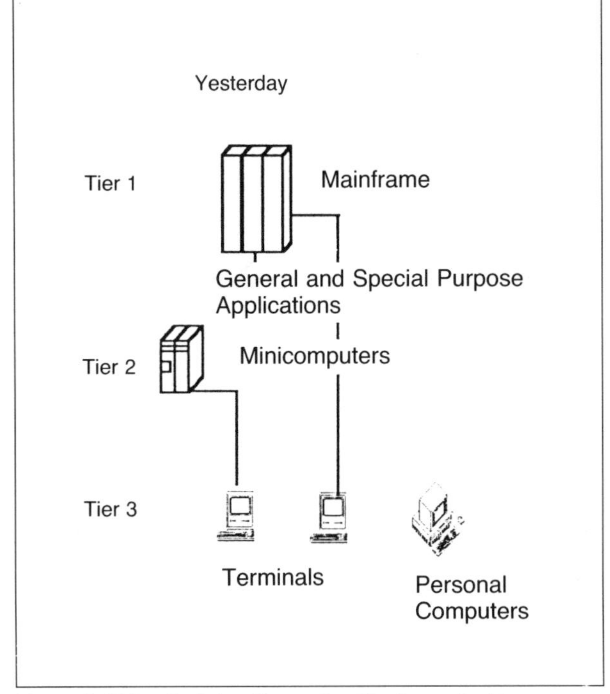

Figure 8-19A. *Evolving Computing Hardware Architectures.*

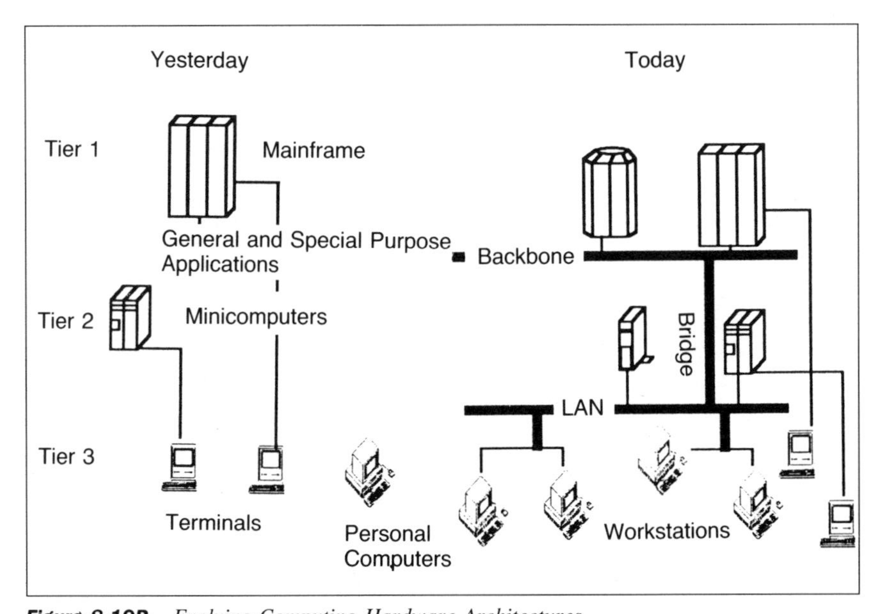

Figure 8-19B. *Evolving Computing Hardware Architectures.*

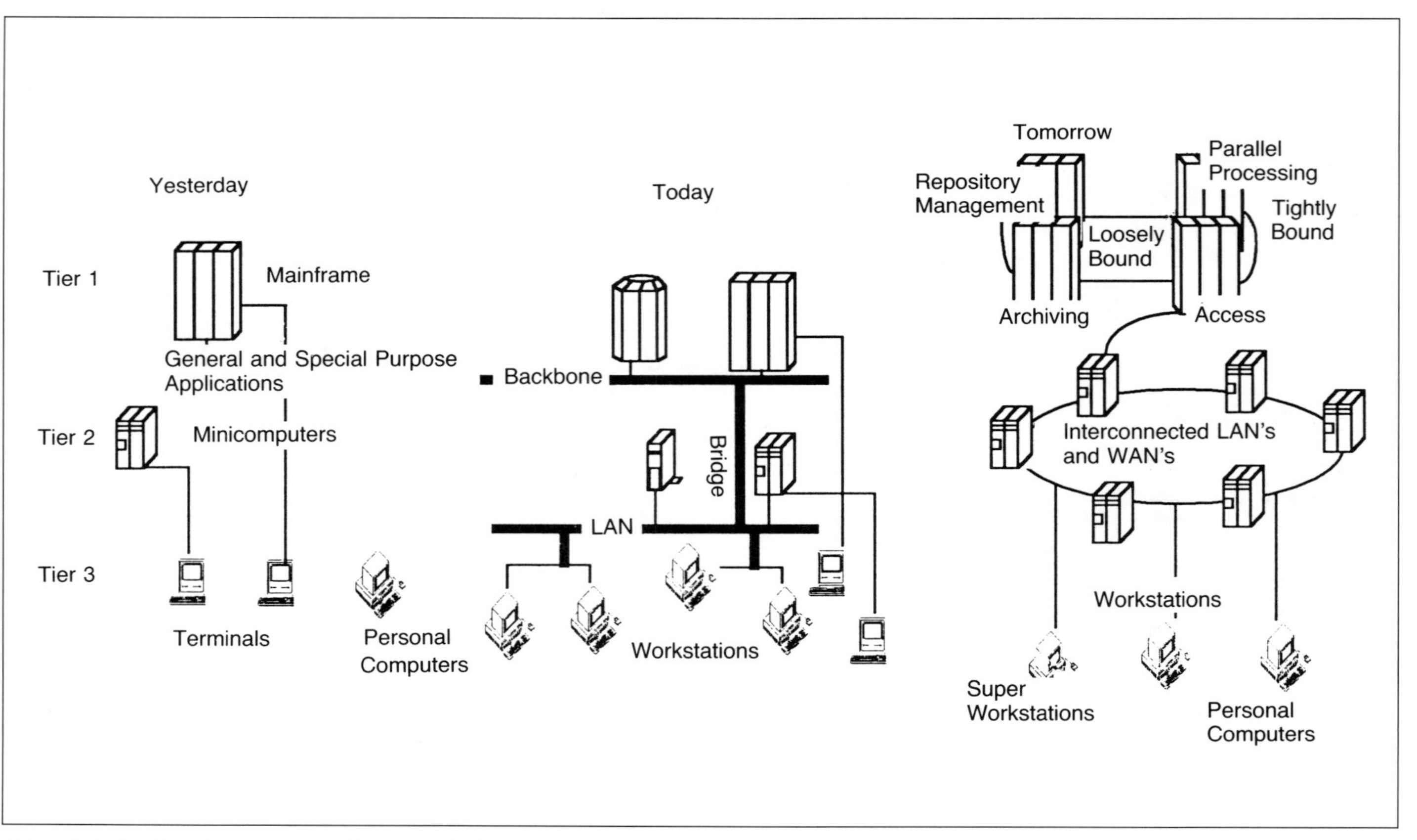

Figure 8-19C. *Evolving Computing Hardware Architectures.*

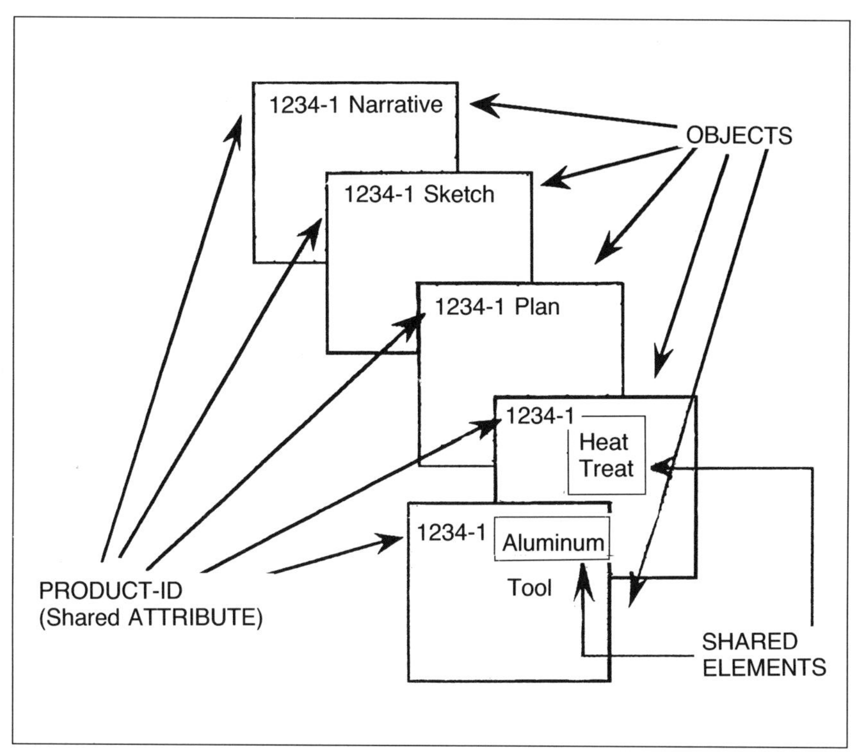

Figure 8-20. *Object Relationships.*

The planning process for CE Design, and its implementation strategy within that process, are described in Chapter 9. From an architectural perspective, there are two computing imperatives to this plan: (1) models and (2) interoperability with a single view. The use of in-line models creates a very powerful construct which the computing industry has been pursuing since its inception. It changes the role of the programmer from managerial process interpreter to tool builder.

Interoperability, achieved by messaging-oriented integration and a single logical view is imperative because it bridges a wide variety of differing levels of legacy systems, new objects, and other types of software into a single stream of activity tailored by the user using a straightforward tool set.

9

IMPLEMENTATION PLANNING FOR CONCURRENT ENGINEERING DESIGN

This chapter focuses on the planning process to implement CE Design. The first eight chapters focused on the "who, what, where and whys" of CE Design. This chapter focuses on "how and when."

A DIFFERENT IMPLEMENTATION PROCESS

While there are many CE Design-related activities under way in industry, some complex manufacturing organizations have begun to consider the many aspects of the implementation process required to effectively implement CE Design. CE Design requires a different implementation process. Previous chapters emphasized different management and technical requirements. However, it is a fair question to ask "Why is a different planning process necessary?" and "What is it?"

The transition from today's organization and its activities to those described in the first eight chapters is a significant issue. Spread throughout the first eight chapters are references to information to be managed, the impact of this level of change on people and the organization, and the various complexities of product, process, and management which must be dealt with. Typically, the implementation of this type of pervasive change takes two approaches.

The Greenfields Implementation Approach

The first change implementation approach is commonly called the "greenfields'' approach. A classic example of this approach is the Saturn division established by General Motors in the U.S. This approach establishes a new management team, staff, and workers at a new facility. A separate team of implementation ''assisters'' of systems, accounting, etc., is also created.

To get true separation from the existing environment, a number of techniques have been used. In some cases, the people put into this situation are not returnable to the existing organization; sometimes this is part of the hiring or transfer process. The perception is that they will become so different, given the culture of the new organizational element, that they would be ''unhappy'' and ''reactionary'' in the existing organizational structure. Sometimes the current leadership is afraid of how this new element's success, or its ideas, might affect them.

In other cases, management has more confidence, but wants to ensure success for the idea first. Then these ''new organization'' people are assimilated back into the old organization to input the experiences and knowledge of what works and doesn't work in the ''new organization.'' These people can be catalysts for change if the inertia of the existing processes does not go on too long and/or become overwhelming to the point that the returnees ''give up.''

In either case, the greenfields approach has significant limitations. It is very expensive and time consuming to set up. If the competitive situation requires ''radical change'' or ''breakthrough'' improvements, then what happens to the existing operations while the ''experimental'' approach is being tried? What happens when that idea is successful? In a complex product manufacturing organization, the real world situation is that the level of change required to have a fully operating CE Design process will require at least several years, even in a greenfields operation. This is due to the technology required and the cumulative rate of individual change necessary. Additional time delay might result in unrecoverable loss of market share. The loss may never be recoverable if a competitor is employing these techniques or has other inherent advantages as described in Chapter 3. In many other competitive shifts of this nature, delayed response has been fatal.

The Transition Approach to Implementation

The second approach to the implementation of CE Design is to transition. In this approach, the present is ''brought into the future'' with improvements along the way. Products are still produced with increasing world-class product and organizational characteristics. This sounds logical and simple. Because transition must deal with the past, a more complex present, and at the same time technological and other constraints, it is perceived to be more difficult than the greenfields approach.

Transition has traditionally been attempted within the same planning approaches used by the organization to plan regular activities with some additional considerations for change impact. Often, because computer systems are involved, the computer project planning approach based on a standard Systems

Development Life Cycle (SysDevCycle) is employed for the changes, computerized or not. This has generally not been very successful because most planning processes have "end-products" or "success points" or "targets." These imply a team effort and significant extra effort. At the "end," after final data conversion and a brief period to eliminate initial problems, the organization can breathe a sigh of relief and return to regular work using the "system" to accomplish that work. However, CE Design's basic tenet is that the nature of the work itself must change.

CE Design has been described as a "manufacturing of information" business process. Chapter 8 generally described the Application, Information, and Automated Infrastructure Support Architectures necessary for full CE Design to be realized. The traditional system development life cycle (SysDevCycle) Information Planning process is particularly inappropriate in the case of significant impact efforts, such as the implementation of CE Design when the emphasis is on business, technical, and managerial "process" changes.

Figure 9-1 depicts the systems development level of effort and cost over the

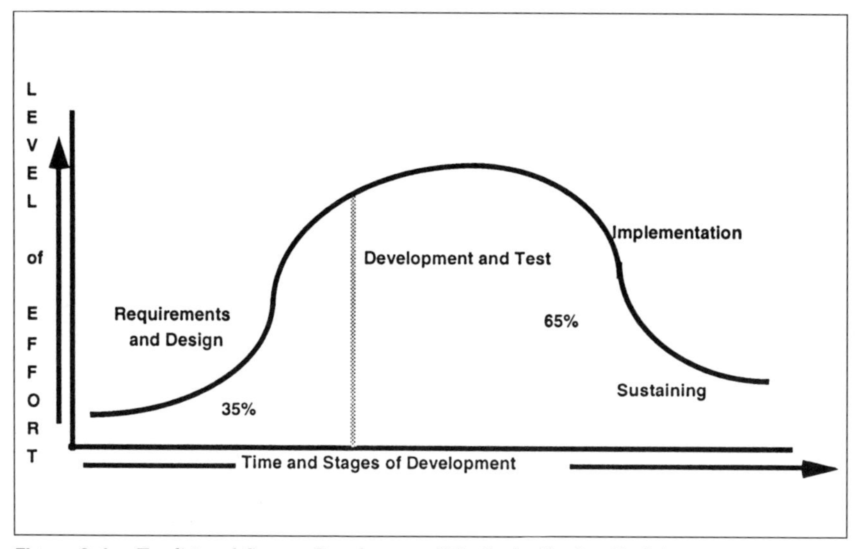

Figure 9-1. *Traditional System Development Life Cycle (SysDevCycle).*

time of the development effort. The traditional SysDevCycle has two important and built-in assumptions. The first is that the system envisioned is going to go through two states—implementation and maintenance (sometimes called sustaining which is a combination of enhancements and maintenance). The perception is that the overall design and most of the detail will not change or will change very little for many years after the first full implementation project.

Many of these systems and their intertwined business, technical, and managerial processes (now called "legacy" or inherited systems) are old. They continue to be subject to slow apparent design drift. Most of these legacy systems are actually Level 1, Business Function-type systems. What the systems group does, as an implementation technique, is reproduce what already is being done manually. Buried in the programming are the methods, forms, and procedures used before. The end of the first implementation cycle, as long as it produces at least 70-80% of the then present manual processes, is close enough to the overall manual system so that the design is also close enough. All that is added are enhancements, legally mandated changes, and gradual improvements. Additionally, the business processes supported are usually administrative in nature and do not change much.

Sometimes, the technology has changed enough for these applications to be "reverse" or "re-engineered." In this software re-engineering approach, the application programs are processed by other programs which restructure the old program's logic statements to make them more understandable and give them "new life." A small and active information industry segment has provided tools for this process. The assumption in the legacy systems arena for some time is that systems go through the life cycle shown in *Figure 9-2*, and that these legacy

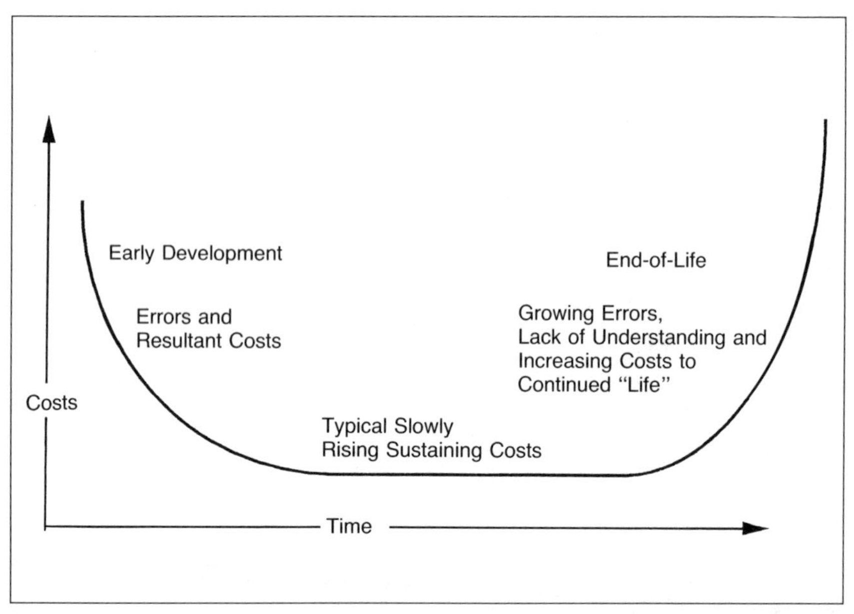

Figure 9-2. *SDLC Cost and Error Cycle.*

systems are now far past their original intended replacement date.

The second SysDevCycle assumption is that a different caliber of person is needed for the initial development project. This is where the systems individual gets to use the current technology and do "new design" and "original programming." Some view the maintenance systems activity as a "training" opportunity and/or an environment for the "average" systems person. Unfortunately, the reverse is true. Currently, it is more difficult to plan and execute in a transition environment (sustaining in the face of the kind of change contemplated for CE Design) than a greenfields environment (traditional SysDevCycle new systems environment). It is also more difficult to execute substantial transition in the information systems area during CE Design's implementation.

The typical SysDevCycle systems planning process uses a "derived futures" approach to address new opportunities. It assumes that if one can just get the requirements of the future understood, as in the case of the typical SysDevCycle, implementing tomorrow is sure to follow. A project like CE Design looks different, but the typical project starts with the basic tenet of imposing tomorrow on today. Just having management tell them "to do it the new way now" is a typical project rallying cry. Unfortunately, there is too much to accomplish. It takes too long. Temporary measures are instituted, technology changes, requirements change (or were never completely understood because they were so different), and eventually the organization rejects the project.

Figure 9-3 depicts how the amount of change between today's processes and

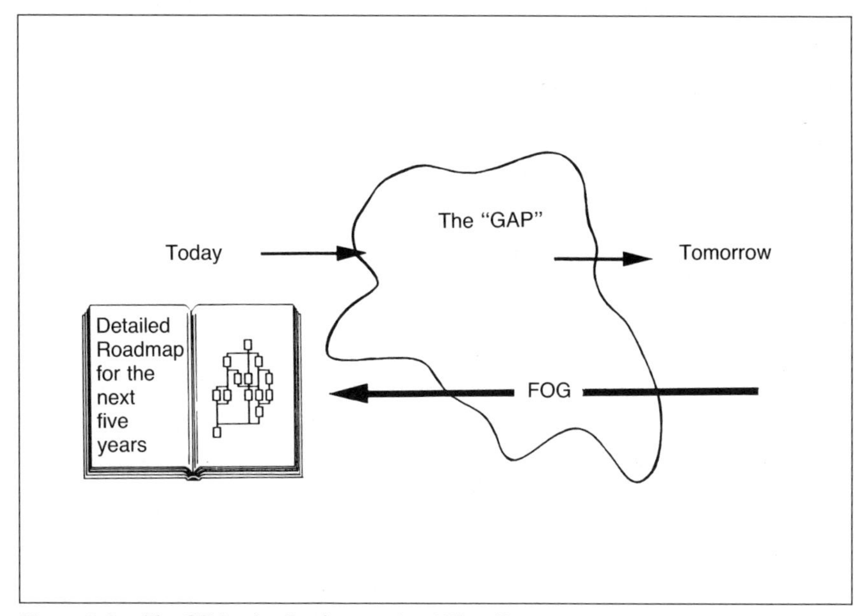

Figure 9-3. *The CE Design Implementation "Gap."*

systems and tomorrow's new processes and systems is too great. It is "fogged" over with too many unknowns and "misknowns." It appears that this type of planning and execution approach is too flawed to repair, at least in the case of the high value opportunity, such as CE Design, in a transition environment. In a greenfields environment, legacy systems and processes are not significant factors except where they interface with upstream or corporate systems for reporting and control purposes.

The final significant factor creating a need for a different implementation process for CE Design is the extendability of the implementation strategies. The business, technical, and managerial processes presently in use in most areas of complex product manufacturing organizations are, as described in the preceding chapters, not extendable in the face of the increasing complexity of product, production process, organization, and business competitive environments. Unfortunately, the "derived future" SysDevCycle approach to the implementation of an integrated single-entity system and its technical and managerial processes and systems is also not extendable in the transition environment. There might be hundreds of individual projects and improvement efforts necessary. Technology and the future vision will drift, and its first implementation might be just 1-5% of the final set of desired processes. CE Design's implementation must be planned to never end if a relative competitive position is to be gained and maintained.

The "Different Implementation Process" Based on Forward Planning

Fortunately, a structured version of what we actually do can be used to address the implementation of CE Design. *This "new" implementation approach's basic assumption is that the complex product manufacturing organization is itself a system of business, technical, and managerial processes.* What we actually do is (1) look at today; (2) try to "envision" where we want to go; and (3) try to move in that direction. This process is simply termed Forward Planning.

Forward Planning is based on several key concepts. The most important is that planning is a closed loop, continuous improvement process for which there is no "end," just transition state changes throughout a continuous improvement process. *Figure 9-4* summarizes various aspects of the Forward Planning concept.

Referring to the numbering of steps in the process illustrated in *Figure 9-4*, the general Forward Planning approach is to (1) establish the process vision for the improvement. This process vision is represented by the process maps in Chapter 7, and the network organization structure mentioned in various chapters. In this case, this book represents the (2) CE Design vision, as supplemented by the many significant efforts in the areas of manufacturability and the many technical aspects of automated infrastructure support which are out of the scope of the book but are described in other manufacturing and computing books. A computing architectural vision (3) to support the process (CE Design) is also established. Chapters 5 through 7 describe a management framework for CE Design processes (2). Chapter 8 describes the vision for the architectures and their automated infrastructure (3). These architectures will reflect the capabilities required to support the CE Design business, technical, and managerial processes which will occur as the present is transitioned into a never-ending chase of the future. These vision statements become the goals of the planning

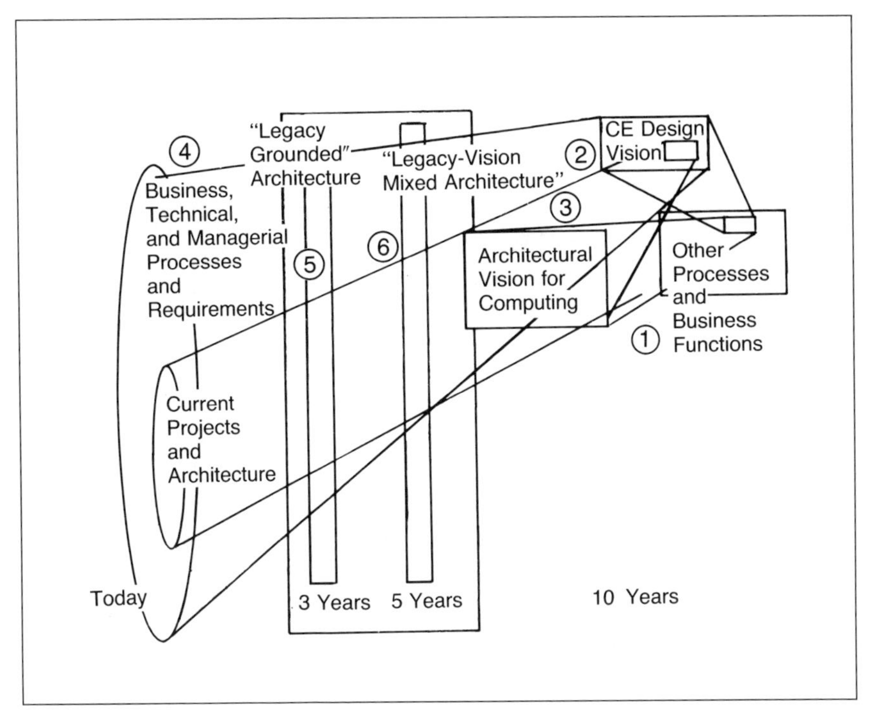

Figure 9-4. *Forward Planning Overview.*

process. Requirements (4) necessary to transition the present toward the future vision are determined. Those capable of being achieved are included in a plan to achieve change, but still based on the present architecture (5), also termed a ''legacy-grounded'' architecture. As this transition is being accomplished, a second intermediate goal, the mixed vision architecture, is determined and planned, and transition takes place. Eventually, after ongoing adjustments, the vision is achieved. However, before (6) is ever completely reached, a new vision has been established and the cycle repeats itself.

In the *Figure 9-5* series, the Forward Planning Process is applied inside a yearly cycle of a paper-based process. This process could be automated using a CE Design automated architecture. As the vision plan is prepared (a), it is used by those establishing user requirements (a1). Commercially available software (a2) and commercially available infrastructure support (a3) information summaries are also updated. The latest application architecture (a4) and automated infrastructure support (a5) plans, based on the current situation at the start of each planning cycle, are updated and prepared. The now updated user requirements are then prepared (b). This process constantly provides users and management with the ''then current'' situation and the near-term plans for architectures and their components. As shown in *Figure 9-4*, the objective of the architectural portion of the *Figure 9-5* planning cycle (d1, d2, d3) is to prepare and monitor compliance with the architectures, current application computing

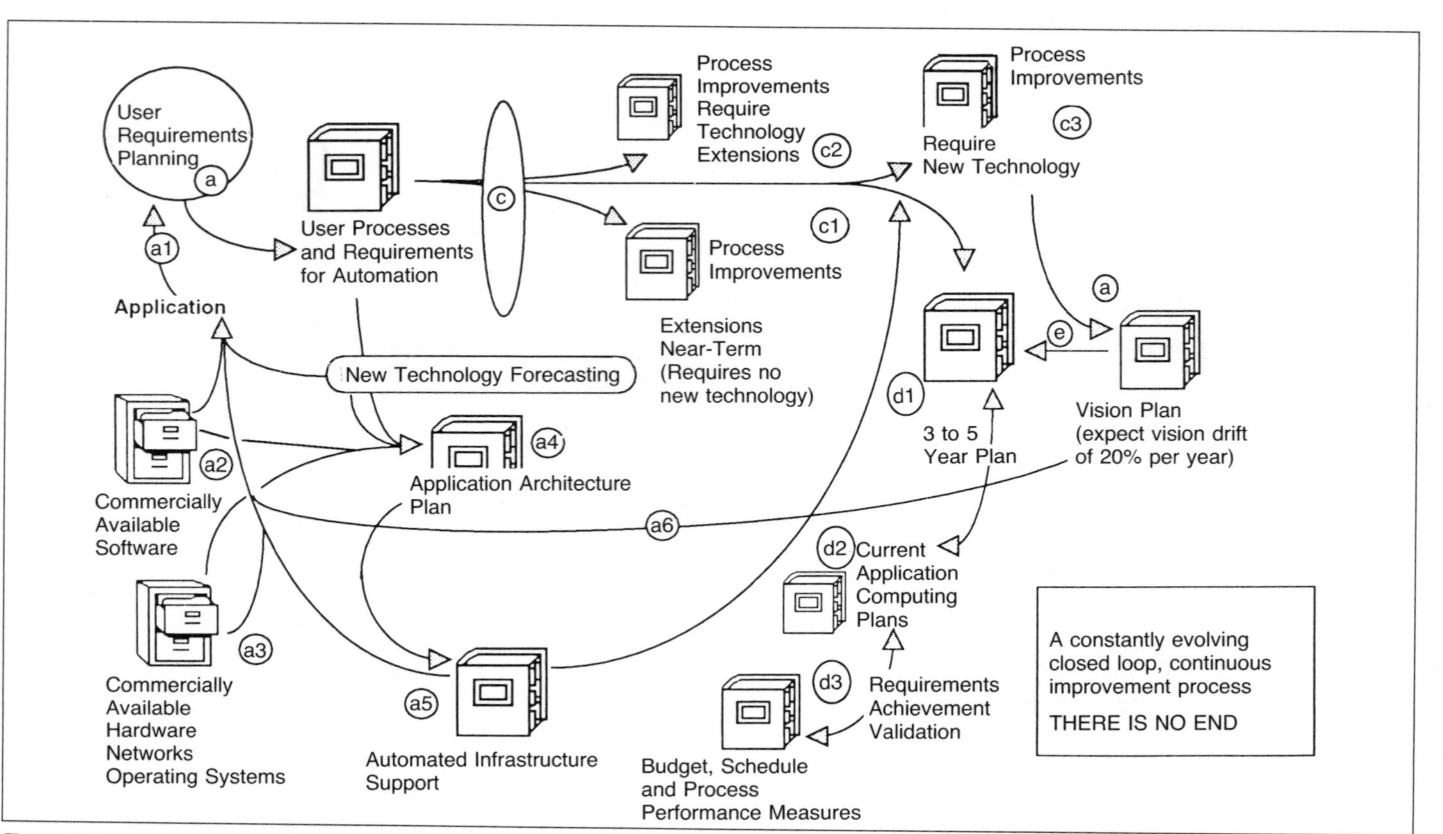

Figure 9-5A. *Forward Planning Document Structure and Process.*

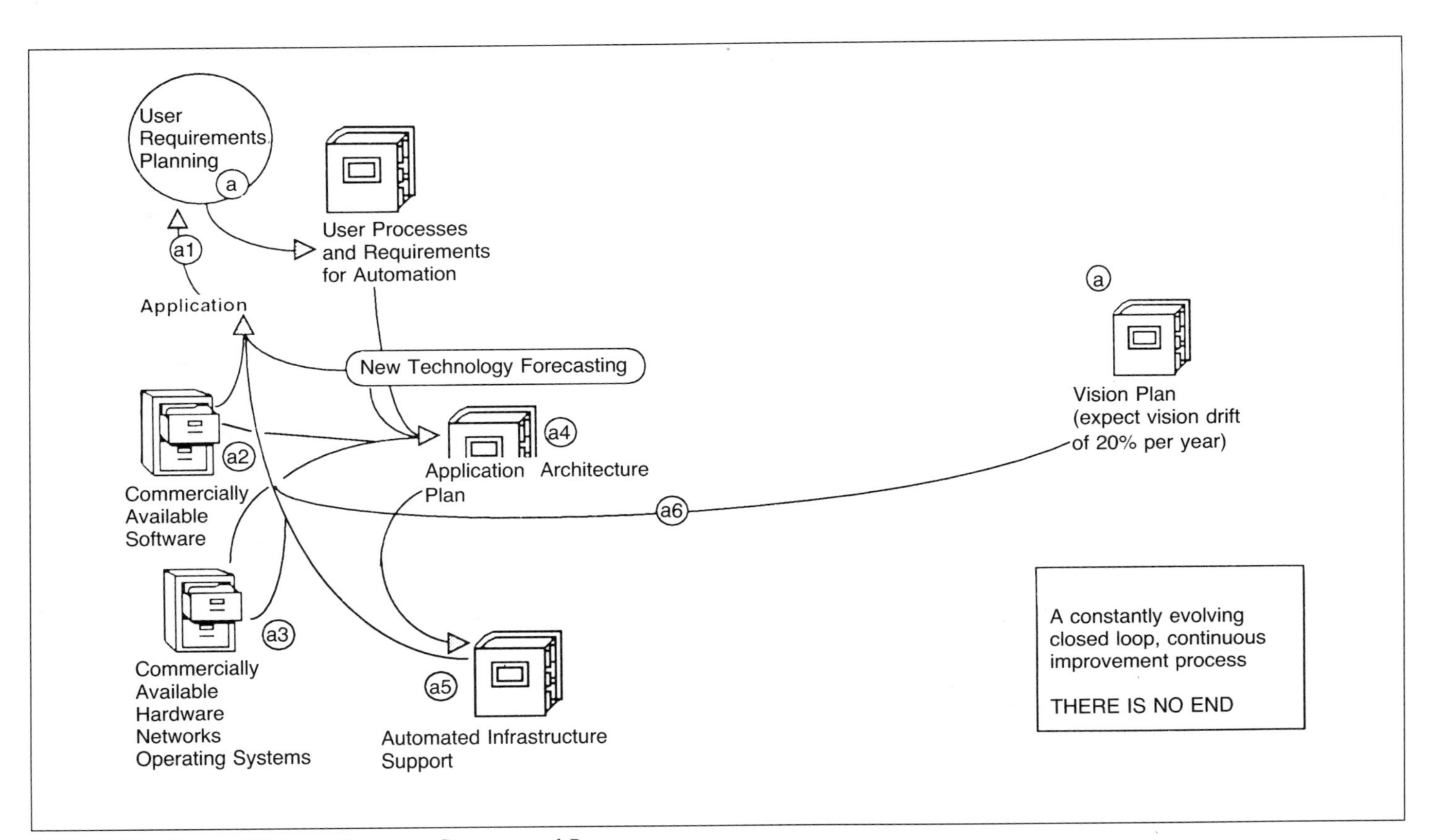

Figure 9-5B. *Forward Planning Document Structure and Process.*

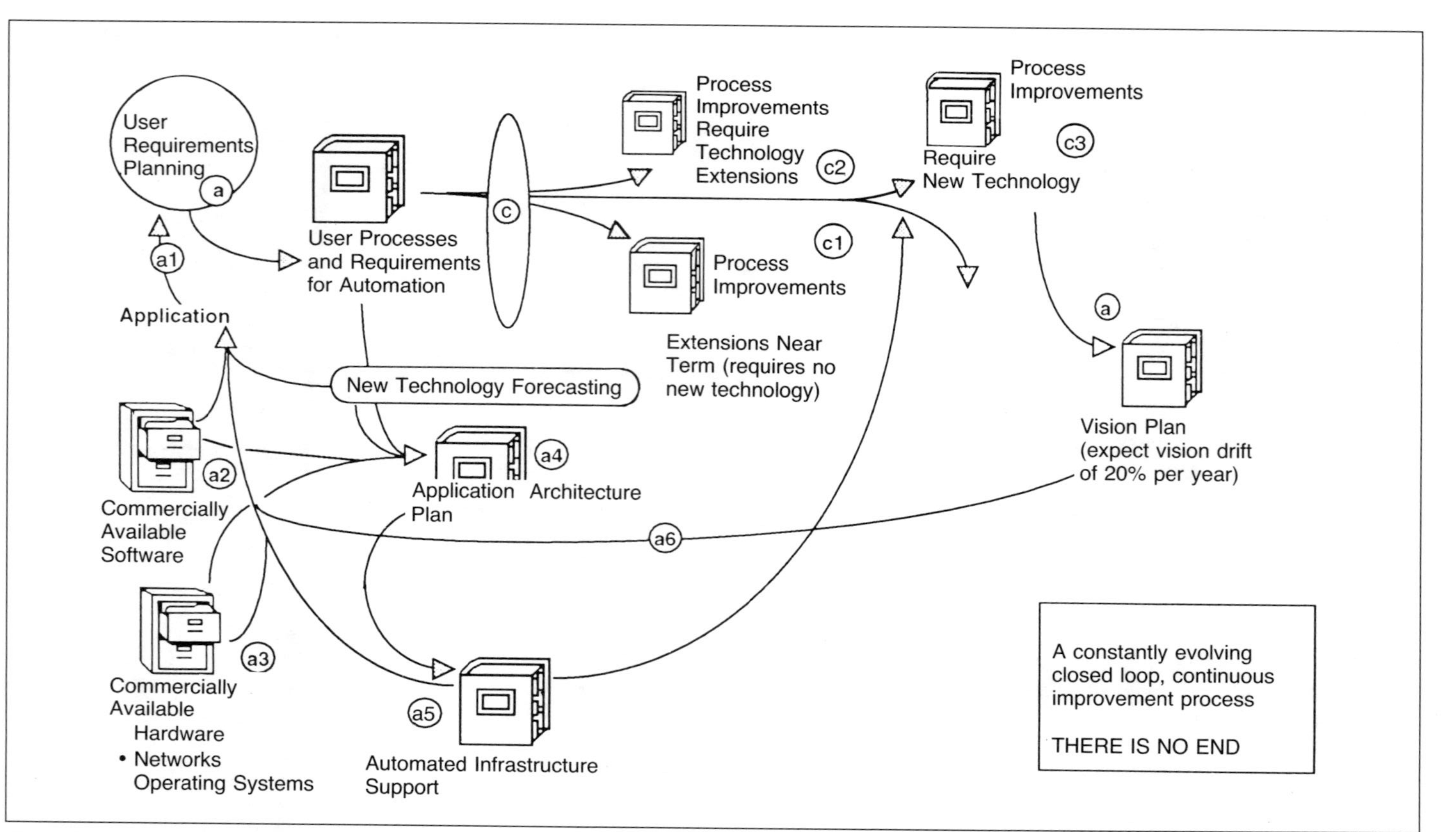

Figure 9-5C. *Forward Planning Document Structure and Process (continued).*

plans, and their corresponding budgets, schedules, and project-related performance measures (what projects generate what benefits, and how to recognize and deliver them).

These application and architectural improvements are generated from a requirements analysis which separates these requirements into near-term extensions of process improvements which can be accomplished now (c1); those process improvements which require, after due consideration, the development of new technology and/or systems to deliver this capability (c2), and those process improvements which require new technology that the organization has decided to wait for (c3).

Figure 9-6A shows the large complex manufacturer with significant comput-

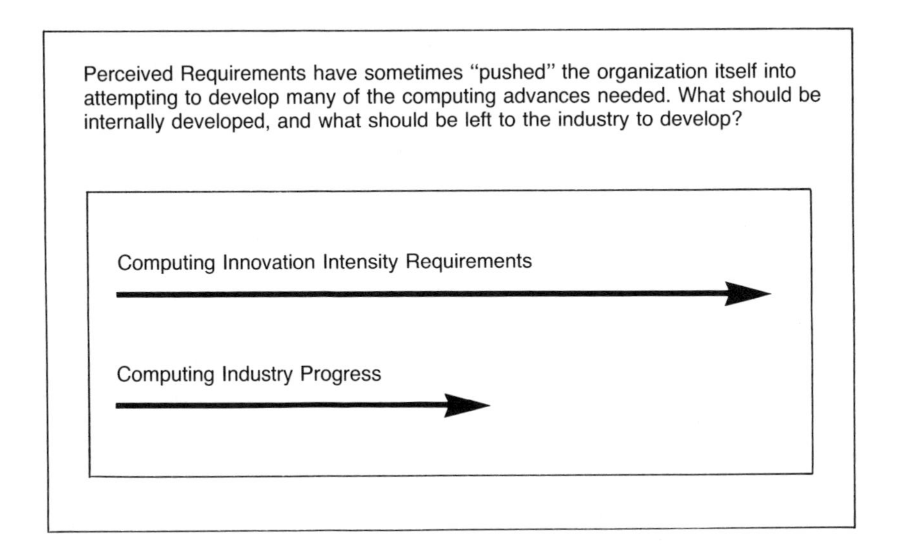

Figure 9-6A. *Pushing Technology.*

ing resources. At times, a manufacturer must decide whether to wait for the required technology to become available, do its own development, or work with a partner. There are, of course, cost, benefit, and other business trade-offs to this analysis.

In addition to some of the consequences in both directions described in Chapter 8 of this "make or wait" decision, there is another planning implication for certain types of computing systems to consider when making decisions about computing. An example is the use of 3D CAD in manufacturing. This type of intensive computing requires new classes of applications and new types of storage mechanisms and management tools.

However, once this technology is used on a major complex product, there is no turning back. All subsequent derived products will use this technology, and the systems to support this approach must continue to grow rapidly in capability.

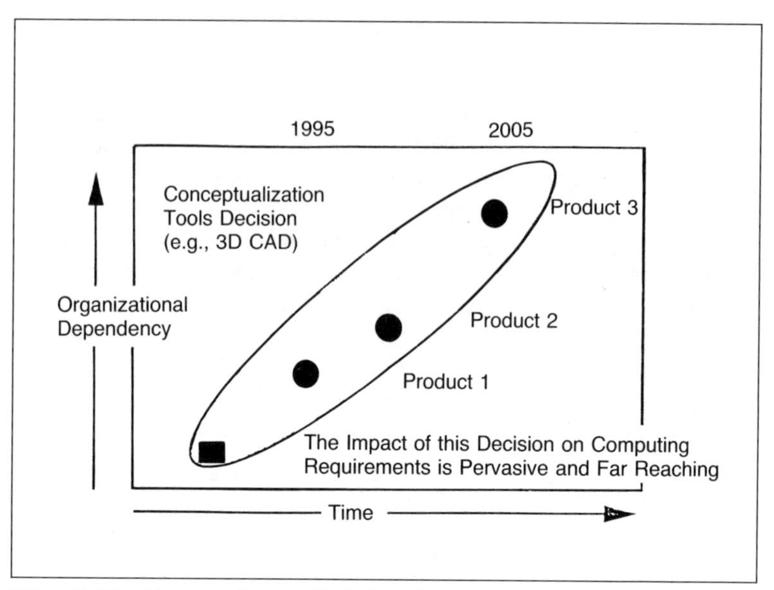

Figure 9-6B. *Conceptualization Tools Decision.*

This "once started" situation is especially prevalent in the area of conceptualization tools. Ramifications of the "no turning back" decision include higher computing storage costs, higher staff and computing costs, license fees, and increased dependence on outside computing software suppliers, and CAD system "experts" retained for their computer CAD skills, not their design or engineering skills. *Figure 9-6B* depicts how increasing commitments and their intrinsic obligations grow over time.

The entire planning process shown in *Figure 9-5* is just another model-driven process. The use of CE Design's process management, design, and manufacturability processes is entirely appropriate to the production of engineered software. Software can be a product which is manufactured, and the manufacturing process can be managed by CE Design and its supporting downstream systems with adjuncts peculiar to software. This CE Design approach to software management can generate the same leverage as CE Design does for conventional process physical systems and other higher level of abstraction systems products. The most important outcome of this change to in-line model management of the software manufacturing process is the return of computing management to traditional management. An experienced CE Design process manager can manage internal or external software systems as any other product development effort would be managed.

Another element of Forward Planning is its ability to permit differentiation between interests at various levels of management. As an element of the discussion of the Process Management Process in Chapter 5, these interest levels were first depicted as layered.

Figure 9-7 and *Figure 9-8* explain the different views or interests which

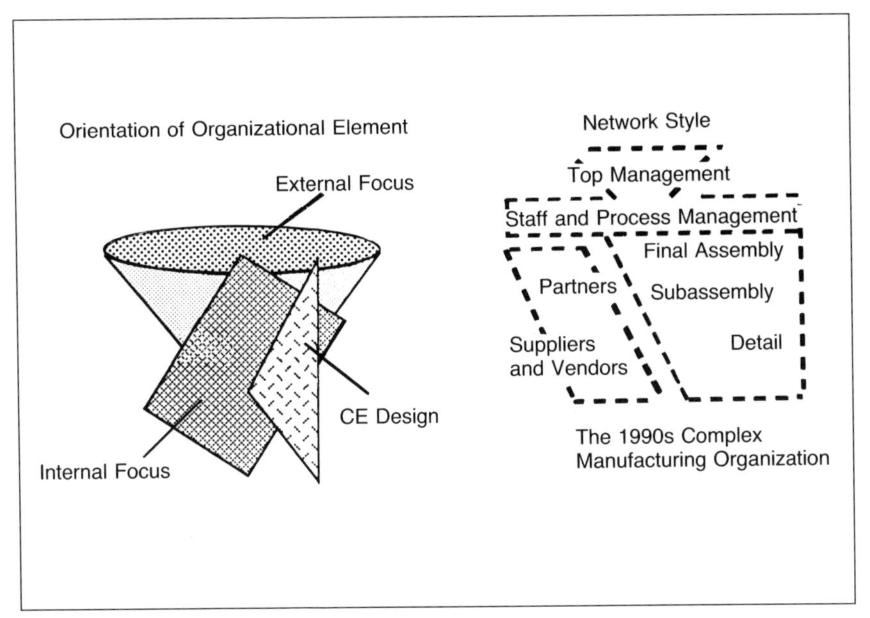

Figure 9-7. *CE Design Within an Overall Interest View.*

different levels of the organization have in the CE Design process and its process and activity components. The concept of process decomposition is also depicted, as is the concept of applying the CE Design process management approach and its management techniques to software which needs to be engineered. In Chapter 6 "levels of abstraction" were introduced. In *Figure 9-8*, these interest areas are expanded into a high-level requirements vision for systems to manage the CE Design process.

In *Figure 9-7*, the executive needs a very broad view outside the firm. The top executives may spend 80% of their time looking at customers, markets, competitors, and other significant external factors which may affect their organization. They are typically supported by staffs who assist in the outside view, as well as in converting this external view to plans and actions inside the organization. The 1990s view of the complex manufacturing organization as a network is compared to the "Field of View" in *Figure 9-7* to represent this external execution focus.

Some elements of CE Design's overall process have an external focus. Examples include new technologies (for products and processes), customer product requirements, and potentially applicable research. When the overall interest view expands to include middle management in the role of final assembly process production managers, the internal focus increases substantially, as does CE Design's. Finally, this vision expands to include business partners (shared design responsibility), suppliers, and vendors (no design responsibility).

In *Figure 9-8*, the relationship between the internal views of various

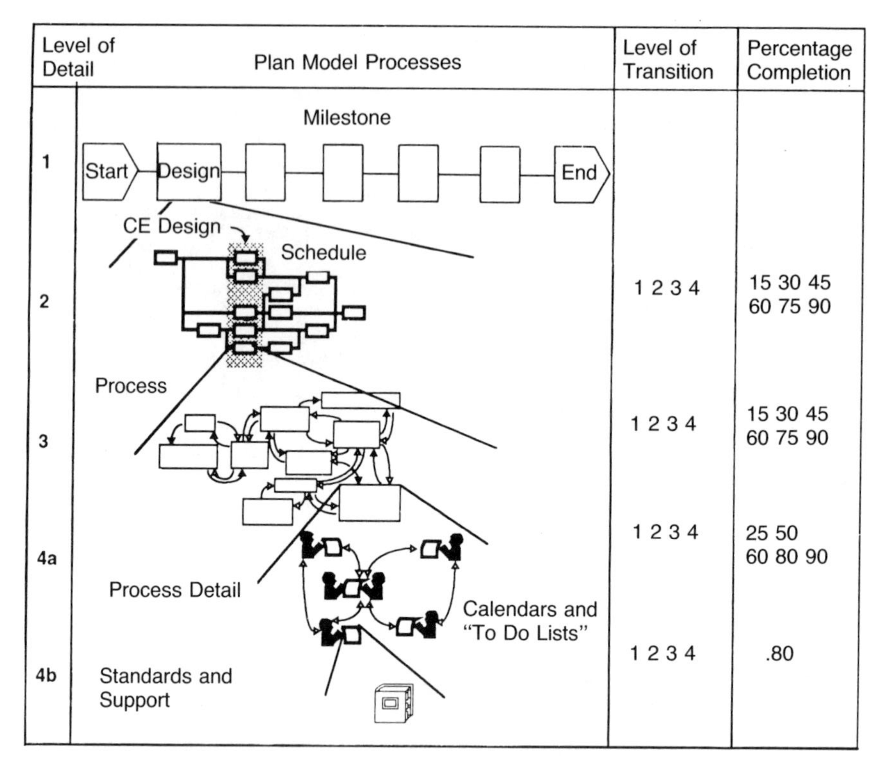

Figure 9-8. *Forward Planning Automation Project Decomposition.*

interested parties is depicted. The Executive (Level 1) is largely interested in goals, objectives, costs, profits, and progress reports about planned major events. As the major events are translated into projects with schedules, the CE Design processes, associated schedules, and the top level major process managers become involved (Level 2). As the individual model of the process is developed, the lead project engineer manages the teams with process management techniques (Level 3). Individual teams deal with project details, checklists, authority concerns, and other design related process and design management issues (Level 4a). Individual engineers use calendars and checklists to complete their activities associated with the process to be executed (Level 4b).

As the organization goes through the multiple transitions planned during the Forward Planning exercises, *Figure 9-5* also could be used to measure progress toward the levels of CE Design. The Level of Transition column and the percent complete column can be used to track progress transition goals, and rules of progress at one level (1, 2, 3, 4a, 4b) before proceeding to another.

During the Forward Planning process, the businesses' goals and objectives also must be considered. *Figure 9-9A* illustrates a decomposition of one of three standard complex manufacturing organization goals as they relate to elements of a CE Design process. Notice that the goals of growth, profitability, and quality

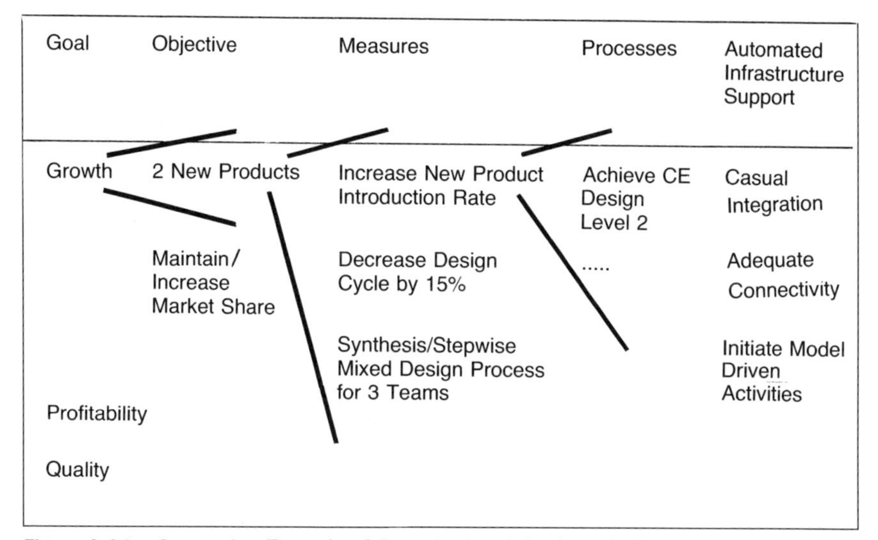

Figure 9-9A. *Integration Example of Organizational Goals and Objectives with Forward Planning.*

can be directly tracked and linked to each project, and measures of success and contribution can also be established. This planning process provides the very important linkage between business and process computing and CE Design Implementation goals; i.e., their justification and rationale. It also sets realistic expectations for management's achievements and contributions, establishing when and in what form the benefits should be expected. In *Figure 9-9A*, the goal of growth is used to illustrate linkage. The objective is to introduce two new products. One needs to meet several measures of success; these include to "decrease design cycle by 15%." For the "increase new product introduction rate" measure, there appears to be a need to reach CE Design implementation Level 2, which has the automated infrastructure support ramifications.

The other important result of this linkage of process computing and CE Design Implementation goals to business goals in the Forward Planning process is *pacing*. At what pace or intensity should the transition to CE Design be pursued? Which improvements should be made first? When process improvement projects are directly linked to business goals, their potential impact can be "rolled up" with other elements to be considered in achieving goals.

In addition to the perceived benefits, an analysis of the risk of achieving the desired result for a particular level of expenditures must be considered. *Figure 9-9B* depicts the range of potential results. The size of the lobes as they reach out toward the improvement target reflect the relative number of projects which, on

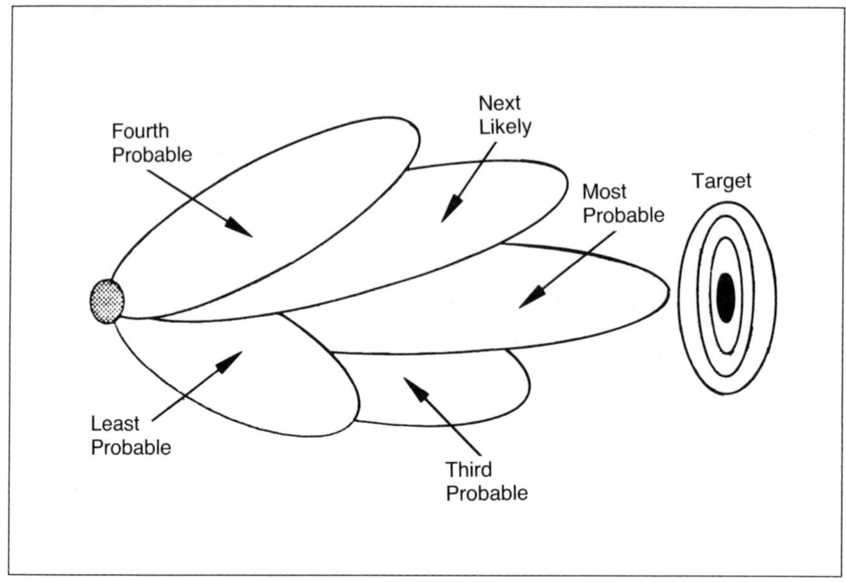

Figure 9-9B. *Computing Expenditure Risk Assessment.*

average, will achieve that type of result. Including a formal risk monitoring process in the Forward Planning process model minimizes the occurrence of undesirable results.

CE DESIGN IMPLEMENTATION SHOULD USE THE FORWARD PLANNING APPROACH

CE Design is a set of interrelated business, technical, and managerial processes. Implementing CE Design in one implementation project is usually not reasonable. One approach which Forward Planning envisions means evolving the organization to the final CE Vision. Forward Planning should be employed for CE Design because both CE Design and a planning process such as Forward Planning support relatively unstructured work and evolving processes.

Figure 9-10 shows each integrated process which makes up CE Design at the Business Function, Business Process, and Process Management levels. It shows that they should be part of the planning process. For each integrated process, an implementation strategy should be developed. This strategy should consider organizational goals and objectives, process improvement objectives, and technology opportunities. The strategy should also include achieving CE Design Levels of Improvement.

Achieving Levels of Improvement as a Focus of CE Design Implementation

Full achievement of CE design capabilities will not be completed overnight.

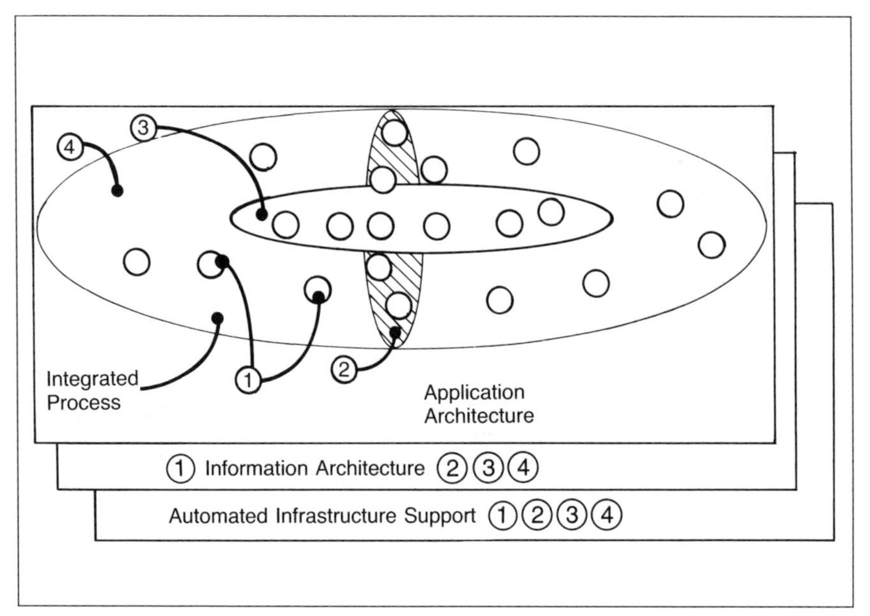

Figure 9-10. *A CE Design Implementation Approach.*

Full achievement could require several years or more. How, then, can the organization progress, improve its competitive position, achieve benefits, lower costs, and not "be there yet?" The concept of Forward Planning has at least two identifiable, intermediate, and achievable objectives which show substantial progress, yet do not waste time, effort, and competitive position. For CE Design, the achievement of levels of improvement tied to Forward Planning transition points has been a workable approach to this progress issue.

Achieving levels of improvement is a central CE Design implementation concept. As shown in *Figure 9-10*, improvements (Level 1) can first occur as isolated improvement points; they should be planned to promote transitions to full CE Design. Improvements to the evolving integrated processes, to the application systems, to the information architecture, and to the automated infrastructure should all be coordinated. Some change not supporting these objectives should be permitted. However, do not focus particular attention on suboptimal improvements that improve a process at the "expense" of another. Analyses and heavy investment improvements of processes and systems not part of the current highest priority improvements should be avoided. Available resources not directed at CE Design priorities should be focused on lower cost process improvements using CQI.

When several of these improvements can be made laterally, then Level 2, process management and Level 3, business process improvements can occur. These next levels are possible because in-line process models can interrelate these improved processes. Finally, when enough of the overall process is improved (at least 70% to ensure design integrity), then the fully integrated process (Level 4) can be instituted. The movement between levels can be as fast as technology, process, and people can allow. This entire process could be termed a "roll up to success" strategy. This strategy is forward-planned as improvements are made, priorities are revised, and new plans developed and completed.

This process improvement strategy should be developed and tracked in three increasingly detailed categories:

1. CE Design,
2. Business Processes, and
3. Business Activity and Supporting Functions.

These categories should have their own measurement processes and success parameters. At the overall CE Design Process Level of Detail, the four implementation levels are summarized in *Figure 9-11.* At Level One, CE

Level	Process Description	Simulation	Percentage Complete	Systems Perspective	Comments
4	Essentially Concurrent Design	Leads to Design	>80%	Concurrent Operations	Significant Concurrency; appears to be simultaneous
3	CE Design Processes in a serial environment	Validates Design	>70%	Single View of data and systems	Electronic Resolution across non-common platforms
2	Concurrent serial operations	Validates Testing	<60%	PDES (formalized interfaces)	Coordination required and "in-line" models of processes executed
1	CE Design tools in a serial operation	Testing Element	15-25%	ad hoc depends on cooperating individuals	Typical starting point

Figure 9-11. *Increasing CE Design Levels for Integration Completion.*

Design-type information tools are in use, but the coordination between groups depends on individual initiative. The use of the evolving Product Data Exchange using STEP (PDES) as a formalized information-sharing mechanism, along with many other open architecture standards, is part of what can aid in making this transition to Level 2 of CE Design using the serial or step-wise refinement design process.

The most crucial improvement is made at Level 2 when coordination becomes formalized. At Level 2, in-line, model-driven processes become active. For the information to be shared across an automated infrastructure, the process management capabilities of Routing and Queuing, Configuration Management, and Resource Management should be active. It is not necessary that all functions be automated when this transition is made, but the remaining manual tasks, plus the cost of the automatic improvements should, even initially, be more than offset by the savings generated by speed, accuracy, and reasonable product quality improvements.

Why are "in-the-loop" model driven processes so important? One of the dilemmas of systems and processes is the matter of process understanding and translation. Increasingly, there is an emphasis placed on developing models of processes. Problematic in this approach is that these models are developed by people who are not experts in process operation. Additionally, they turn the models over to others for software development. This "multiple hand-off" creates misunderstandings. As the process description is passed from person to person, each "hand-off" creates an error-making opportunity. Also, the model of the process which is developed rapidly becomes obsolete as changes to the real-world process are made. As a result, an expensive and time-consuming process model maintenance activity must be instituted. Finally, the process model may not prove useful. Gaining a representation of a process via a model is only useful in this context if it can change. Because process changes are targeted in Forward Planning, the models of other processes may be useless until late. In-line models resolve this dilemma by being actively executed and changed as the process is executed.

As shown in *Figure 9-12*, these in-line process models should be developed in no more than six levels, with five degrees of increasing granularity. This example has five. Most important, the person actively engaged in executing the process should directly build the model of the process to be executed as a natural element of the job. This is why the Process Management "in-line," model driven, process capability is so important, and should receive such a high priority so early in a CE Design implementation strategy. When models are built as they are being used, they are inherently accurate, up-to-date, and without ambiguity because they are driving the execution of actual work. This simple concept is very powerful. Because there is a complete model to examine, the manufacturer returns to the basic systems development concept of "process pull," when technology and Implementation Strategy are considered. As described in Chapter 8, it moves the system developer out of the model development process, and allows system development to become another standard business process managed activity set. It permits simulation of proposed process improvements. It provides a base model from which evaluated process models and variants for particular conditions may be developed. It allows the automated infrastructure and application architecture continuous improvement efforts to proceed using various levels of granularity while actually executing models.

Prior process improvements can provide the basis for proceeding to Level 3 of the four levels of CE Design Integration and Evolution as described in *Figure 9-10*. The importance of achieving the Level 2 to Level 3 transition capability—simultaneously managing differing levels of process granularity within in-line process models as in *Figure 9-12*—cannot be overemphasized. This linking of process elements provides, for the first time, the ability to begin to demonstrate the single flow and single view of CE Design information. This linking also

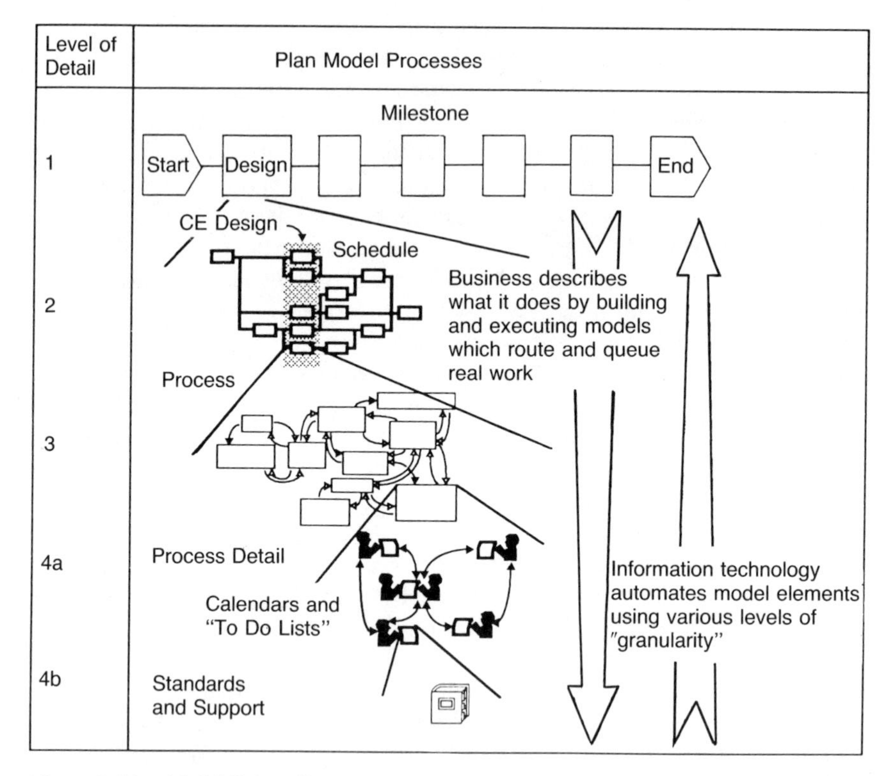

Figure 9-12. *Model Driven Processes.*

provides the environment in which Level 3 can be achieved.

Level 3 provides a single view of data, systems, and processes. In a typical transitional situation, the process model has been built and improvement projects instituted to first build interfaces between various automated and manual systems and procedures at various levels of granularity. Each "granularized" process element is surrounded by an object oriented information envelope and message-oriented architecture which captures defined inputs and outputs. Implemented message handlers and object-oriented "envelopes" enable the passage of this information to other process elements.

The Routing and Queuing Model monitors progress, reports status, schedules resources, and performs other tasks associated with Process Management. Usually, at least one element of this transitioning process is a "legacy system" which has been encapsulated, and purged of previously needed "redundant" activities. These redundant activities reflect the usual necessity for each system to be self-sufficient. Each single-entity system has to be designed using an approach which did not consider any other system. Many times it has to carry, manipulate, and report data across redundant or similar reporting processes.

The power of the "in-line" model can be seen in this mixed systems situation. Improvements can be made as the process continues. Many improvements are invisible to the process owner or active user; many may only need to be present to turn off redundant and competing system functions.

In *Figure 9-13*, Level 3 CE Design operation permits any level of the

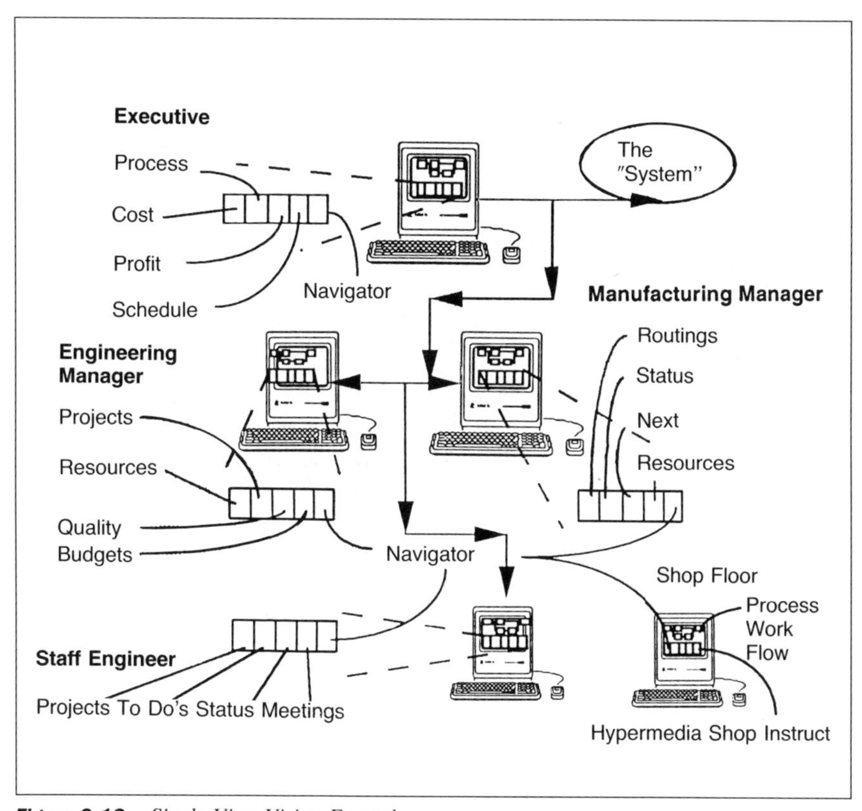

Figure 9-13. *Single-View Vision Example.*

complex product manufacturing operation to view and manipulate information, at any level of detail, through a workstation without regard to where the process is executed or the information is located. The executive can view the highest level of granularity CE Design process model and monitor overall cost, projected profits, schedules, and other related processes, such as marketing, sales, and competitive analysis, which would benefit from the use of the CE Design planning, implementation, and information architecture and automated infrastructure support approaches. Because of the potential complexity of the total process model-driven information architecture, a "navigator," or intelligent guide to information of interest, is also provided at each level of the organization.

Provided within each organization unit are appropriate levels of granularity for viewing executing processes and navigation tools. Engineering management, for example, might focus on project status, resource utilization, and queuing and routing problems (by exception). Manufacturing management might focus on schedules, incorporation schedules, process/routing statuses, or budget variances. Finally, the action users and builders of these models might monitor calendars, project status, queued work, or upcoming events.

The third level of CE Design evolutionary development, to Level 4 operation, is achieving essentially concurrent operations. In this operation, quality of the component systems has reached the point where several people can be utilizing the same conceptual tools, working on the same problem, and using the same information. So it appears they are simultaneously updating the same portion of the information "envelope." For example, in *Figure 9-14*, the object spool is

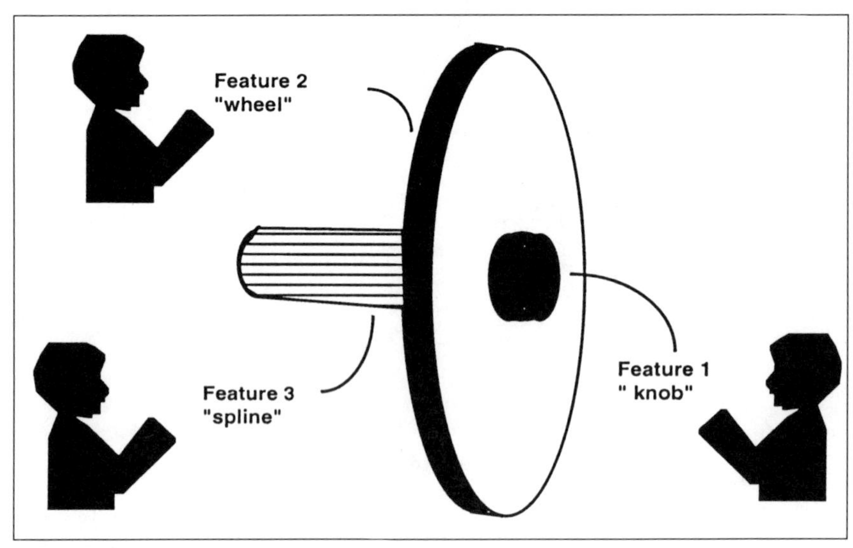

Figure 9-14. *Apparent Concurrency on 3D CAD.*

being simultaneously modified by several people. This is possible because the granularity of the configuration control of the graphical object is at the FEATURE level, as is the support provided by the appropriate CE Design Automated Infrastructure.

Figure 9-15 summarizes these major breakpoints from the automated system builder's perspective. The Level 2 process breakpoint of in-line process models is discussed in Chapter 8. The application architecture and automated infrastructure support work necessary to provide the "granularity" and "single-view" capabilities depend on evolving technologies and the treatment of systems as just another business, managerial, and business process, which is managed and measured as the rest of the organization's engineering and design processes. Software also can be manufactured using this implementation strategy and approach.

Level	Point Description	Comments
4	Users of Business Process Models operate in a *concurrent* mode of operation using "in the loop" simulation of the design and its operation	In collaborative style, using different elements of a design in a simultnaseous fashion such that the pieces are re-configured as made available
3	Software construction process drives up the process model decomposition	Supports various levels of "granularities" of objects within electronic "envelopes" across a single view of the information
2	Business Process Model Drives Routing, Queuing, and Process Management Tasks	• Models are not really useful until they are used • Can be forced by scanning in documents until Messaging E-mail, and the majority of Routing and Queuing capability can be assembled and constructed
1	CE Design Tools Become Useful	Laying the groundwork via "seduction"

Figure 9-15. *Breakthrough Points for CE Design with Automated Infrastructure Support Evolution.*

In *Figure 9-16*, there is a direct relationship between the "in-line" process models, especially those "nested" with higher level representations as in *Figure 9-8*, and the product differentiation separating the product in-design areas. There are a number of product breakdown approaches in design. These types of product differentiation include WBSs and decomposition. Product differentiation occurs when the product breakdown is planned around reusable components and systems.

In *Figure 9-16*, a sample "lifter" product is broken down in a differentiated fashion. Each level is equated to an equivalent level of process model detail. At the full product level, an executive level process model can manage the entire product process. Individual major subassemblies are managed using a process model matched to the needs of a product element. Smaller assemblies have their subprocess model as well. These submodels operate as integrated elements of the higher level process models. As the submodels are executed, teams coordinate their efforts using the various design approaches and "design for's ...." An example of their individual activities and the acceptable variation which can occur in these nested processes is represented by the idea of using cement in the leg base elements of the lowest differentiated element of the sample "lifter" product model.

As the CE Design Implementation Plan progresses through Levels of Improvement, much work in improving the application system architecture, information architecture, and their automated infrastructure support must be occurring. *Figure 9-11* showed Level 1, 2, 3, and 4 improvement projects in these areas occuring roughly simultaneously with the process improvements themselves. In addition, process models that guide the work can be constantly updated to reflect changing and improving circumstances. This update can occur as fast as is appropriate, and does not depend on computing resources for these updates. Integration proceeds at its own pace, as can process and architecture

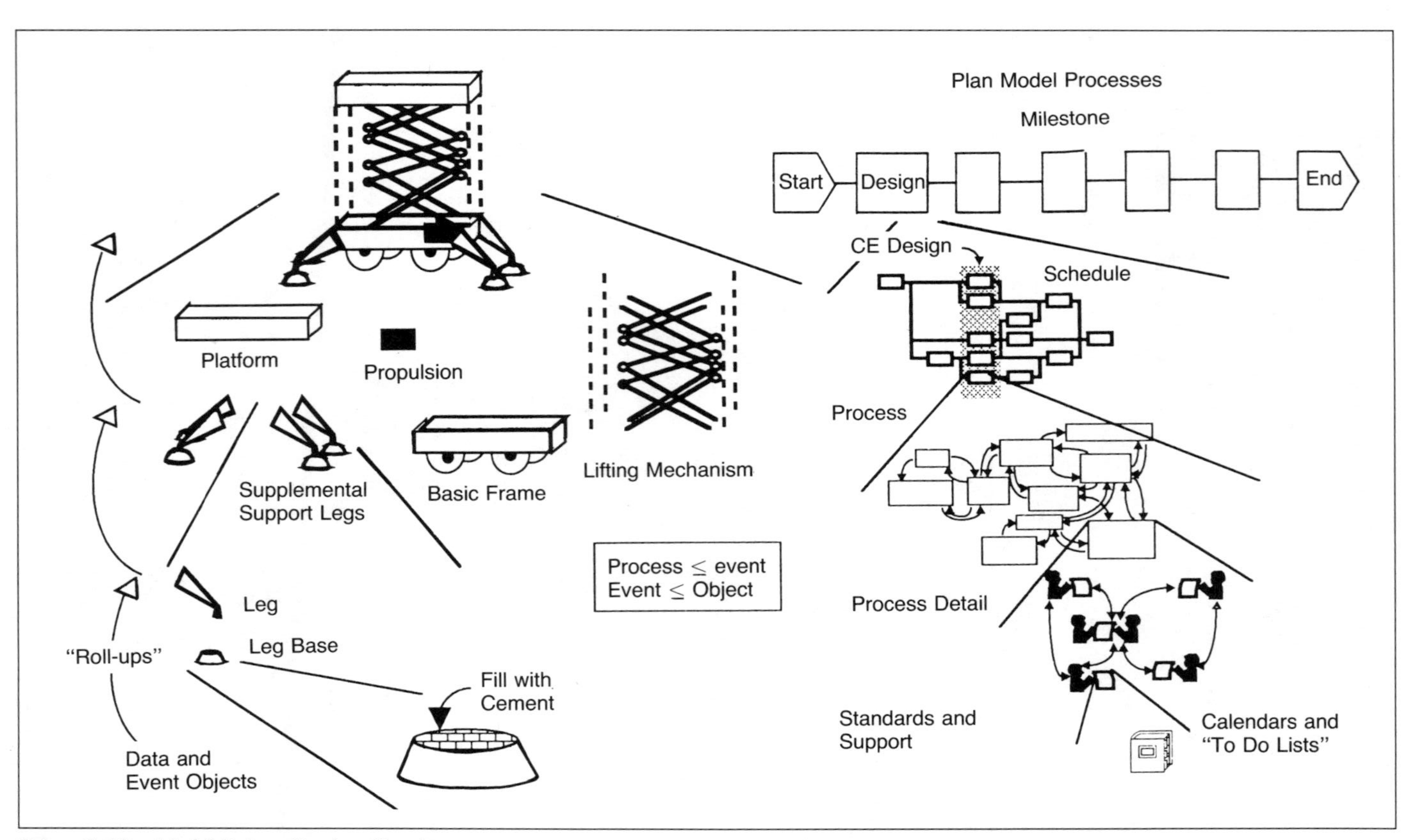

Figure 9-16. *Model-Driven Product Development.*

change. This highly flexible and multiple-dimensioned change environment is at the heart of the CE Design Implementation strategy.

Figure 9-17 tracks the three general types of integration (casual, exchange, operation) and some specific "hows" of each type. Casual integration is a prerequisite for Level 1 CE Design improvements. Formalized Exchange is a prerequisite for Level 2, and Operational Integration for Level 3.

Type	Casual Integration	Exchange	Operation
Principal Characteristic	Common Access From Same Device	Exchange of Information Across Interface without Manual Intervention	"Single View" of Information
How	"Windows" of Inconsistent User Interface Operations	"Windows" into Consistent User Interface	"Windows" into Consistent User Interface
	Re-enter Data	Automated Data Exchange Standards	Consistent Infomration Management & Integration
	Multi-system Access	Multi-system / Platform Access	Interoperability via OO Messaging Architecture
	Between Windows	Between Applications	Across Hetereogeneous Platforms
	Uses Facilities of Workstation to Move Data	• Uses Communications Facilities (Intra- & Inter-application Systems) • Uses Dictionaries & Directories as Roadmaps to Information	Object Oriented Message Architecture Treats All Information Types with Common Approach

Figure 9-17. *Integration Levels.*

Transition to Level 4, fully integrated CE Design, requires changes in the processes themselves, the application architecture, and automated infrastructure support, which are not derivative of existing and current capabilities in most complex product manufacturing organizations. *Figure 9-12* and *Figure 9-15*, as well as *Figure 9-18*, all describe the movement from Level 3 to Level 4 from various perspectives.

The most important differences at this point are no longer process support improvements. One might characterize the evolution of the first three levels to be preparation for fully capable collaborative, concurrent CE Design. This high-leverage process is the objective of the overall implementation plan for CE Design. Once the support mechanisms are in place, these powerful business, technical, and managerial processes can be implemented. During the implementation planning process for this portion of the Forward Planned evolution, particular emphasis should be placed on the relationship between the people and the processes.

Figure 9-19 shows that CE Design builds a "support envelope" for each individual in the organization. Thus, he or she is constantly touched unobtru-

Feature of Application Architecture	Automated Infrastrucutre Support
Intellectual Capitalization	Reusability , Software Manufacturing,* and Storage and Retrieval of Information (not data sets)
Full CE Design *Concurrency*	Interoperability (same process, multiple platforms, Slot level configuration Control) via OO Message architecture
Concurrent Process / Object Availability	Connectivity (high speed responsive access to various elements (slots) of the same object) [high speed = inside user awareness]
Electronic Information "Envelope"	Distributed Information Repository (Meta data management) using a confederacy approach, where objects are volunteered and not mandated)
Electronic Information "Envelope" (from a process perspective)	Process Management via "*In-line* Models" with varying granularity
Models	Drive Milestone, Schedule, and Processes

* = may not be necessary

Figure 9-18. *Level-Four Integration with Increased Capabilities.*

sively by process flows, process management, and the automated infrastructure support necessary for the individual, group, and organizational unit to function. This is the big change. Preparing individuals to be part of such an environment should receive significant attention. There are several matrix teaming objectives which should be emphasized during this planning process. They probably need formal training support. These are summarized in *Figure 9-20*.

One of the teaming objectives is empowerment. The collaborative, concurrent design process has one important difference from the step-wise refinement design process. In step-wise refinement, personal specialization results in people being responsible for task completion within an overall process. They are not directly responsible for any results. In the collaborative, concurrent design process, each individual or design group takes responsibility for a product or a portion of a product and its manufacturing completion and success. This difference must be reflected not only in the design process, but in the administrative atmosphere, personnel policies, and personal evaluation criteria surrounding the individual. Emphasis is on surrounding the individual with CE Design automated support as depicted in *Figure 9-19* and revised personnel policies shown in *Figure 9-20*.

INTEGRATION OF IMPROVEMENT PROJECTS

Several of the implementation concepts, and of CE Design, can be utilized throughout the complex product manufacturing organization. The "in-line"

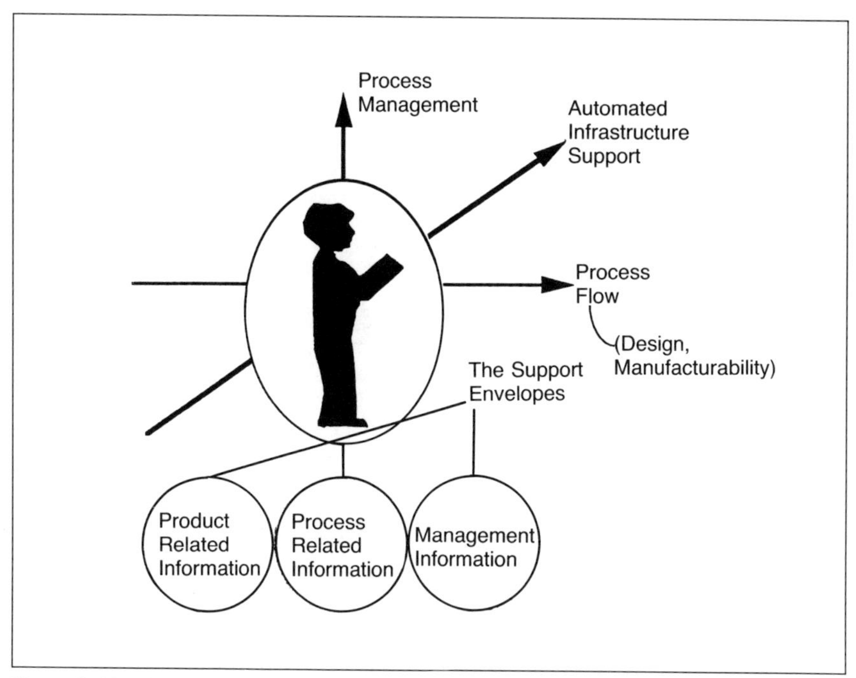

Figure 9-19. *Level 4 CE Design and the Individual.*

process model of Level 2 is the first of these. When this capability has propagated itself through a good portion of an organization (which will take time), then the forward planning process can be integrated with the processes it intends to improve. As shown in the *Figure 9-21* series, planning process models can now cross-reference, or even include the "to be affected" processes. Additionally, active models of the preferred, or vision process are maintained and compared with the now current process. Progress is measured, and priorities established through analysis, simulation, and comparison. These can also be compared to the organizational goals and objectives in *Figure 9-7* for the purposes of more accurately selecting, measuring, and defining the success parameters of these objectives.

Not all organizations must go completely through all four levels of implementation. Appendix A is intended to provide the complex product manufacturing organization with insight into:

- Guideline 1–Is your organization a prime candidate for CE Design?
- Guideline 2–At what level is the organization currently operating?

Some organizations achieve spectacular success by applying elements of the CE Design concept. For example, complex computer-on-a-chip processor manufacturers use various techniques to continue evolving their products. Some of these processes operate at Levels 3 and 4, but are highly focused on just a few elements of the overall CE Design process's scope. These same techniques may not be pervasive throughout engineering, or the organization, however, and may

Area of Emphasis	Objectives
Collaboration	Mix team members and roles to balance skills and ability to contribute; set up training programs to develop interpersonal skills; involve managers in training
Empowerment	Adjust personnel policies and personnel evàluation and compensation review processes to emphasize personal responsibility for results
Product Orientation	Focus on responsibility for product results and not on task completion
Shared Responsibility	Adjust relationships with suppliers and vendors to reflect shared responsibility in results

Figure 9-20. *Matrix Teaming Objectives.*

suffer from granularity concerns. For these organizations, applying known process techniques, using this same Forward Planning implementation approach, may provide a "fast track" to their successful propagation.

In any case, is it over when the organization reaches Level 4, or full CE Design operation? Of course not! *In fact, improvement has just begun.* At this point, aggressive pursuit of the other characteristics of world-class manufacturing, which are enabled by CE Design, can be completed. Processes can continue to be improved. *The power of CE Design can be fully applied to all products and processes.*

A Sample CE Design Implementation Plan

The challenge facing CE Design implementation is that it cannot be implemented "in a vacuum" in a transition environment. In the *Figure 9-21* series, a sample CE Design Implementation Plan for an individual improvement is depicted. This sample set of integrated plans has four general subject areas:

1. Corporate plans;
2. Product plans;
3. Process improvement and performance measurement plans, and
4. Executable in-line process models.

In *Figure 9-21A*, several goals have been established at the corporate, or the complex product manufacturing organization level. These goals include "quality in everything we do," "increased profitability for ourselves and for our shareholders," and "empowerment of our work force." These are all important

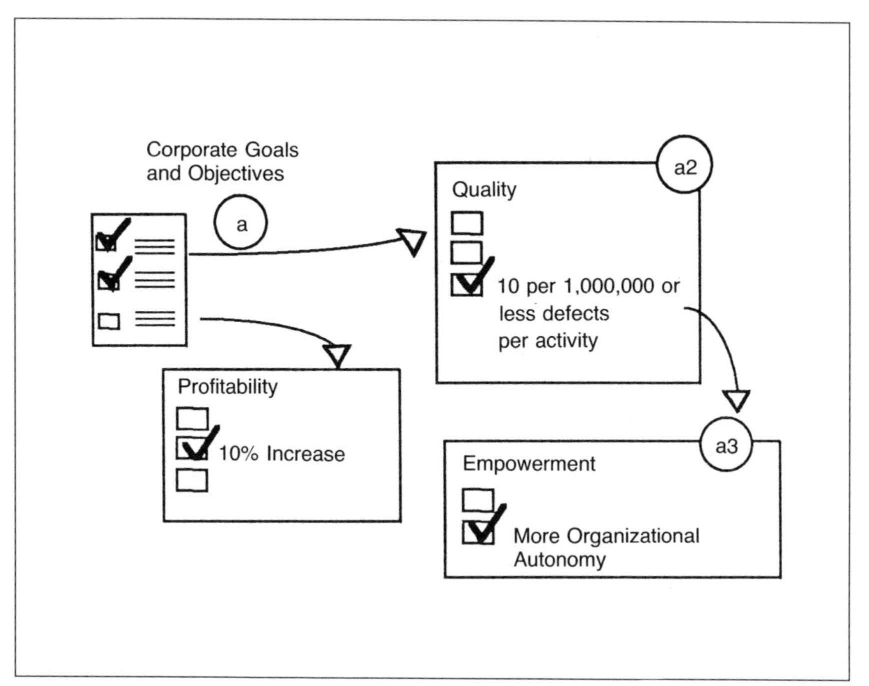

Figure 9-21A. *Sample CE Design Implementation Plan for Selective Improvement in a Level Improvement Plan (a,a2,a3).*

goals. They appear as the (a) group in *Figure 9-21A*. Quality should pervade the organization, and design quality is a very important component of that goal. Profitability means the ability to invest in new products and processes, and provide adequate pay, benefits, and working conditions. Empowerment is particularly powerful. It includes pushing greater decision-making power down to lower levels of the organization, and providing more authority and ownership of the work to those doing the work. *Figure 9-22* is a sample of such a mission and goals statement for a maker of powered lifters.

These goals must be translated into actionable objectives. In *Figure 9-22*, each goal has been assigned several objectives the organization is to achieve in addition to the necessary and vital goal of producing and selling product. These objectives are part of the (a) grouping in *Figure 9-21A*. For this example, one of the *Quality* goals is to reduce 20 defects per one million activities to 10 defects per one million activities. For *Empowerment*, the objectives include organizational changes in two major product groups to foster increased local autonomy. The *Profitability* objectives include a 10% improvement in profitability in the mid-range lifter product group area, in spite of heavy competition and some price cutting by competitors.

For this example, the organization decided that to achieve the profitability objective, they needed to introduce a new, lower-cost-to-build-and-service mid-range lifter. This product planning decision was made with significant

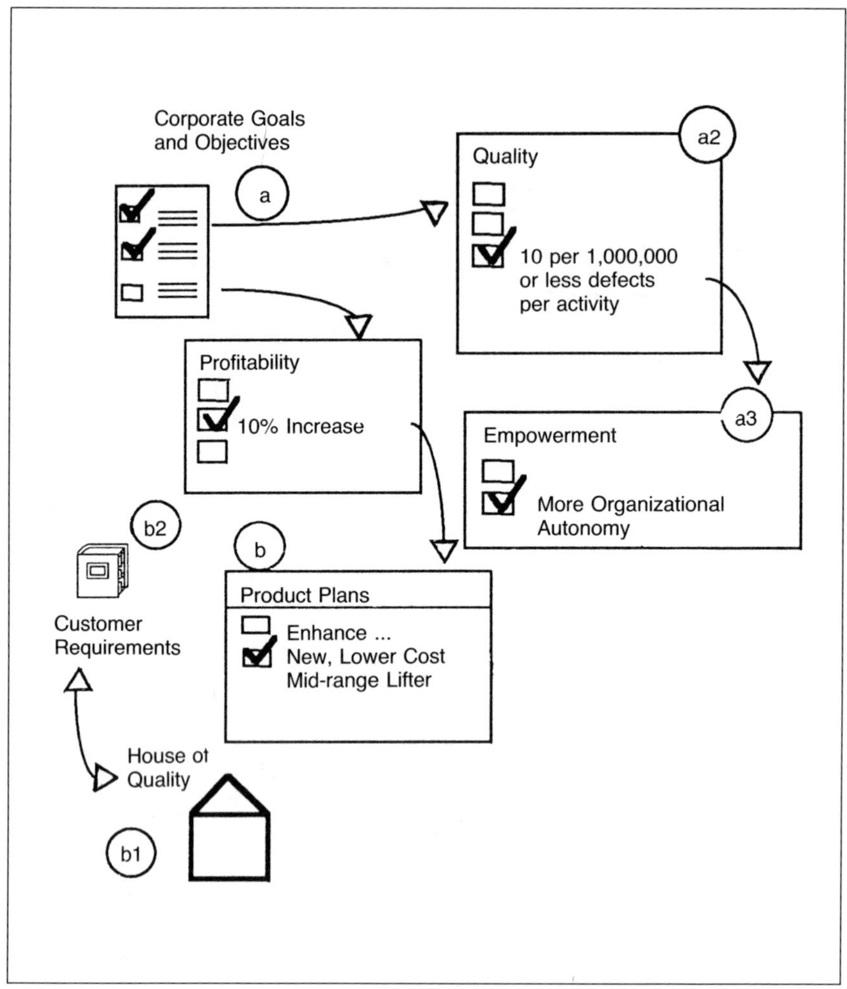

Figure 9-21B. *Sample CE Design Implementation Plan for Selective Improvement in a Level Improvement Plan (continued with b,b2,b3 added).*

customer and potential customer input. Product-related plans and actions are shown in the (b) group of *Figure 9-21B*.

The House-of-Quality technique was used in *Figure 9-21C*, to identify (1) lower cost opportunities for purchase and support; (2) the ability to lift greater weight to the same height; and (3) much lower maintenance costs as the key customer concerns for this type of product. Engineering analysis recognized only a new design could meet these objectives. In addition, the profitability objective appeared to be possible through better design and other improvements gained as the organization was simultaneously proceeding through a multiple year CE Design implementation plan—a strategic initiative for the organization.

While these product plans were being established, these product plans were

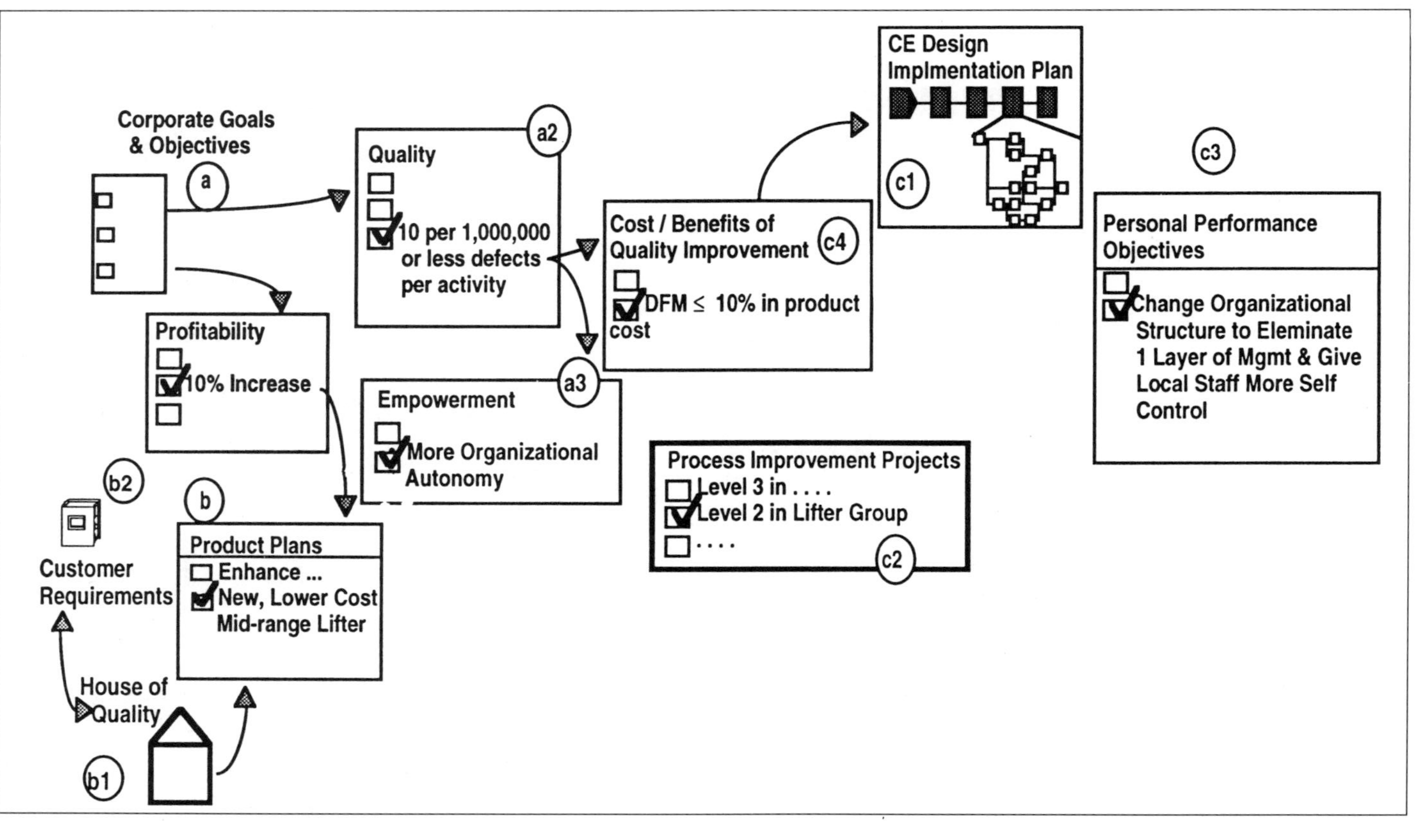

Figure 9-21C. *Sample CE Design Implementation Plan for Selective Improvement in a Level Improvement Plan (continued with c1,c2,c3,c4 added).*

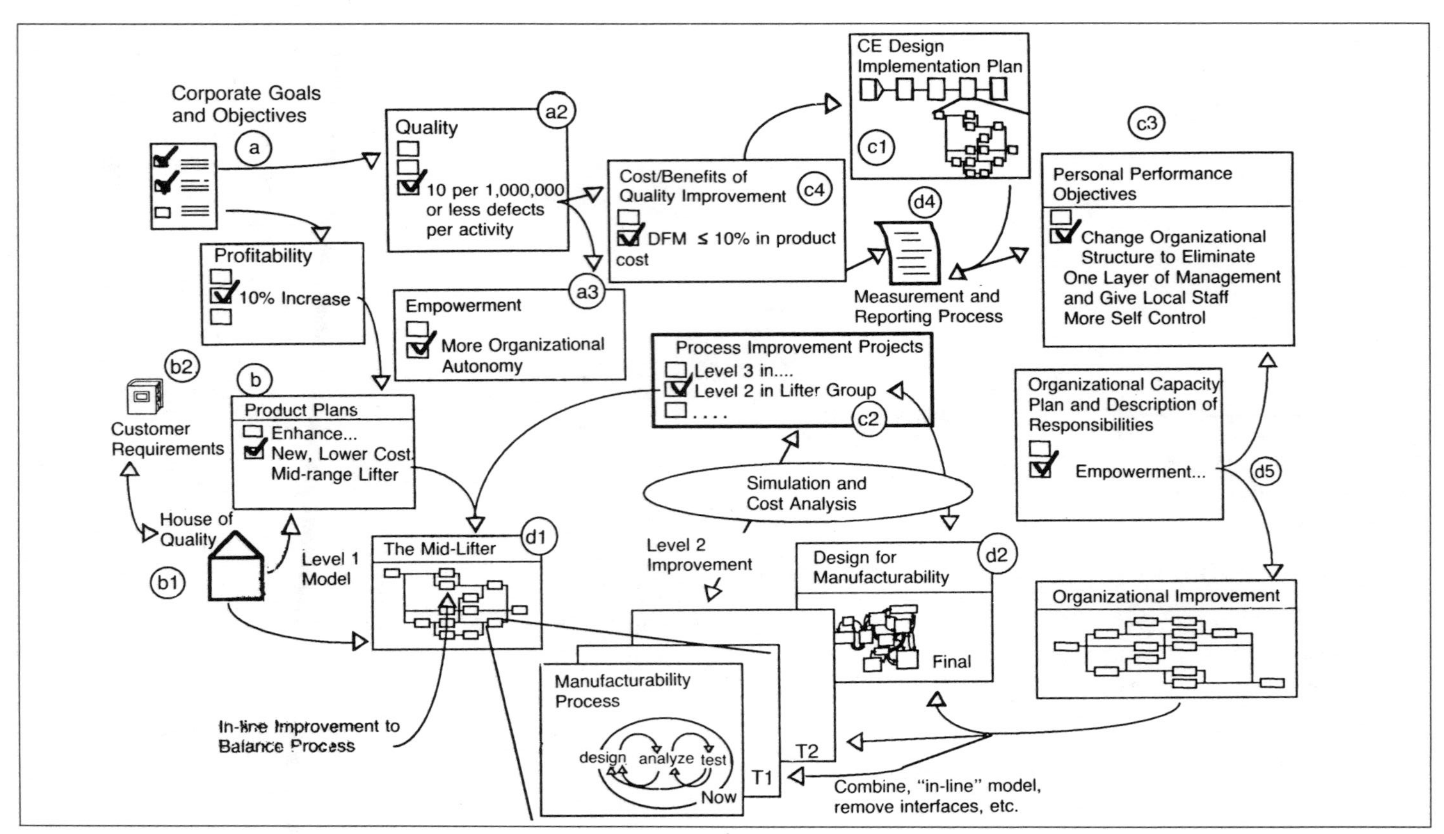

Figure 9-21D. *Sample CE Design Implementation Plan for Selective Improvement in a Level Improvement Plan (continued with d2 through d5 added).*

Goal	Objectives
(1) Selling and producing product	Continue Present Operations at Present Rates with Agreed to Revenue Objectives
(2) Profitability	10% improvement in profitability in the mid-range lifter product group area
Compertitive Advantage Related Goals	
(3) Quality	From20 defects per one million activities to 10 defects per one million activities
(4) Empowerment	Organizational changes in Mid-Lifter product groups to foster increased local autonomy
Strategic Initiatives	
(5) Continue to Implement CE Design	Reach Level 3 Capability in Mid-Range Lifter Product Group

Figure 9-22. *Objectives Statements.*

established, a revision to the CE Design Implementation plans occurred. The revision process used forward planning concepts, considered any new technological opportunities, and carefully considered the cost benefit opportunities (savings, avoidance, revenue enhancement) possible. This revised plan was the product of the Forward Planning document structure and planning process, shown in *Figure 9-5* series.

Figure 9-23 shows that customer requirements always drive product decisions. Internal organizational goals also drive product decisions. The impact of these choices and their integrated process improvement and product improvement plans is shown in *Figure 9-21D.* The relationships between process improvements, organizational improvements, and product improvements are shown.

Figure 9-24 illustrates a sample CE Design planning model set. The objectives and highest level in-line process planning and execution models can be used to actually improve processes to achieve the quality improvement (20 to 10 defects...) and profitability (10% increase...) objectives. These Level 1 plans have been selectively, for this sample, decomposed into Levels 2 and 3 in-line process planning and execution models. These lower level plans are part of the (c) group of *Figure 9-21*. Remember, these plans are executed, and not just reported against, via CE Design's routing and queuing workflow management facility (in-line model execution).

Describing *Figure 9-25* further, engineering management, after receiving general authorization to proceed on the new mid-range lifter, prepares a

Customer Requests	Reason for Request	Internal Objective	Design Response
(1) Lower cost of purchase and support process cost	Competition attempting to enter market	Reduce Product Cost by 10%	Using Design for Manufacturability Techniques, Reduce part count, component cost
(2) The ability to lift greater weight to the same height with new OSHA safety regulations	Size and weight of equipment plus personnel rising for 3 story buildings	Increase composite weight by 100 pounds	Increase weight lifting capacity of motor by 60 lbs., decrease weight of platform by 40 lbs. using new materials and style of floor mesh
(3) Much lower maintenance costs	Propulsion system requires too much maintenance	Reduce part count on motor	Go to fuel injection, remove carburetor and choke

QFD Process; available from American Supplier Institute

Figure 9-23. *Requirements and Internal Objectives Drive Product Decisions.*

top-level work statement. Authorized sets of teams with a new product WBS and appropriate matrix design teams, in collaboration with the team leaders, prepare Level 1 and Level 2 in-line design process milestone schedules and detailed schedules, and review their design process models consistent with their current CE Design Strategic Initiative Implementation status.

These now particularized process models contain, as an integral part of the model, activities which represent the incorporation of further improvements (from *Figure 9-23*, and as represented by the (d) series of *Figure 9-21*). In *Figure 9-21D*, this Level 2 process is further decomposed into selective Level 3 activity models to demonstrate the improvement transition states in the design for manufacturability (DFM) portion of the manufacturability process for the new mid-range lifter. These improvements are focused on improving the relationships and information passages between various DFM activities, and improving and leveling the granularity of this portion of the overall process.

As these product and process improvements and plans were being pursued, improvements in the functioning of the organization were also being pursued. As noted in *Figure 9-21D*, employee empowerment was also established as a high-level goal. This goal led to its objectives (organizational changes), which

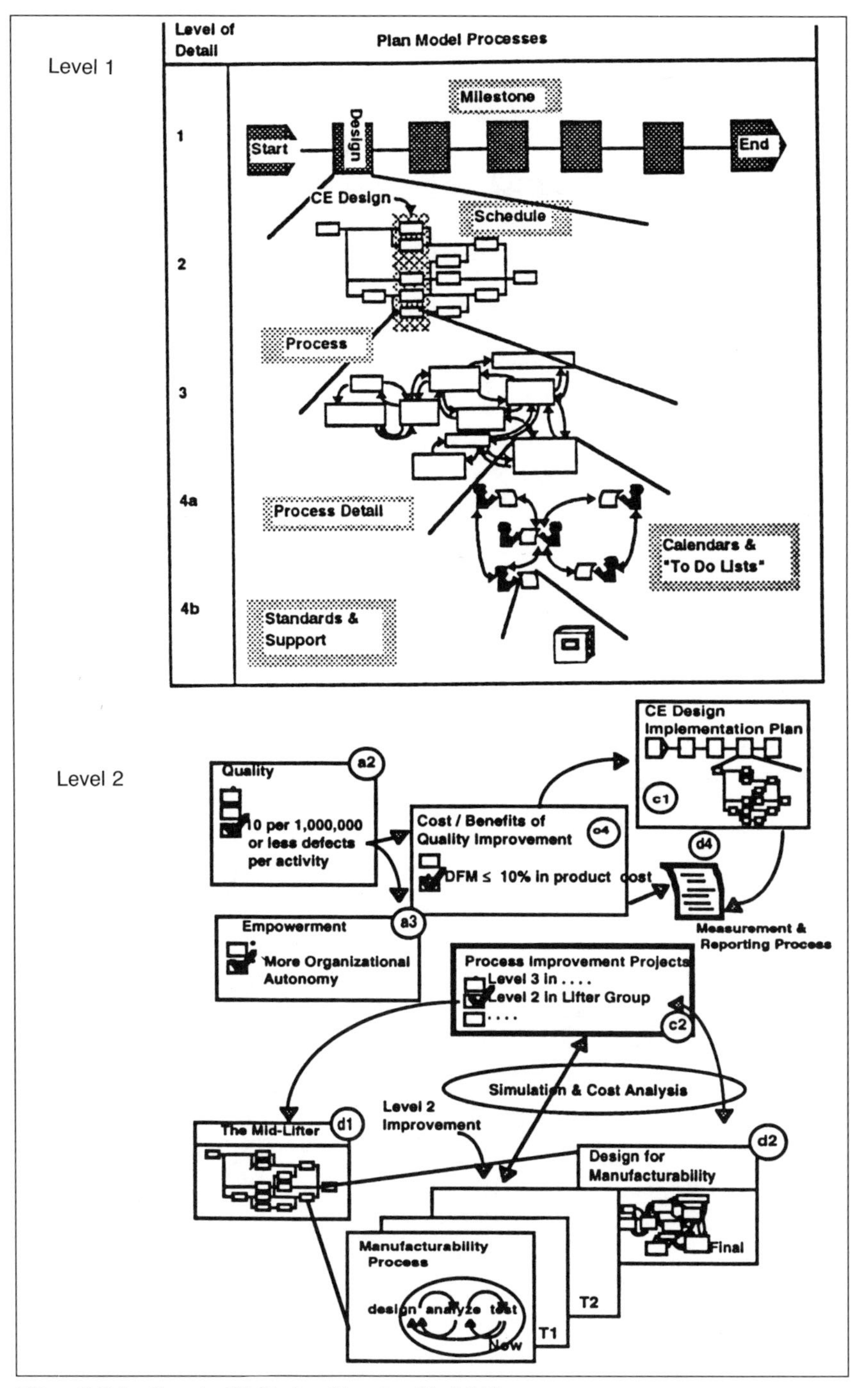

Figure 9-24. *Sample CE Design Planning Model Element.*

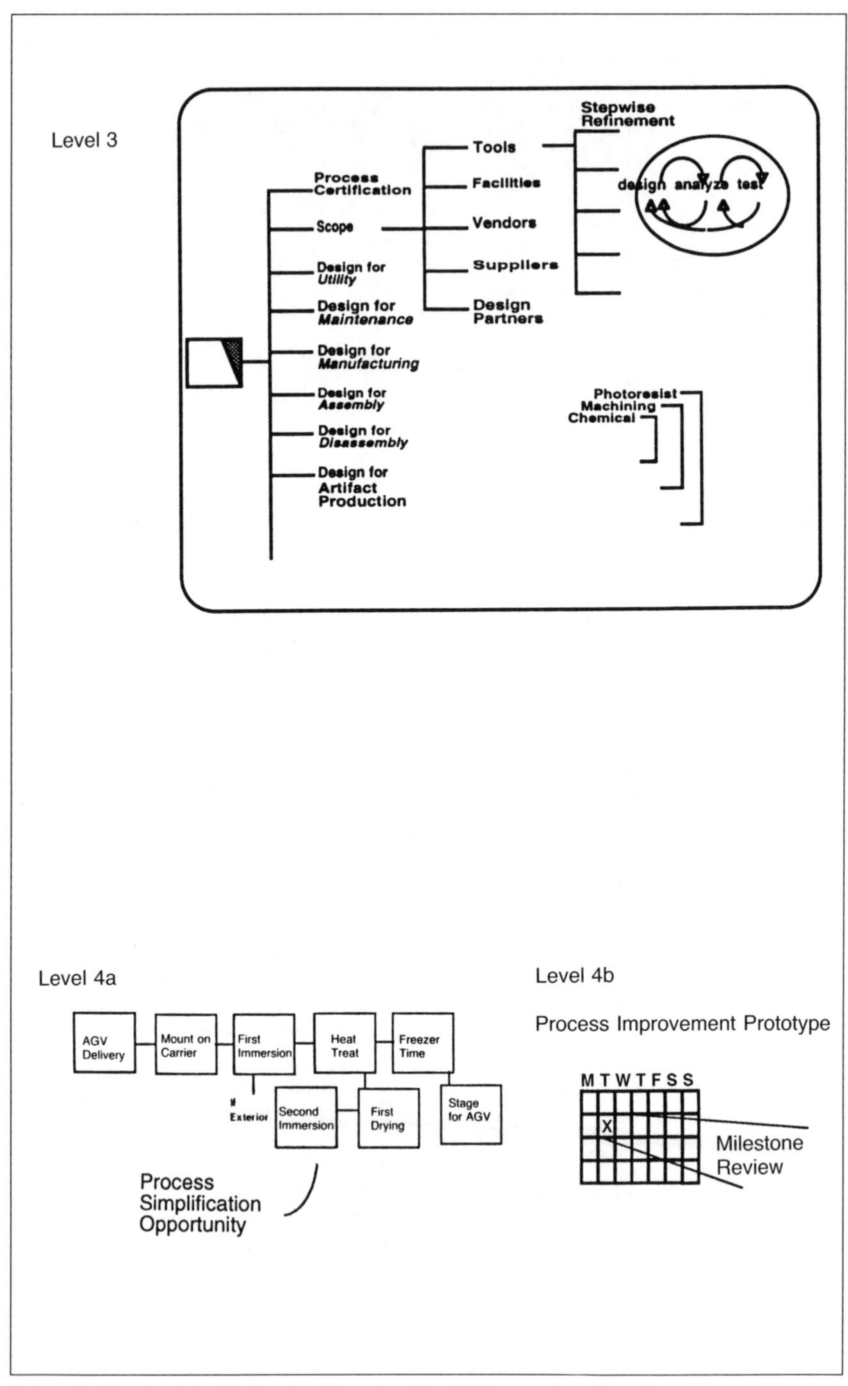

Figure 9-24. *Sample CE Design Planning Model Element (continued).*

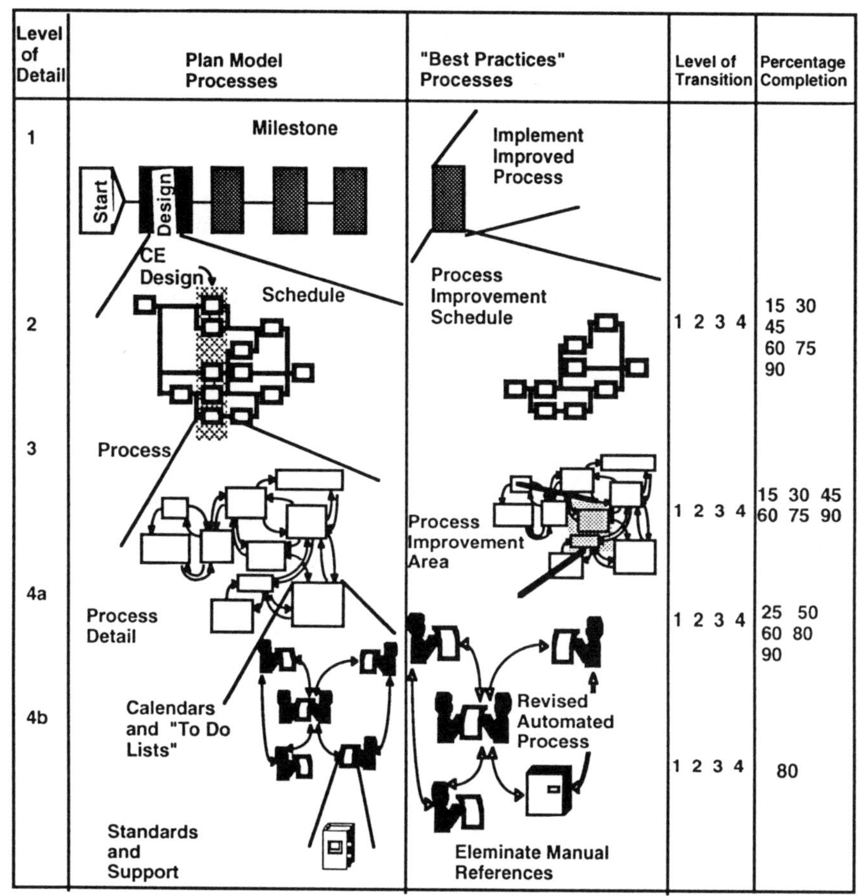

Figure 9-25. *Forward Planning Process Improvement Project Integration.*

are to be achieved in the mid-range lifter group (*Figures 9-21, 9-22,* and *9-23*). These objectives were linked to the personal performance objectives set by the manager of the mid-range lifter group. The manager agreed the movement to CE Design Level 3 should be accompanied by the redeployment of a layer of quality control personnel because the process owners (workers) were now inspecting their own work for quality. This change in responsibility was aided by an in-line process transition model. It included a partial phaseout of these checks before the process owners took full control. These organizational change process models were linked to those at Level 3, for executing the new mid-range lifter and its integrated granularity process improvement effort.

When the revised quality objectives were set, they were not set on an arbitrary basis. A cost/benefit analysis established the approximate projected cost of these changes and their benefits. The proposed objectives, their projects, and their cost benefit analysis were considered along with others, and their higher ROI and lower risk assessment caused them to be selected for this year. These cost analysis objectives were added to the project accounting measurement and reporting process which monitors in-line model activities using ABC costing techniques as indicated in *Figure 9-21*. Simulation was used to forecast improvement impact and cost benefit timing. Results were rolled up to the major objective level to report on progress.

The integrated planning and execution process just described must be duplicated across all projects, goals, and objectives. While it may seem involved, this type of planning has resulted in greatly improved complex product manufacturing organizations over the past few years. Success can breed success.

Appendix

Self-evaluation Guides

Glossary

APPENDIX A

SELF-EVALUATION GUIDES

This evaluation table assists the reader in assessing the urgency of applying Concurrent Engineering Design (CE Design) to their organization. There are no "right" or "wrong" answers, except if these answers are difficult to answer. The answers to these questions should be known by most people in the organization who have some experience with the firm.

TABLE 1. CE DESIGN APPROPRIATENESS

Question	Answer	Evaluation Guidelines
1. Do you have multiple products? If so, does the part count per product exceed an average of 1000?		• Multiple programs, each with multiple products, probably mean complexity in management processes. • Part counts close to, or over 1000 usually mean complexity and variety.
2. What percentage of the product is managed under an interference management scheme?		• If interference analysis, or worrying about the erroneous intersection of different parts designed by different individuals, exceeds 25% of the part count, then complexity is present.
3. What percentage of the products have options, and what percentage have customer special orders?		• A combination of options and special orders from customers, especially special orders unique to a customer on 20% or more of the products, means substantial complexity.
4. What is the duration of production runs for typical major products?		• Multi-year product runs mean substantial concern about the configuration control of products through high change rates if these long run products are also the option/special products of question.
5. What is the rate of change per product per week?		• Constant product change indicates configuration control issues and complexity of design incorporation into production products.
6. What percentage of design changes end up as inline rework, offline rework, or retrofit after production?		• This type of late incorporation is very costly. If the percentage is over 10%, then design to manufacturing coordination using CE Design is required.

TABLE 1. CE DESIGN APPROPRIATENESS. (CONTINUED)

Question	Answer	Evaluation Guidelines
7. What percentage of the products require multiple design teams?		• As the number of design teams goes up, the need for coordination and communication between them increases exponentially, creating the need for CE Design process management.
8. What percentage of the products require after the sale configuration tracking for regulatory, maintenance, parts, or general support reasons?		• Continuing to maintain configuration control over delivered products increases concerns for design artifact (CAD, parts lists, etc.) storage, currentness, and correctness.
9. What effectivity schemes are used for products and their designs?		• If the effectivity scheme requires more than date or serial number effectivity, then complexity increases rapidly.
10. What percentage of today's parts has been prepared digitally?		• A high rate of digital parts production means computing architecture and management issues for Engineering and Design, as well as for manufacturing.
11. Would you characterize the organization as being organized by profit center, program (or matrixed), or by Managed Business Unit (MBU)?		• Being organized along these modes means that the product is probably complex enough, or the processes complex enough, for the organization to attempt to use organizational structure as a complexity reduction technique.
12. Has Concurrent Engineering (CE) been used on an end product already?		• Using CE means that the organization has already recognized some elements of the situation, or more, and is communicating with production manufacturing already. Once combined with design and process management, these beginnings can be even more powerful.
13. When you used CE, did you reorganize or set up a separate organization for the new product?		• This separation indicates concerns about the ability of the organization to change; a requirement of completing the transition to CE Design, but also a recognition of the complexities of CE Design implementation.

This evaluation table assists in evaluating the organization or firm's present awareness, or progress toward the implementation of several of the key concepts and capabilities required to successfully achieve CE Design. The intent is not to establish organizational strength or weakness, but to progress toward the vision of CE Design.

TABLE 2. CE DESIGN IMPLEMENTATION EVALUATION

Level of Operation Table	Key Area →	Firm reorganized around managerial, technical, and business processes.	Engineering, design, and production manuf. teams organized around collaborative or concurrent processes.	Design process is concurrent and collaborative.	Org. culture, org. structure, and personal evaluation process established around teams and Managed Business Units.	Product Configuration control is accomplished via effectivity and other integrated marker systems.
Level	**Status**					
5	perceived to be 80% or accomplished.					
4	transitioning to "new" environment.					
3	planned changes.					
2	aware and concerned.					
1	not generally aware.					

EVALUATING THE RESULTS OF THE SELF-ASSESSMENT

There is not a "good" or "bad" score after using these tables. The first table indicates the organization "need" for CE Design. The more complex the circumstances of the organization, the more CE Design is needed. The need for CE Design also is determined by the frequency of product change and the speed of the change. If all of the three factors are present in a significant manner, then CE Design is extremely important to the organization.

If the first table indicated that CE Design is appropriate to the organization but the second table shows that the organization is just starting, (i.e. the choices regarding status are at either the 1 or 2 level), then the plans, the process engineering, and the establishment of the tools and computing architectures should be begun. This book is intended to act as a manager's guide to the direction to take the organization once the journey has begun.

If the second table indicates that the organization has significant programs underway, and if the CE Design journey is practically complete, the management's focus should be on pacing the progress of the organization consistent with the levels of CE Design "progress." This "progress" is outlined in Chapter 8 and Chapter 9. The manager's planning task is to coordinate business needs, technological and team progress, and budget to move the organization, with the right mix of these elements, toward full CE Design.

CE Design Glossary

ABC - Activity Based Casting; a different approach to assigning business costs based on activities, either human or otherwise, as compared to the traditional approach associated with burdened overhead hours.

activity - an event with an identifiable start and stop which transforms input(s) to positive output(s).

algorithms - a repeatable set of procedural steps; many times expressed in a mathematically oriented fashion.

allocation - in accounting, the activity of assigning a pool or set of costs to various accounts based on a pre-established set of logic.

ambiguity - in logic, a statement that can be interpreted or successfully executed in more than one fashion.

applicability - in effectivity, the set of end products affected by effectivity markers.

application software - software supporting the execution of business, technical, and managerial processes and not the functioning of the computing equipment itself.

approvals - in process management, a formal agreement to proceed; usually requiring a person's signature.

architecture - in computing, a pre-planned configuration of computing hardware, software, and networks providing intended services.

archival - information stored for later retrieval but in another than immediately usable form.

artifact - an end-product or byproduct of an activity.

attributes - a type of product characteristic basic to the product.

authority - in process management, the person or group responsible for an activity or process and produced end products; this person or group can generate approvals or may seek the concurrence of others before final approval.

authorization - the act of approval.

autonomous - in process management, the ability to act without the need to pre-authorize such activity.

bandwidth - in computing, the amount of information transmittable during a period of time.

batch - a pre-set number of products processed as a set.

best practices - using a set of industry process examples, mapping the best parts of the process examples into a process representing the "best of the best."

bottleneck - in process management, a point in a process where throughput is less than other parts of the process.

breakpoint - in process improvements, a point in time when the improvements, taken together, have improved the overall process so that a significant overall benefit can now be achieved.

breakthrough - a type of improvement requiring "investment," but whose return is orders of magnitude larger than the investment.

CALS. - (Computer-aided Acquisition and Logistics Support) A DoD initiative to accelerate the transition from present paper intensive, non-integrated weapon system design, manufacturing, and support processes to a highly automated, integrated mode of operation (from CALS Vision, Draft 1.21, November 1989, US Department of transportation, Research and Special Programs Administration, Transportation Systems Center, Cambridge, MA 02142).

catalyst - in a process, an agent aiding in the execution of input to output transformation without being consumed in the transformation.

cells - in manufacturing, a logically grouped set of equipment devoted to a particular process.

Change Control. - An activity or procedure within an enterprise managing the incorporation of engineering change into the product definition.

charting - a quality improvement technique in which the results of process monitoring are displayed in tabular form.

CIM - Computer Integrated Manufacturing.

CIM-OSA - (Computer Integrated Manufacturing-Open Systems Architecture) A European initiative to develop a reference architecture for information systems.

collaborative, concurrent - a CE design process focusing on simultaneous design.

commitment - the decision to proceed.

complex - a body of knowledge no one human can comprehend at once (or in a short period of time).

compliance - consistent with pre-established direction.

composites - a manufactured component assembled from resins and fabric.

conceptualization - the first of three stages of intellectual idea or concept development.

concurrence - in process management, arranging for another to agree before finishing.

concurrent - the execution of more than one activity in what appears to be a simultaneous fashion.

confederation - in process management, an approach to management where the pattern of control rests not in a higher authority (such as in a federated style of control), but in the consensus of the group; where all or a majority must agree before anything is actually undertaken.

configuration - a particular, specific set of items.

constraints - restrictions imposed in a particular situation.

context - the entire environment of a decision.

continuous - in manufacturing, a process involving products that cannot be individually counted.

cooperative - performed in conjunction with another.

coordination - the act of cooperative communications, used, for example, to obtain a concurrence.

core competencies - a certain set of processes which uniquely distinguish the organization from any other and provides competitive advantage.

craftsman - a highly skilled individual who has "mastered" the art of a particular type of work.

culture - in organizations, the emergent characteristics the organization as a whole displays to outside observers.

custom - a product feature unique to a single customer or so unusual that it is not a regularly sold product feature.

cycle time - the time required to complete the entire process and be able to re-start the process.

data - in computing, an individual instance of information requiring software for its human interpretations to be reasonably possible.

database - a storage management software system.

date - a type of effectivity based on the calendar.

decomposition - an approach reducing complexity to smaller, more under-standable components.

defect - any type of error.

deliverable - an end product for a process.

deployment - the commitment of resources.

derivative - a type of product based on, but evolved from, another product.

detection - one of the four stages of quality (prevention, detection, internal failure, external failure).

deviation - a variation from a pre-established standard.

differentiated - an approach used to reduce complexity based on general systems theory.

disassemble - in product design, the ability to take the product apart in such a manner that it can also be reassembled, or put back together, in a fully usable, as before, state.

disciplines - various types of work assignments, each based on a distinct body of knowledge.

discrete - in manufacturing, a process type having countable products.

distributed computing - a type of computing using more than one processing element to complete a useful piece of computing service.

drafting - the preparation of an engineering drawing.

drawing - a dimensionally accurate set of documents, including a pictorial representation of at least a portion of the product to be manufactured and included additional explanatory information.

effectivity - a type of marker system used to conditionally associate elements of a whole.

embedded software - computing programs assisting in operating a physical product.

empowerment - a management approach permitting more decision making by employees.

encapsulated - the act of including something in another larger set.

enhancement - adding value to an end product by adding new features or attributes of interest.

entity - A set of real-world objects (people, places, things, events, etc.) with characteristics in common and within the scope of a project. Each entity has a name that is a singular noun or noun phrase describing the object it represents.

event - a time-based occurrence.

exception - a non-standard occurrence.

explosion - the activity of breaking a whole product into its set of part numbers.

failure - the occurrence of an error resulting in product non-performance.

feature - a type of product characteristic which is either optional or of special interest.

finite - a countable number.

flow control - in process management, the management technique of using continuous process, feedback loop oriented management logic.

framework - in computing, a general architecture.

functionality - in computing, software capabilities providing services to the software's user.

generative - in processes, using the design to directly produce a process for the execution or production of the design.

geometry - in computing, the computer generates an image of the product to be produced from its data, or geometric constructs.

granularity - the level of detail.

graphics - in computing, the image on the display.

handwriting - in engineering, the notes added to a drawing.

hardware - in computing, the actual physical devices.

heterogeneous - in computing, different vendor's computing hardware combined in the same interconnected network.

hierarchically - **(1)** a top-down control technique, or **(2)** an approach to structuring information with the index at the top of the hierarchy.

holistic - considering as a whole.

IGES - Initial Graphics Exchange Specification. An ASME/ANSI Y14.26M-

1989 Standard. A standard for three-dimensional physical models.

incorporation - in production manufacturing, the point at which a change is to be included in the product.

index/indexed - in computing, the establishment of a separate, quicker to access data file containing a few key data items, each of which "points" to a larger file record containing the remainder of the information.

indicators - pieces of information not directly measuring process performance but implying how performance is proceeding.

indirect - in cost accounting, all activities not directly associated with producing the product.

information - Data which has meaning by virtue of the fact that an agreement exists, via a data definition or protocol, to interpret the data. Data having a context.

information envelope - a descriptive term to describe how information can be stored so as to be associated with other information in a less rigid form than that of a database.

infrastructure - in computing, the underlying or base elements of the architecture.

innovation - an improvement creating additional value in a product.

instances - a particular separately identified item in a class of such items.

intellectual capitalization - the process of transforming information into something of enduring value.

interfaces - in computing, the point of data exchange between different software systems.

interoperability - in computing, the ability of a software program or a system of programs to operate on more than one manufacturers' hardware.

introduction - in manufacturing, the process of beginning change incorporation.

iteration - in design, the approach of design-test-analyze being repeated over and over.

knowledge - information which is useful, adds value, and is reusable.

legacy system - a computing system impeding the introduction of process improvements.

levels of abstraction - the mental process of substituting more general purpose concepts for more specific objects or concepts. See extended discussion in Chapter 3.

maintainability - the degree to which a product can be easily kept in excellent operating condition.

manufacturability - one of the key CE Design processes.

marker - an effectivity information element, usually in **mnemonic** form.

mechanism - in architecture, the actual instantiation of the general concept called for in an architectural policy or principle.

messaging - an interface technique.

methodology - a predetermined set of procedural steps.

metric - in process, an accurate measurement of process performance.

milestone - in scheduling, a high-level, important event in a project.

minicomputers - a type of general purpose computer with relative power between PCs and mainframes.

mnemonic - in effectivity, the use of an abbreviation, or short easy to remember representation of some other item. For example, "USA" can be used to represent "United States of America."

mockup - in design, either a physical representation of the assembled end product or the computerized equivalent of a physical representation.

model - A description of something or the description of all possible interpretations of something. A representation or abstraction of an object, function, or event. Random House unabridged 1974 defines model as "a system of things and relations satisfying a set of axioms, so that the axioms can be interpreted as true statements." Also, from the same source, "a representation, generally in miniature, to show the construction or serve as a copy of something."

navigator - in computing, a user-interface aid for making one's way through the maze of possible actions which can be taken in a complex computing environment.

neural networks - in computing, a type of logic circuit which simulating the activity of brain neurons. Typically used in evaluating different patterns of information.

object - in computing, the use of "real world" representations for information. An "object" contains the programming and the data necessary to represent this "real world" item. A discipline and thought process applied to computing begun in the field of simulation in the 1970s.

objectives - targets for the organization to achieve, as in "goals and objectives."

object-oriented - in computing, the use of objects in design and programming.

offerings - in manufacturing, options and custom features made available for customers to order.

omission - a type of error in which something is not done, in contrast to **commission**, in which an error is actually committed.

open systems - in computing, the label given to the computing industry trend of enabling applications **interoperability**.

originator - in process management, the individual or group who initially introduced the information or idea into the process or product; many times this individual or group also has authorization over change through process control techniques.

outsourcing - in manufacturing, using another organization to make a part of the final end product.

overhead - an accounting term describing activities not directly relating to actual production.

ownership - in computing, the group of business people who use the computing and contribute value to the products of the organization through the computing.

packaging - in design, the part of the product acting as a platform or container for the real product. In computing, for example, the metal box in which the computer circuits rests is a "package."

paradigm - as in "paradigm shift," a change of such significance that it results in a redefinition of the competitive position of firms in a market.

parameterization - in design, the use of values, usually mathematically expressed, which can generate the geometry of a part or product.

partnering - in design, sharing responsibility for the product, including legal liability.

permission - a state of version control.

phantoms - in production manufacturing, a "fake" part number or assembly introduced into the structured bill of materials for the convenience of production manufacturing, usually to reduce the complexity of assembly by introducing smaller steps of product build-up.

pilot - in implementation and planning, the execution of a "small" or "test" set of sample items through a new or changed process in order to finalize any errors or shortcomings before committing the entire set of items to the process.

pointers - in computing, the ability to provide short hand indicators of more complete information accumulated in the storage systems of the computer.

portability - in computing, the ability to move a program from computer to computer, usually of different manufacture, with little or no change.

prevention - the most desirable stage of quality control, where errors are not allowed to happen.

prime - in aerospace and defense, the organization which contracts directly with the government for an end product or service on behalf of itself, its partners, and its suppliers and vendors.

proceduralized - in processes, the establishment of a proscribed set of activities before the execution of the first activity. In computing, the capture of these proscribed activities inside the computer program.

process - a set of activities transforming inputs into products.

procurement - in manufacturing, the process used to acquire something.

product - the result of the completion of a set of activities in a process which transforms the inputs to the process into a set of end results, or "product."

Product Data Exchange Standard - Standard (under development) for communicating a complete product model with sufficient information content to be interpretable directly by advanced CAD/CAM applications such as generative process planning and CAD directed inspection. The expression of what a product is, ideally, with sufficient clarity to be able to manufacture the product. PDES version 1 became an international standard called STEP. PDES now stands for Product Data Exchange using Step.

Product Definition Management (PDM) . The set of data elements completely defining a product for a certain context, a subset of product data are now contained as in check-in/out systems on a computer also called PDM Systems.

production manufacturing - that part of the overall manufacturing organization transforming raw materials into finished, physical manifestations or end products.

productivity - the proficiency with which an activity is performed.

profiles - in computing, sets of standards, which together are implementable; for example, the US government has issued a set of profiles of standards for use in procuring computing.

program - in manufacturing, a set of similar products under a single product manager.

promise, available to - in the control of production in manufacturing, the difference between the number of products for sale in a category of products not previously committed to another customer and the number of these products actually in inventory and in work-in-process capable of being shipped when the customer requests. Cycle time reduction makes this number larger because the time to respond to orders is smaller, a very important capability to offer prospective customers.

properties - in design, aspects of the products evident only when the product is actually produced, or the product's actions are simulated.

queue - in processes, the area in which opportunities are staged, usually in priority of arrival, to be transformed by the process.

realization - in design, the third stage of intellectual development of an idea; a conceptualization becomes a visualization, which is turned into a physical manifestation of the idea, or "realization."

regulation - a specific set of directions from a regulatory body.

regulatory - an outside organization, usually governmental, constraining the activities of the organization.

repeatable - to execute the same activity again with the same inputs and the same outputs.

repetitive - to execute the same activity again.

restructure - in business process change, the activities associated with modifying organizations and their activities to accomplish the desired change.

scalable - in computing architecture, the ability of a computing capability to operate as equally effectively when used on a personal computer as on a mainframe.

schedule - a representation of activities to be performed presented within the context of time, usually to enable supervisors, managers, and executives to monitor these events and the resources consumed by these activities.

schematics - the representation of electronic circuits in a wiring diagram form.

serial - in design, the execution of a sequence of activities, where one activity must complete before the next can begin.

signature, electronic - in CE Design, the ability to obtain the approval of an individual by acquiring and storing the electronic equivalent of the individual's actual signature.

simulation - in design and computing, the execution of a model of the process such that it appears that the process is being exercised when in fact it is not.

simultaneous; apparently - within the context of CE Design, the execution of activities against the same element of the design in computing such that it appears that both sets of activities are manipulating the same information under configuration control.

sketch - in design, an initial, not to scale or finalized image of a visualization.

specialization - in business, the tendency to become an expert in a small area of the organization's overall activities.

specifications - in design, those characteristics of the product provided by the customer.

stabilized processes - in business, a repeatable set of activities; the same inputs will results in the same outputs, with the same level of quality.

standardization - in computing, the process of achieving agreement on an area of computing; for example, agreement on the format of a message between two organizations, established by an independent group and accepted and used by these organizations.

states - in a change environment, the physical manifestations of this change go through stages or identifiable "states" of transition, these "states" usually have differentiating features to distinguish the current "state" from previous and next "states."

stepwise refinement - in design, a design process using an iterative, or sequential set of activities to accomplish the goal of product design.

storage - in computing, the portion of the computing hardware devoted to accumulating digital information for recall. This recall is possible even after the power has been turned off, and back on again. Also called "persistent storage" to differentiate between this longer-term capability and the capability to temporarily accumulate information lost when electrical power is lost or the next computing task is initiated.

subassembly - in manufacturing, a set of individual parts put together and exhibiting new or emergent properties not available before.

supercomputers - a class of computers with unusually high computing capabilities, typically in the area of mathematically oriented computing.

supervision - in organizations, individuals who directly manage product value enhancing activities and the people who are part of the value adding activities.

sustainability - the degree to which a product can be maintained in high state of use over a significant period of time.

synthesis - in product design, an approach which is intended to be inclusive, or which considers the entire product and its desired operating characteristics at every stage of the design process. This term can be associated with "holistic" design, as well as with design by simulation.

teams - in CE Design, the basis of the organizational strategy is groups of people who have an allegiance to both the overall organization as well as to the group or "team."

throughput - in production manufacturing, the rate at which products are produced. This term can be used in computing to describe the aggregate amount of transactions performed and can be used to describe the rate of smaller events completing in processes.

tiered - in computing, the approach to computing networking where different types of machines are afforded different roles; these roles are based on the relative power of the computers.

tolerance - in the context of product quality, this is the degree to which accuracy to a design norm is required.

tooling- that part of production manufacturing concerned with developing, manufacturing and maintaining the implements to make the product; e.g., a screwdriver and a stamping form for a part are "tools."

transaction - a single discrete event which can add, delete or change information in a larger group of information. Usually, in computing, this term describes traditional computing application program activity; e.g., a transaction in accounting is, for example, a purchase or sale event.

tree - a technique displaying the work breakdown structure or differentiated product. Many times associated with the structured bill of materials, but actually useful throughout the overall manufacturing process.

variant - an effectivity technique controlling design ideas which are changed from a previous version but which have not yet been formally approved for inclusion as an accepted design.

version - an effectivity control technique traditionally used in software, documents or drawings. Versions can be displayed numerically, where 12.3 would be the equivalent of the 12th major revision, and the third iteration of minor changes to the 12th revision.

visualization - one of the three stages of intellectual development of an idea. In this stage, the individual prepares a display of the idea sufficient to communicate it to others.

waterfall chart- a type of planning chart used to schedule, among other items, the final assembly of complex products. It depicts the schedule in GANTT chart format with the sequenced events associated with assembly displayed in such as manner as the look "staggered" on the page or appear to resemble a "waterfall."

windowing- a computing technique permitting the simultaneous display of several different computer program's human interfaces at once on the same display device.

work breakdown structure - a technique separating large, complex problems and their personal work assignments into smaller, more manageable pieces. In this book, CE Design recommends using the differentiation approach.

workstation - a type of desktop computer connected to other desktop computers, or into a network interconnecting other computers, and which operates as an integral part of the network. To be distinguished from a personal computer, which may connect into networks, but is focused mainly on tasks not requiring such interconnections.

Index

INDEX

Q

R

S

T

V

W